Cambridge
International AS and A Level Mathematics

Pure Mathematics 2 and 3

Sophie Goldie
Series Editor: Roger Porkess

Questions from the Cambridge International AS and A Level Mathematics papers are reproduced by permission of University of Cambridge International Examinations.

Questions from the MEI AS and A Level Mathematics papers are reproduced by permission of OCR.

We are grateful to the following companies, institutions and individuals who have given permission to reproduce photographs in this book.

Photo credits: page 2 © Tony Waltham / Robert Harding / Rex Features; page 51 left © Mariusz Blach – Fotolia; page 51 right © viappy – Fotolia; page 62 © Phil Cole/ALLSPORT/Getty Images; page 74 © Imagestate Media (John Foxx); page 104 © Ray Woodbridge / Alamy; page 154 © VIJAY MATHUR/X01849/Reuters/Corbis; page 208 © Krzysztof Szpil – Fotolia; page 247 © erikdegraaf – Fotolia

All designated trademarks and brands are protected by their respective trademarks.

Every effort has been made to trace and acknowledge ownership of copyright. The publishers will be glad to make suitable arrangements with any copyright holders whom it has not been possible to contact.

®IGCSE is the registered trademark of University of Cambridge International Examinations.

Hachette UK's policy is to use papers that are natural, renewable and recyclable products and made from wood grown in sustainable forests. The logging and manufacturing processes are expected to conform to the environmental regulations of the country of origin.

Orders: please contact Bookpoint Ltd, 130 Milton Park, Abingdon, Oxon OX14 4SB. Telephone: (44) 01235 827720. Fax: (44) 01235 400454. Lines are open 9.00–5.00, Monday to Saturday, with a 24-hour message answering service. Visit our website at www.hoddereducation.com

Much of the material in this book was published originally as part of the MEI Structured Mathematics series. It has been carefully adapted for the Cambridge International AS and A Level Mathematics syllabus.

The original MEI author team for Pure Mathematics comprised Catherine Berry, Bob Francis, Val Hanrahan, Terry Heard, David Martin, Jean Matthews, Bernard Murphy, Roger Porkess and Peter Secker.

Copyright in this format © Roger Porkess and Sophie Goldie, 2012

First published in 2012 by
Hodder Education, an Hachette UK company,
338 Euston Road
London NW1 3BH

Impression number   5 4 3 2
Year   2016 2015 2014

All rights reserved. Apart from any use permitted under UK copyright law, no part of this publication may be reproduced or transmitted in any form or by any means, electronic or mechanical, including photocopying and recording, or held within any information storage and retrieval system, without permission in writing from the publisher or under licence from the Copyright Licensing Agency Limited. Further details of such licences (for reprographic reproduction) may be obtained from the Copyright Licensing Agency Limited, Saffron House, 6–10 Kirby Street, London EC1N 8TS.

Cover photo © Irochka – Fotolia
Illustrations by Pantek Media, Maidstone, Kent
Typeset in Minion by Pantek Media, Maidstone, Kent
Printed in Dubai

A catalogue record for this title is available from the British Library

ISBN 978 1444 14646 2

Contents

Key to symbols in this book vi
Introduction vii
The Cambridge International AS and A Level Mathematics syllabus viii

P2 Pure Mathematics 2 — 1

Chapter 1 Algebra — 2
Operations with polynomials — 3
Solution of polynomial equations — 8
The modulus function — 17

Chapter 2 Logarithms and exponentials — 23
Logarithms — 23
Exponential functions — 28
Modelling curves — 30
The natural logarithm function — 39
The exponential function — 43

Chapter 3 Trigonometry — 51
Reciprocal trigonometrical functions — 52
Compound-angle formulae — 55
Double-angle formulae — 61
The forms $r\cos(\theta \pm \alpha)$, $r\sin(\theta \pm \alpha)$ — 66
The general solutions of trigonometrical equations — 75

Chapter 4 Differentiation — 78
The product rule — 78
The quotient rule — 80
Differentiating natural logarithms and exponentials — 85
Differentiating trigonometrical functions — 92
Differentiating functions defined implicitly — 97
Parametric equations — 104
Parametric differentiation — 108

Chapter 5 — Integration — 117

- Integrals involving the exponential function — 117
- Integrals involving the natural logarithm function — 117
- Integrals involving trigonometrical functions — 124
- Numerical integration — 128

Chapter 6 — Numerical solution of equations — 136

- Interval estimation – change-of-sign methods — 137
- Fixed-point iteration — 142

P3 Pure Mathematics 3 — 153

Chapter 7 — Further algebra — 154

- The general binomial expansion — 155
- Review of algebraic fractions — 164
- Partial fractions — 166
- Using partial fractions with the binomial expansion — 173

Chapter 8 — Further integration — 177

- Integration by substitution — 178
- Integrals involving exponentials and natural logarithms — 183
- Integrals involving trigonometrical functions — 187
- The use of partial fractions in integration — 190
- Integration by parts — 194
- General integration — 204

Chapter 9 — Differential equations — 208

- Forming differential equations from rates of change — 209
- Solving differential equations — 214

Chapter 10 — Vectors — 227

- The vector equation of a line — 227
- The intersection of two lines — 234
- The angle between two lines — 240
- The perpendicular distance from a point to a line — 244
- The vector equation of a plane — 247
- The intersection of a line and a plane — 252
- The distance of a point from a plane — 254
- The angle between a line and a plane — 256
- The intersection of two planes — 262

Chapter 11	**Complex numbers**	**271**
	The growth of the number system	271
	Working with complex numbers	273
	Representing complex numbers geometrically	281
	Sets of points in an Argand diagram	284
	The modulus–argument form of complex numbers	287
	Sets of points using the polar form	293
	Working with complex numbers in polar form	296
	Complex exponents	299
	Complex numbers and equations	302
	Answers	**309**
	Index	**341**

Key to symbols in this book

? This symbol means that you may want to discuss a point with your teacher. If you are working on your own there are answers in the back of the book. It is important, however, that you have a go at answering the questions before looking up the answers if you are to understand the mathematics fully.

p This symbol invites you to join in a discussion about proof. The answers to these questions are given in the back of the book.

! This is a warning sign. It is used where a common mistake, misunderstanding or tricky point is being described.

This is the ICT icon. It indicates where you could use a graphic calculator or a computer. Graphic calculators and computers are not permitted in any of the examinations for the Cambridge International AS and A Level Mathematics 9709 syllabus, however, so these activities are optional.

b This symbol and a dotted line down the right-hand side of the page indicate material that you are likely to have met before. You need to be familiar with the material before you move on to develop it further.

e This symbol and a dotted line down the right-hand side of the page indicate material which is beyond the syllabus for the unit but which is included for completeness.

Introduction

This is part of a series of books for the University of Cambridge International Examinations syllabus for Cambridge International AS and A Level Mathematics 9709. It follows on from *Pure Mathematics 1* and completes the pure mathematics required for AS and A level. The series also contains a book for each of mechanics and statistics.

These books are based on the highly successful series for the Mathematics in Education and Industry (MEI) syllabus in the UK but they have been redesigned for Cambridge international students; where appropriate, new material has been written and the exercises contain many past Cambridge examination questions. An overview of the units making up the Cambridge international syllabus is given in the diagram on the next page.

Throughout the series the emphasis is on understanding the mathematics as well as routine calculations. The various exercises provide plenty of scope for practising basic techniques, they also contain many typical examination questions.

An important feature of this series is the electronic support. There is an accompanying disc containing two types of Personal Tutor presentation: examination-style questions, in which the solutions are written out, step by step, with an accompanying verbal explanation, and test-yourself questions; these are multiple-choice with explanations of the mistakes that lead to the wrong answers as well as full solutions for the correct ones. In addition, extensive online support is available via the MEI website, www.mei.org.uk.

The books are written on the assumption that students have covered and understood the work in the Cambridge IGCSE® syllabus. However, some of the early material is designed to provide an overlap and this is designated 'Background'. There are also places where the books show how the ideas can be taken further or where fundamental underpinning work is explored and such work is marked as 'Extension'.

The original MEI author team would like to thank Sophie Goldie who has carried out the extensive task of presenting their work in a suitable form for Cambridge international students and for her many original contributions. They would also like to thank University of Cambridge International Examinations for their detailed advice in preparing the books and for permission to use many past examination questions.

Roger Porkess
Series Editor

The Cambridge International AS and A Level Mathematics syllabus

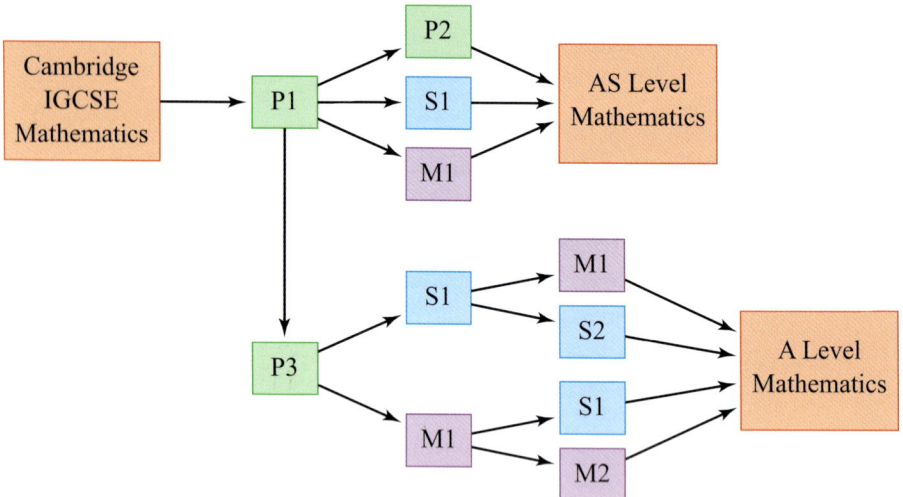

Pure Mathematics 2

Algebra

No, it [1729] is a very interesting number. It is the smallest number expressible as the sum of two cubes in two different ways.

Srinivasa Ramanujan

A brilliant mathematician, Ramanujan was largely self-taught, being too poor to afford a university education. He left India at the age of 26 to work with G.H. Hardy in Cambridge on number theory, but fell ill in the English climate and died six years later in 1920. On one occasion when Hardy visited him in hospital, Ramanujan asked about the registration number of the taxi he came in. Hardy replied that it was 1729, an uninteresting number; Ramanujan's instant response is quoted above.

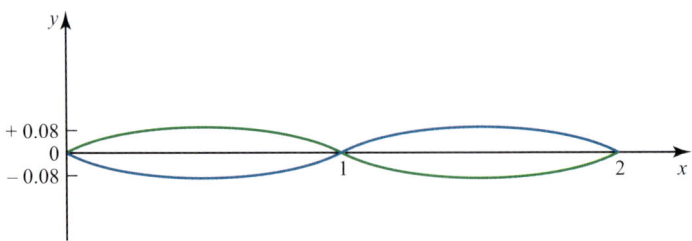

The photograph shows the Tamar Railway Bridge. The spans of this bridge, drawn to the same horizontal and vertical scales, are illustrated on the graph as two curves, one green, the other blue.

 How would you set about trying to fit equations to these two curves?

You will already have met quadratic expressions, like $x^2 - 5x + 6$, and solved quadratic equations, such as $x^2 - 5x + 6 = 0$. Quadratic expressions have the form $ax^2 + bx + c$ where x is a variable, a, b and c are constants and a is not equal to zero. This work is covered in *Pure Mathematics 1* Chapter 1.

An expression of the form $ax^3 + bx^2 + cx + d$, which includes a term in x^3, is called a *cubic* in x. Examples of cubic expressions are

$$2x^3 + 3x^2 - 2x + 11, \qquad 3y^3 - 1 \qquad \text{and} \qquad 4z^3 - 2z.$$

Similarly a *quartic* expression in x, like $x^4 - 4x^3 + 6x^2 - 4x + 1$, contains a term in x^4; a *quintic* expression contains a term in x^5 and so on.

All these expressions are called *polynomials*. The *order* of a polynomial is the highest power of the variable it contains. So a quadratic is a polynomial of order 2, a cubic is a polynomial of order 3 and $3x^8 + 5x^4 + 6x$ is a polynomial of order 8 (an *octic*).

Notice that a polynomial does not contain terms involving $\sqrt{x}$, $\frac{1}{x}$, etc. Apart from the constant term, all the others are multiples of x raised to a positive integer power.

Operations with polynomials

Addition of polynomials

Polynomials are added by adding like terms, for example, you add the coefficients of x^3 together (i.e. the numbers multiplying x^3), the coefficients of x^2 together, the coefficients of x together and the numbers together. You may find it easiest to set this out in columns.

EXAMPLE 1.1 Add $(5x^4 - 3x^3 - 2x)$ to $(7x^4 + 5x^3 + 3x^2 - 2)$.

SOLUTION

$$\begin{array}{rrrrr} 5x^4 & 3x^3 & & -2x & \\ +\,(7x^4 & +5x^3 & +3x^2 & & -2) \\ \hline 12x^4 & +2x^3 & +3x^2 & -2x & -2 \end{array}$$

Note

This may alternatively be set out as follows:

$$(5x^4 - 3x^3 - 2x) + (7x^4 + 5x^3 + 3x^2 - 2) = (5 + 7)x^4 + (-3 + 5)x^3 + 3x^2 - 2x - 2$$
$$= 12x^4 + 2x^3 + 3x^2 - 2x - 2$$

Subtraction of polynomials

Similarly polynomials are subtracted by subtracting like terms.

EXAMPLE 1.2

Simplify $(5x^4 - 3x^3 - 2x) - (7x^4 + 5x^3 + 3x^2 - 2)$.

SOLUTION

$5x^4$	$-3x^3$		$-2x$	
$-(7x^4$	$+5x^3$	$+3x^2$		$-2)$
$-2x^4$	$-8x^3$	$-3x^2$	$-2x$	$+2$

⚠ Be careful of the signs when subtracting. You may find it easier to change the signs on the bottom line and then go on as if you were adding.

> *Note*
>
> This, too, may be set out alternatively, as follows:
>
> $(5x^4 - 3x^3 - 2x) - (7x^4 + 5x^3 + 3x^2 - 2) = (5-7)x^4 + (-3-5)x^3 - 3x^2 - 2x + 2$
> $= -2x^4 - 8x^3 - 3x^2 - 2x + 2$

ⓑ Multiplication of polynomials

When you multiply two polynomials, you multiply each term of the one by each term of the other, and all the resulting terms are added. Remember that when you multiply powers of x, you add the indices: $x^5 \times x^7 = x^{12}$.

EXAMPLE 1.3

Multiply $(x^3 + 3x - 2)$ by $(x^2 - 2x - 4)$.

SOLUTION

Arranging this in columns, so that it looks like an arithmetical long multiplication calculation you get:

		x^3		$+3x$	-2
×			x^2	$-2x$	-4
Multiply top line by x^2		x^5	$+3x^3$	$-2x^2$	
Multiply top line by $-2x$	$-2x^4$		$-6x^2$	$+4x$	
Multiply top line by -4		$-4x^3$		$-12x$	$+8$
Add	x^5	$-2x^4$	$-x^3$	$-8x^2$	$-8x$	$+8$

> *Note*
>
> Alternatively:
>
> $(x^3 + 3x - 2) \times (x^2 - 2x - 4) = x^3(x^2 - 2x - 4) + 3x(x^2 - 2x - 4) - 2(x^2 - 2x - 4)$
> $= x^5 - 2x^4 - 4x^3 + 3x^3 - 6x^2 - 12x - 2x^2 + 4x + 8$
> $= x^5 - 2x^4 + (-4+3)x^3 + (-6-2)x^2 + (-12+4)x + 8$
> $= x^5 - 2x^4 - x^3 - 8x^2 - 8x + 8$

Division of polynomials

Division of polynomials is usually set out rather like arithmetical long division.

EXAMPLE 1.4 Divide $2x^3 - 3x^2 + x - 6$ by $x - 2$.

> If the dividend is missing a term, leave a blank space. For example, write $x^3 + 2x + 5$ as $x^3 \quad + 2x + 5$. Another way to write it is $x^3 + 0x^2 + 2x + 5$.

SOLUTION

Method 1

$$\begin{array}{r} 2x^2 \\ x-2 \overline{\smash{\big)}\, 2x^3 - 3x^2 + x - 6} \\ 2x^3 - 4x^2 \end{array}$$

> Found by dividing $2x^3$ (the first term in $2x^3 - 3x^2 + x - 6$) by x (the first term in $x - 2$).

> $2x^2(x-2)$

Now subtract $2x^3 - 4x^2$ from $2x^3 - 3x^2$, bring down the next term (i.e. x) and repeat the method above:

> $x^2 \div x$

$$\begin{array}{r} 2x^2 + x \\ x-2 \overline{\smash{\big)}\, 2x^3 - 3x^2 + x - 6} \\ 2x^3 - 4x^2 \\ x^2 + x \\ x^2 - 2x \end{array}$$

> $x(x-2)$

Continuing gives:

> This is the answer. It is called the *quotient*.

$$\begin{array}{r} 2x^2 + x + 3 \\ x-2 \overline{\smash{\big)}\, 2x^3 - 3x^2 + x - 6} \\ 2x^3 - 4x^2 \\ x^2 + x \\ x^2 - 2x \\ 3x - 6 \\ 3x - 6 \\ \hline 0 \end{array}$$

> The final remainder of zero means that $x - 2$ divides exactly into $2x^3 - 3x^2 + x - 6$.

Thus $(2x^3 - 3x^2 + x - 6) \div (x - 2) = (2x^2 + x + 3)$.

Method 2

Alternatively this may be set out as follows if you know that there is no remainder.

Let $(2x^3 - 3x^2 + x - 6) \div (x - 2) = ax^2 + bx + c$

> The polynomial here must be of order 2 because $2x^3 \div x$ will give an x^2 term.

Multiplying both sides by $(x - 2)$ gives

$(2x^3 - 3x^2 + x - 6) = (ax^2 + bx + c)(x - 2)$

Multiplying out the expression on the right

$2x^3 - 3x^2 + x - 6 \equiv ax^3 + (b - 2a)x^2 + (c - 2b)x - 2c$

> The identity sign is used here to emphasise that this is an identity and true for all values of x.

Comparing coefficients of x^3

$$2 = a$$

Comparing coefficients of x^2

$$-3 = b - 2a$$
$$-3 = b - 4$$
$$\Rightarrow b = 1$$

Comparing coefficients of x

$$1 = c - 2b$$
$$1 = c - 2$$
$$\Rightarrow c = 3$$

Checking the constant term

$$-6 = -2c \text{ (which agrees with } c = 3).$$

So $ax^2 + bx + c$ is $2x^2 + x + 3$

i.e. $(2x^3 - 3x^2 + x - 6) \div (x - 2) \equiv 2x^2 + x + 3$.

Method 3

With practice you may be able to do this method 'by inspection'. The steps in this would be as follows.

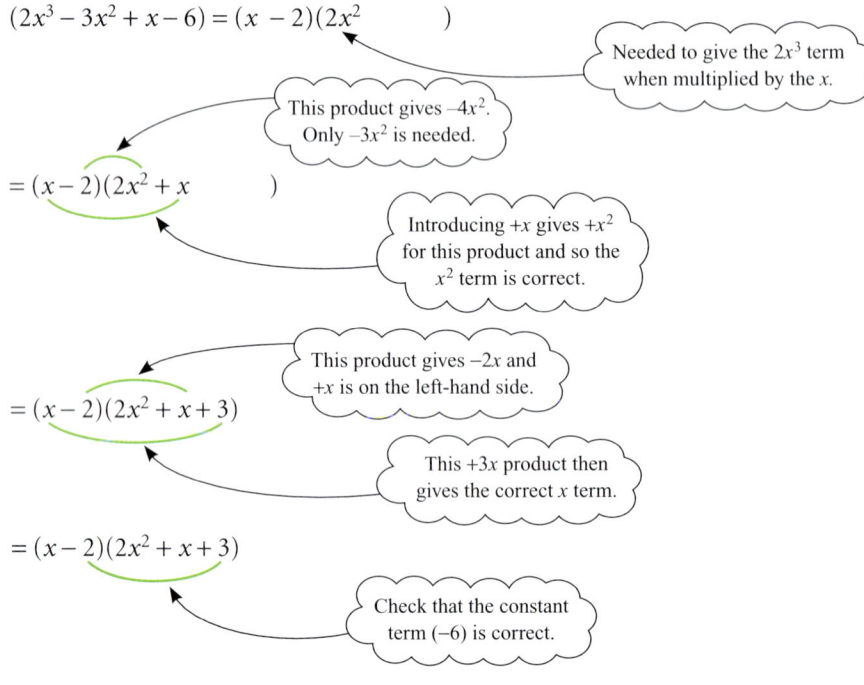

So $(2x^3 - 3x^2 + x - 6) \div (x - 2) \equiv 2x^2 + x + 3$.

A *quotient* is the result of a division. So, in the example above the quotient is $2x^2 + x + 3$.

EXERCISE 1A

1. State the orders of the following polynomials.

   (i) $x^3 + 3x^2 - 4x$ (ii) x^{12} (iii) $2 + 6x^2 + 3x^7 - 8x^5$

2. Add $(x^3 + x^2 + 3x - 2)$ to $(x^3 - x^2 - 3x - 2)$.

3. Add $(x^3 - x)$, $(3x^2 + 2x + 1)$ and $(x^4 + 3x^3 + 3x^2 + 3x)$.

4. Subtract $(3x^2 + 2x + 1)$ from $(x^3 + 5x^2 + 7x + 8)$.

5. Subtract $(x^3 - 4x^2 - 8x - 9)$ from $(x^3 - 5x^2 + 7x + 9)$.

6. Subtract $(x^5 - x^4 - 2x^3 - 2x^2 + 4x - 4)$ from $(x^5 + x^4 - 2x^3 - 2x^2 + 4x + 4)$.

7. Multiply $(x^3 + 3x^2 + 3x + 1)$ by $(x + 1)$.

8. Multiply $(x^3 + 2x^2 - x - 2)$ by $(x - 2)$.

9. Multiply $(x^2 + 2x - 3)$ by $(x^2 - 2x - 3)$.

10. Multiply $(x^{10} + x^9 + x^8 + x^7 + x^6 + x^5 + x^4 + x^3 + x^2 + x^1 + 1)$ by $(x - 1)$.

11. Simplify $(x^2 + 1)(x - 1) - (x^2 - 1)(x - 1)$.

12. Simplify $(x^2 + 1)(x^2 + 4) - (x^2 - 1)(x^2 - 4)$.

13. Simplify $(x + 1)^2 + (x + 3)^2 - 2(x + 1)(x + 3)$.

14. Simplify $(x^2 + 1)(x + 3) - (x^2 + 3)(x + 1)$.

15. Simplify $(x^2 - 2x + 1)^2 - (x + 1)^4$.

16. Divide $(x^3 - 3x^2 - x + 3)$ by $(x - 1)$.

17. Find the quotient when $(x^3 + x^2 - 6x)$ is divided by $(x - 2)$.

18. Divide $(2x^3 - x^2 - 5x + 10)$ by $(x + 2)$.

19. Find the quotient when $(x^4 + x^2 - 2)$ is divided by $(x - 1)$.

20. Divide $(2x^3 - 10x^2 + 3x - 15)$ by $(x - 5)$.

21. Find the quotient when $(x^4 + 5x^3 + 6x^2 + 5x + 15)$ is divided by $(x + 3)$.

22. Divide $(2x^4 + 5x^3 + 4x^2 + x)$ by $(2x + 1)$.

23. Find the quotient when $(4x^4 + 4x^3 - x^2 + 7x - 4)$ is divided by $(2x - 1)$.

24. Divide $(2x^4 + 2x^3 + 5x^2 + 2x + 3)$ by $(x^2 + 1)$.

25. Find the quotient when $(x^4 + 3x^3 - 8x^2 - 27x - 9)$ is divided by $(x^2 - 9)$.

26. Divide $(x^4 + x^3 + 4x^2 + 4x)$ by $(x^2 + x)$.

27. Find the quotient when $(2x^4 - 5x^3 - 16x^2 - 6x)$ is divided by $(2x^2 + 3x)$.

28. Divide $(x^4 + 3x^3 + x^2 - 2)$ by $(x^2 + x + 1)$.

Solution of polynomial equations

You have already met the formula

$$x = \frac{-b \pm \sqrt{b^2 - 4ac}}{2a}$$

for the solution of the quadratic equation $ax^2 + bx + c = 0$.

Unfortunately there is no such simple formula for the solution of a cubic equation, or indeed for any higher power polynomial equation. So you have to use one (or more) of three possible methods.

- Spotting one or more roots.
- Finding where the graph of the expression cuts the x axis.
- A numerical method.

EXAMPLE 1.5

Solve the equation $4x^3 - 8x^2 - x + 2 = 0$.

SOLUTION

Start by plotting the curve whose equation is $y = 4x^3 - 8x^2 - x + 2$. (You may also find it helpful at this stage to display it on a graphic calculator or computer.)

x	−1	0	1	2	3
y	−9	+2	−3	0	35

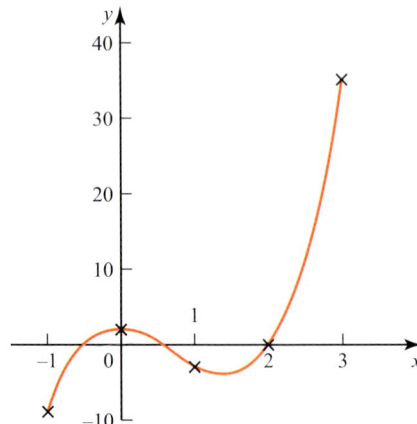

Figure 1.1

Figure 1.1 shows that one root is $x = 2$ and that there are two others. One is between $x = -1$ and $x = 0$ and the other is between $x = 0$ and $x = 1$.

Try $x = -\frac{1}{2}$.

Substituting $x = -\frac{1}{2}$ in $y = 4x^3 - 8x^2 - x + 2$ gives
$$y = 4 \times \left(-\frac{1}{8}\right) - 8 \times \frac{1}{4} - \left(-\frac{1}{2}\right) + 2$$
$$y = 0$$

So in fact the graph crosses the x axis at $x = -\frac{1}{2}$ and this is a root also.

Similarly, substituting $x = +\frac{1}{2}$ in $y = 4x^3 - 8x^2 - x + 2$ gives
$$y = 4 \times \frac{1}{8} - 8 \times \frac{1}{4} - \frac{1}{2} + 2$$
$$y = 0$$

and so the third root is $x = \frac{1}{2}$.

The solution is $x = -\frac{1}{2}, \frac{1}{2}$ or 2.

This example worked out nicely, but many equations do not have roots which are whole numbers or simple fractions. In those cases you can find an approximate answer by drawing a graph. To be more accurate, you will need to use a *numerical* method, which will allow you to get progressively closer to the answer, homing in on it. Such methods are covered in Chapter 6.

The factor theorem

The equation $4x^3 - 8x^2 - x + 2 = 0$ has roots that are whole numbers or fractions. This means that it could, in fact, have been factorised.

$$4x^3 - 8x^2 - x + 2 = (2x + 1)(2x - 1)(x - 2) = 0$$

Few polynomial equations can be factorised, but when one can, the solution follows immediately.

Since $(2x + 1)(2x - 1)(x - 2) = 0$

it follows that either $2x + 1 = 0 \Rightarrow x = -\frac{1}{2}$
or $2x - 1 = 0 \Rightarrow x = \frac{1}{2}$
or $x - 2 = 0 \Rightarrow x = 2$

and so $x = -\frac{1}{2}, \frac{1}{2}$ or 2.

This illustrates an important result, known as the *factor theorem*, which may be stated as follows.

If $(x - a)$ is a factor of the polynomial f(x), then f(a) = 0 and $x = a$ is a root of the equation f(x) = 0. Conversely if f(a) = 0, then $(x - a)$ is a factor of f(x).

EXAMPLE 1.6 Given that $f(x) = x^3 - 6x^2 + 11x - 6$:

(i) find $f(0)$, $f(1)$, $f(2)$, $f(3)$ and $f(4)$

(ii) factorise $x^3 - 6x^2 + 11x - 6$

(iii) solve the equation $x^3 - 6x^2 + 11x - 6 = 0$

(iv) sketch the curve whose equation is $f(x) = x^3 - 6x^2 + 11x - 6$.

SOLUTION

(i) $f(0) = 0^3 - 6 \times 0^2 + 11 \times 0 - 6 = -6$
$f(1) = 1^3 - 6 \times 1^2 + 11 \times 1 - 6 = 0$
$f(2) = 2^3 - 6 \times 2^2 + 11 \times 2 - 6 = 0$
$f(3) = 3^3 - 6 \times 3^2 + 11 \times 3 - 6 = 0$
$f(4) = 4^3 - 6 \times 4^2 + 11 \times 4 - 6 = 6$

(ii) Since $f(1)$, $f(2)$ and $f(3)$ all equal 0, it follows that $(x-1)$, $(x-2)$ and $(x-3)$ are all factors. This tells you that

$$x^3 - 6x^2 + 11x - 6 = (x-1)(x-2)(x-3) \times \text{constant}$$

By checking the coefficient of the term in x^3, you can see that the constant must be 1, and so

$$x^3 - 6x^2 + 11x - 6 = (x-1)(x-2)(x-3)$$

(iii) $x = 1$, 2 or 3

(iv)

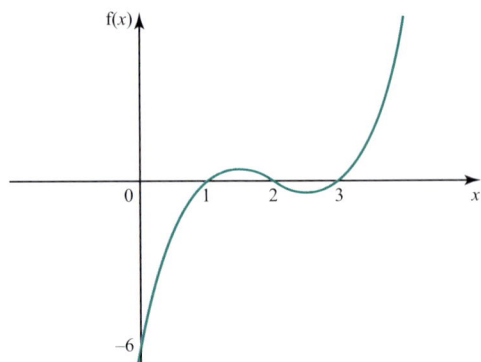

Figure 1.2

In the previous example, all three factors came out of the working, but this will not always happen. If not, it is often possible to find one factor (or more) by 'spotting' it, or by sketching the curve. You can then make the job of searching for further factors much easier by dividing the polynomial by the factor(s) you have found: you will then be dealing with a lower order polynomial.

EXAMPLE 1.7 Given that $f(x) = x^3 - x^2 - 3x + 2$:

(i) show that $(x - 2)$ is a factor

(ii) solve the equation $f(x) = 0$.

SOLUTION

(i) To show that $(x - 2)$ is a factor, it is necessary to show that $f(2) = 0$.

$$f(2) = 2^3 - 2^2 - 3 \times 2 + 2$$
$$= 8 - 4 - 6 + 2$$
$$= 0$$

Therefore $(x - 2)$ is a factor of $x^3 - x^2 - 3x + 2$.

(ii) Since $(x - 2)$ is a factor you divide $f(x)$ by $(x - 2)$.

$$\begin{array}{r}
x^2 + x - 1 \\
x - 2 \overline{\smash{)}\, x^3 - x^2 - 3x + 2} \\
\underline{x^3 - 2x^2} \\
x^2 - 3x \\
\underline{x^2 - 2x} \\
-x + 2 \\
\underline{-x + 2} \\
0
\end{array}$$

So $f(x) = 0$ becomes $(x - 2)(x^2 + x - 1) = 0$,

$\Rightarrow$ either $x - 2 = 0$ or $x^2 + x - 1 = 0$.

Using the quadratic formula on $x^2 + x - 1 = 0$ gives

$$x = \frac{-1 \pm \sqrt{1 - 4 \times 1 \times (-1)}}{2}$$
$$= \frac{1 \pm \sqrt{5}}{2}$$
$$= -1.618 \text{ or } 0.618 \text{ (to 3 d.p.)}$$

So the complete solution is $x = -1.618, 0.618$ or 2.

Spotting a root of a polynomial equation

Most polynomial equations do not have integer (or fraction) solutions. It is only a few special cases that work out nicely.

To check whether an integer root exists for any equation, look at the constant term. Decide what whole numbers divide into it and test them.

EXAMPLE 1.8 Spot an integer root of the equation $x^3 - 3x^2 + 2x - 6 = 0$.

SOLUTION

The constant term is -6 and this is divisible by $-1, +1, -2, +2, -3, +3, -6$ and $+6$. So the only possible factors are $(x \pm 1), (x \pm 2), (x \pm 3)$ and $(x \pm 6)$. This limits the search somewhat.

$f(1) = -6$ No; $f(-1) = -12$ No;
$f(2) = -6$ No; $f(-2) = -30$ No;
$f(3) = 0$ Yes; $f(-3) = -66$ No;
$f(6) = 114$ No; $f(-6) = -342$ No.

$x = 3$ is an integer root of the equation.

EXAMPLE 1.9 Is there an integer root of the equation $x^3 - 3x^2 + 2x - 5 = 0$?

SOLUTION

The only possible factors are $(x \pm 1)$ and $(x \pm 5)$.

$f(1) = -5$ No; $f(-1) = -11$ No;
$f(5) = 55$ No; $f(-5) = -215$ No.

There is no integer root.

The remainder theorem

Using the long division method, any polynomial can be divided by another polynomial of lesser order, but sometimes there will be a remainder. Look at $(x^3 + 2x^2 - 3x - 7) \div (x - 2)$.

$$
\begin{array}{r}
x^2 + 4x + 5 \\
x - 2 \overline{\smash{\big)}\, x^3 + 2x^2 - 3x - 7} \\
\underline{x^3 - 2x^2} \\
4x^2 - 3x \\
\underline{4x^2 - 8x} \\
5x - 7 \\
\underline{5x - 10} \\
3
\end{array}
$$

The quotient is $x^2 + 4x + 5$ and the remainder is 3.

You can write this as

$$x^3 + 2x^2 - 3x - 7 = (x - 2)(x^2 + 4x + 5) + 3$$

At this point it is convenient to call the polynomial $x^3 + 2x^2 - 3x - 7 = f(x)$.

So f$(x) = (x - 2)(x^2 + 4x + 5) + 3$.   ①

Substituting $x = 2$ into both sides of ① gives f$(2) = 3$.

So f(2) is the remainder when f(x) is divided by $(x - 2)$.

This result can be generalised to give the *remainder theorem*. It states that for a polynomial, f(x),

f(a) is the remainder when f(x) is divided by $(x - a)$.

f$(x) = (x - a)$g$(x) +$ f(a) (the remainder theorem)

EXAMPLE 1.10 Find the remainder when $2x^3 - 3x + 5$ is divided by $x + 1$.

SOLUTION

The remainder is found by substituting $x = -1$ in $2x^3 - 3x + 5$.

$$2 \times (-1)^3 - 3 \times (-1) + 5$$
$$= -2 + 3 + 5$$
$$= 6$$

So the remainder is 6.

EXAMPLE 1.11 When $x^2 - 6x + a$ is divided by $x - 3$, the remainder is 2. Find the value of a.

SOLUTION

The remainder is found by substituting $x = 3$ in $x^2 - 6x + a$.

$$3^2 - 6 \times 3 + a = 2$$
$$9 - 18 + a = 2$$
$$-9 + a = 2$$
$$a = 11$$

When you are dividing by a linear expression any remainder will be a constant; dividing by a quadratic expression may give a linear remainder.

? A polynomial is divided by another of degree n.

What can you say about the remainder?

When dividing by polynomials of order 2 or more, the remainder is usually found most easily by actually doing the long division.

EXAMPLE 1.12 Find the remainder when $2x^4 - 3x^3 + 4$ is divided by $x^2 + 1$.

SOLUTION

$$
\begin{array}{r}
2x^2 - 3x - 2 \\
x^2 + 1 \overline{\smash{)}\, 2x^4 - 3x^3 + 4}\\
2x^4 + 2x^2 \\
\hline
-3x^3 - 2x^2 \\
-3x^3 - 3x \\
\hline
-2x^2 + 3x + 4\\
-2x^2 - 2\\
\hline
3x + 6
\end{array}
$$

The remainder is $3x + 6$.

 In a division such as the one in Example 1.12, it is important to keep a separate column for each power of x and this means that sometimes it is necessary to leave gaps, as in the example above. In arithmetic, zeros are placed in the gaps. For example, 2 thousand and 3 is written 2003.

EXERCISE 1B

1 Given that $f(x) = x^3 + 2x^2 - 9x - 18$:

   (i)   find $f(-3)$, $f(-2)$, $f(-1)$, $f(0)$, $f(1)$, $f(2)$ and $f(3)$
   (ii) factorise $f(x)$
   (iii) solve the equation $f(x) = 0$
   (iv) sketch the curve with the equation $y = f(x)$.

2 The polynomial $p(x)$ is given by $p(x) = x^3 - 4x$.

   (i)   Find the values of $p(-3)$, $p(-2)$, $p(-1)$, $p(0)$, $p(1)$, $p(2)$ and $p(3)$.
   (ii) Factorise $p(x)$.
   (iii) Solve the equation $p(x) = 0$.
   (iv) Sketch the curve with the equation $y = p(x)$.

3 You are given that $f(x) = x^3 - 19x + 30$.

   (i)   Calculate $f(0)$ and $f(3)$. Hence write down a factor of $f(x)$.
   (ii) Find p and q such that $f(x) \equiv (x - 2)(x^2 + px + q)$.
   (iii) Solve the equation $x^3 - 19x + 30 = 0$.
   (iv) Without further calculation draw a sketch of $y = f(x)$.

[MEI]

4 **(i)** Show that $x - 3$ is a factor of $x^3 - 5x^2 - 2x + 24$.
   (ii) Solve the equation $x^3 - 5x^2 - 2x + 24 = 0$.
   (iii) Sketch the curve with the equation $y = x^3 - 5x^2 - 2x + 24$.

5 **(i)** Show that $x = 2$ is a root of the equation $x^4 - 5x^2 + 2x = 0$ and write down another integer root.
   (ii) Find the other two roots of the equation $x^4 - 5x^2 + 2x = 0$.
   (iii) Sketch the curve with the equation $y = x^4 - 5x^2 + 2x$.

6 **(i)** The polynomial $p(x) = x^3 - 6x^2 + 9x + k$ has a factor $x - 4$. Find the value of k.
   (ii) Find the other factors of the polynomial.
   (iii) Sketch the curve with the equation $y = p(x)$.

7 The diagram shows the curve with the equation $y = (x + a)(x - b)^2$ where a and b are positive integers.

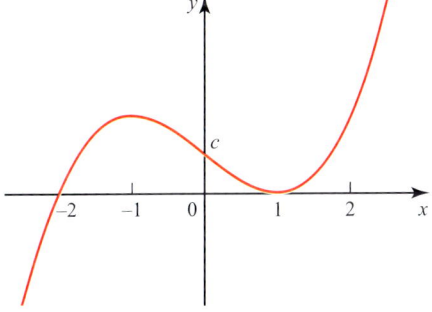

   (i) Write down the values of a and b, and also of c, given that the curve crosses the y axis at $(0, c)$.
   (ii) Solve the equation $(x + a)(x - b)^2 = c$ using the values of a, b and c you found in part **(i)**.

8 The function $f(x)$ is given by $f(x) = x^4 - 3x^2 - 4$ for real values of x.

   (i) By treating $f(x)$ as a quadratic in x^2, factorise it in the form $(x^2 + \ldots)(x^2 + \ldots)$.
   (ii) Complete the factorisation as far as possible.
   (iii) How many real roots has the equation $f(x) = 0$? What are they?

9 **(i)** Show that $x - 2$ is not a factor of $2x^3 + 5x^2 - 7x - 3$.
   (ii) Find the quotient and the remainder when $2x^3 + 5x^2 - 7x - 3$ is divided by $x - 2$.

10 The equation $f(x) = x^3 - 4x^2 + x + 6 = 0$ has three integer roots.

   (i) List the eight values of a for which it is sensible to check whether $f(a) = 0$ and check each of them.
   (ii) Solve $f(x) = 0$.

11 Factorise, as far as possible, the following expressions.

   (i) $x^3 - x^2 - 4x + 4$ given that $(x - 1)$ is a factor.
   (ii) $x^3 + 1$ given that $(x + 1)$ is a factor.
   (iii) $x^3 + x - 10$ given that $(x - 2)$ is a factor.
   (iv) $x^3 + x^2 + x + 6$ given that $(x + 2)$ is a factor.

12 (i) Show that neither $x = 1$ nor $x = -1$ is a root of $x^4 - 2x^3 + 3x^2 - 8 = 0$.
   (ii) Find the quotient and the remainder when $x^4 - 2x^3 + 3x^2 - 8$ is divided by
   (a) $(x - 1)$ (b) $(x + 1)$ (c) $(x^2 - 1)$.

13 When $2x^3 + 3x^2 + kx - 6$ is divided by $x + 1$ the remainder is 7.
   Find the value of k.

14 When $x^3 + px^2 + p^2x - 36$ is divided by $x - 3$ the remainder is 21.
   Find a possible value of p.

15 When $x^3 + ax^2 + bx + 8$ is divided by $x - 3$ the remainder is 2 and when it is divided by $x + 1$ the remainder is -2.
   Find a and b and hence obtain the remainder on dividing by $x - 2$.

16 When $f(x) = 2x^3 + ax^2 + bx + 6$ is divided by $x - 1$ there is no remainder and when $f(x)$ is divided by $x + 1$ the remainder is 10.
   Find a and b and hence solve the equation $f(x) = 0$.

17 The cubic polynomial $ax^3 + bx^2 - 3x - 2$, where a and b are constants, is denoted by $p(x)$. It is given that $(x - 1)$ and $(x + 2)$ are factors of $p(x)$.
   (i) Find the values of a and b.
   (ii) When a and b have these values, find the other linear factor of $p(x)$.
   [Cambridge International AS & A Level Mathematics 9709, Paper 2 Q4 June 2006]

18 The polynomial $2x^3 + 7x^2 + ax + b$, where a and b are constants, is denoted by $p(x)$. It is given that $(x + 1)$ is a factor of $p(x)$, and that when $p(x)$ is divided by $(x + 2)$ the remainder is 5. Find the values of a and b.
   [Cambridge International AS & A Level Mathematics 9709, Paper 2 Q4 June 2008]

19 The polynomial $2x^3 - x^2 + ax - 6$, where a is a constant, is denoted by $p(x)$.
   It is given that $(x + 2)$ is a factor of $p(x)$.
   (i) Find the value of a.
   (ii) When a has this value, factorise $p(x)$ completely.
   [Cambridge International AS & A Level Mathematics 9709, Paper 2 Q2 November 2008]

20 The polynomial $x^3 + ax^2 + bx + 6$, where a and b are constants, is denoted by $p(x)$. It is given that $(x - 2)$ is a factor of $p(x)$, and that when $p(x)$ is divided by $(x - 1)$ the remainder is 4.
   (i) Find the values of a and b.
   (ii) When a and b have these values, find the other two linear factors of $p(x)$.
   [Cambridge International AS & A Level Mathematics 9709, Paper 2 Q6 June 2009]

21 The polynomial $x^3 - 2x + a$, where a is a constant, is denoted by $p(x)$.
   It is given that $(x + 2)$ is a factor of $p(x)$.
   (i) Find the value of a.
   (ii) When a has this value, find the quadratic factor of $p(x)$.
   [Cambridge International AS & A Level Mathematics 9709, Paper 3 Q2 June 2007]

The modulus function

Look at the graph of $y = f(x)$, where $f(x) = x$.

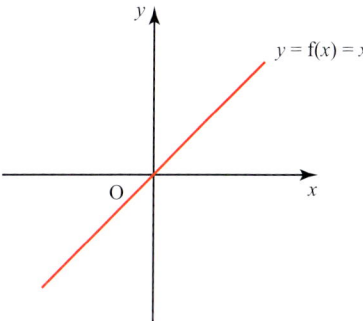

Figure 1.3

The function $f(x)$ is positive when x is positive and negative when x is negative.

Now look at the graph of $y = g(x)$, where $g(x) = |x|$.

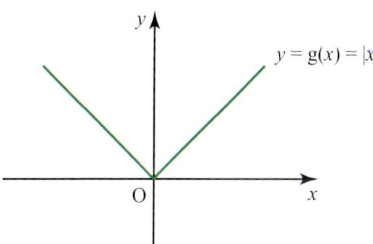

Figure 1.4

The function $g(x)$ is called the *modulus* of x. $g(x)$ always takes the positive numerical value of x. For example, when $x = -2$, $g(x) = 2$, so $g(x)$ is always positive. The modulus is also called the *magnitude* of the quantity.

Another way of writing the modulus function $g(x)$ is

$g(x) = x$ for $x \geq 0$
$g(x) = -x$ for $x < 0$.

❓ What is the value of $g(3)$ and $g(-3)$?

What is the value of $|3 + 3|$, $|3 - 3|$, $|3| + |3|$ and $|3| + |-3|$?

The graph of $y = g(x)$ can be obtained from the graph of $y = f(x)$ by replacing values where $f(x)$ is negative by the equivalent positive values. This is the equivalent of reflecting that part of the line in the x axis.

EXAMPLE 1.13   Sketch the graphs of the following on separate axes.

(i) $y = 1 - x$

(ii) $y = |1 - x|$

(iii) $y = 2 + |1 - x|$

SOLUTION

(i) $y = 1 - x$ is the straight line through $(0, 1)$ and $(1, 0)$.

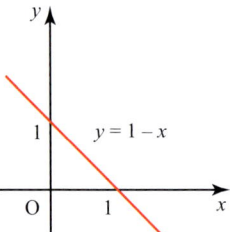

Figure 1.5

(ii) $y = |1 - x|$ is obtained by reflecting the part of the line for $x > 1$ in the x axis.

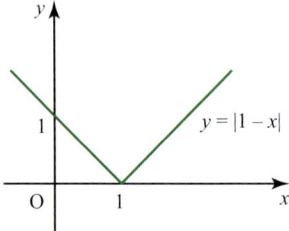

Figure 1.6

(iii) $y = 2 + |1 - x|$ is obtained from the previous graph by applying the translation $\begin{pmatrix} 0 \\ 2 \end{pmatrix}$.

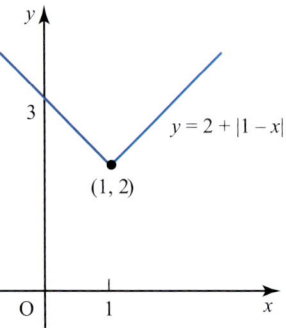

Figure 1.7

Inequalities involving the modulus sign

You will often meet inequalities involving the modulus sign.

❓ Look back at the graph of $y = |x|$ in figure 1.4.

How does this show that $|x| < 2$ is equivalent to $-2 < x < 2$?

Here is a summary of some useful rules.

Rule	Example
$|x| = |-x|$	$|3| = |-3|$
$|a - b| = |b - a|$	$|8 - 5| = |5 - 8| = +3$
$|x|^2 = x^2$	$|-3|^2 = (-3)^2$
$|a| = |b| \Leftrightarrow a^2 = b^2$	$|-3| = |3| \Leftrightarrow (-3)^2 = 3^2$
$|x| \leq a \Leftrightarrow -a \leq x \leq a$	$|x| \leq 3 \Leftrightarrow -3 \leq x \leq 3$
$|x| > a \Leftrightarrow x < -a$ or $x > a$	$|x| > 3 \Leftrightarrow x < -3$ or $x > 3$

EXAMPLE 1.14

Solve the following.

(i) $|x + 3| \leq 4$

(ii) $|2x - 1| > 9$

(iii) $5 - |x - 2| > 1$

SOLUTION

(i) $|x + 3| \leq 4 \Leftrightarrow -4 \leq x + 3 \leq 4$
$\Leftrightarrow -7 \leq x \leq 1$

(ii) $|2x - 1| > 9 \Leftrightarrow 2x - 1 < -9$ or $2x - 1 > 9$
$\Leftrightarrow 2x < -8$ or $2x > 10$
$\Leftrightarrow x < -4$ or $x > 5$

(iii) $5 - |x - 2| > 1 \Leftrightarrow 4 > |x - 2|$
$\Leftrightarrow |x - 2| < 4$
$\Leftrightarrow -4 < x - 2 < 4$
$\Leftrightarrow -2 < x < 6$

Note

The solution to part **(ii)** represents two separate intervals on the number line, so cannot be written as a single inequality.

EXAMPLE 1.15 Express the inequality $-2 < x < 6$ in the form $|x - a| < b$, where a and b are to be found.

SOLUTION

$|x - a| < b \iff -b < x - a < b$
$\iff a - b < x < a + b$

Comparing this with $-2 < x < 6$ gives

$a - b = -2$
$a + b = 6$.

Solving these simultaneously gives $a = 2$, $b = 4$, so $|x - 2| < 4$.

EXAMPLE 1.16 Solve $2x < |x - 3|$.

SOLUTION

It helps to sketch a graph of $y = 2x$ and $y = |x - 3|$.

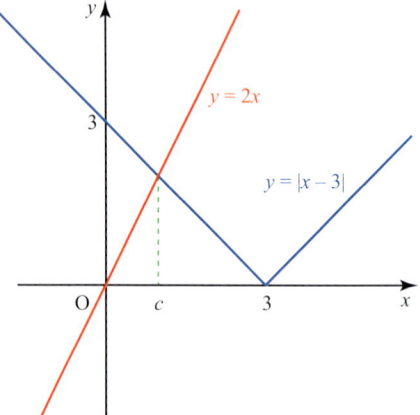

Figure 1.8

You can see that the graph of $y = 2x$ is below $y = |x - 3|$ for $x < c$.

You can find the critical region by solving $2x < -(x - 3)$.

$2x < -(x - 3)$
$2x < -x + 3$
$3x < 3$
$x < 1$

c is at the intersection of the lines $y = 2x$ and $y = -(x - 3)$.

EXAMPLE 1.17

(i) Solve $|2x-1|=|x-2|$.
(ii) Solve $|2x-1|<|x-2|$.

SOLUTION

(i) Sketching a graph of $y=|2x-1|$ and $y=|x-2|$ shows that the equation is true for two values of x.

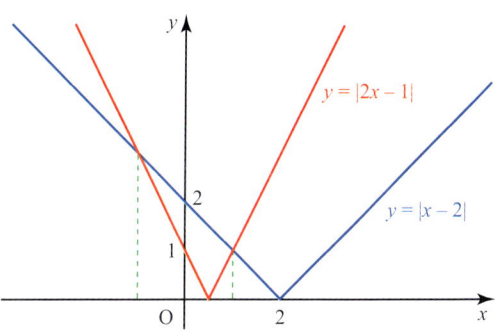

Figure 1.9

You can find these values by solving $|2x-1|=|x-2|$.

One method is to use the fact that $|a|=|b| \Leftrightarrow a^2=b^2$.

$$|2x-1|=|x-2|$$
Squaring: $(2x-1)^2 = (x-2)^2$
Expanding: $4x^2 - 4x + 1 = x^2 - 4x + 4$
Rearranging: $3x^2 - 3 = 0$
$\Rightarrow$ $x^2 - 1 = 0$
Factorising: $(x-1)(x+1) = 0$

So the solution is $x=-1$ or $x=1$.

(ii) When $|2x-1|<|x-2|$, $y=|2x-1|$ (drawn in red) is below $y=|x-2|$ (drawn in blue) on the graph. So the solution to the inequality is $-1 < x < 1$.

EXERCISE 1C

1 Solve the following equations.

(i) $|x+4|=5$
(ii) $|x-3|=4$
(iii) $|3-x|=4$
(iv) $|4x-1|=7$
(v) $|2x+1|=5$
(vi) $|8-2x|=6$
(vii) $|2x+1|=|x+5|$
(vii) $|4x-1|=|9-x|$
(ix) $|3x-2|=|4-x|$

2 Solve the following inequalities.

(i) $|x+3|<5$
(ii) $|x-2| \leqslant 2$
(iii) $|x-5|>6$
(iv) $|x+1| \geqslant 2$
(v) $|2x-3|<7$
(vi) $|3x-2| \leqslant 4$

3 Express each of the following inequalities in the form $|x-a| < b$, where a and b are to be found.

(i) $-1 < x < 3$
(ii) $2 < x < 8$
(iii) $-2 < x < 4$
(iv) $-1 < x < 6$
(v) $9.9 < x < 10.1$
(vi) $0.5 < x < 7.5$

4 Sketch each of the following graphs on a separate set of axes.

(i) $y = |x+2|$
(ii) $y = |2x-3|$
(iii) $y = |x+2| - 2$
(iv) $y = |x| + 1$
(v) $y = |2x+5| - 4$
(vi) $y = 3 + |x-2|$

5 Solve the following inequalities.

(i) $|x+3| < |x-4|$
(ii) $|x-5| > |x-2|$
(iii) $|2x-1| \leq |2x+3|$
(iv) $|2x| \leq |x+3|$
(v) $|2x| > |x+3|$
(vi) $|2x+5| \geq |x-1|$

6 Solve the inequality $|x| > |3x-2|$.

[Cambridge International AS & A Level Mathematics 9709, Paper 2 Q1 June 2005]

7 Solve the inequality $2x > |x-1|$.

[Cambridge International AS & A Level Mathematics 9709, Paper 3 Q2 June 2006]

8 Given that a is a positive constant, solve the inequality $|x-3a| > |x-a|$.

[Cambridge International AS & A Level Mathematics 9709, Paper 3 Q1 November 2005]

KEY POINTS

1 A polynomial in x has terms in positive integer powers of x and may also have a constant term.

2 The order of a polynomial in x is the highest power of x which appears in the polynomial.

3 The factor theorem states that if $(x-a)$ is a factor of a polynomial f(x) then f(a) = 0 and $x = a$ is a root of the equation f(x) = 0.
Conversely if f(a) = 0, then $x - a$ is a factor of f(x).

4 The remainder theorem states that f(a) is the remainder when the polynomial f(x) is divided by $(x-a)$.

5 The modulus of x, written $|x|$, means the positive value of x.

6 The modulus function is

$$|x| = x, \quad \text{for } x \geq 0$$
$$|x| = -x, \quad \text{for } x < 0.$$

2 Logarithms and exponentials

> **Normally speaking it may be said that the forces of a capitalist society, if left unchecked, tend to make the rich richer and the poor poorer and thus increase the gap between them.**
>
> *Jawaharlal Nehru*

This cube has volume of 500 cm³.

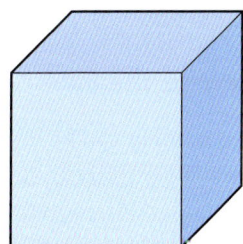

❓ How would you calculate the length of its side, correct to the nearest millimetre, without using the cube root button on your calculator?

Logarithms

You can think of multiplication in two ways. Look, for example, at 81×243, which is $3^4 \times 3^5$. You can work out the product using the numbers or you can work it out by adding the powers of a common base – in this case base 3.

Multiplying the numbers: $81 \times 243 = 19\,683$

Adding the powers of the base 3: $4 + 5 = 9$ and $3^9 = 19\,683$

Another name for a power is a *logarithm*. Since $81 = 3^4$, you can say that the logarithm to the base 3 of 81 is 4. The word logarithm is often abbreviated to log and the statement would be written $\log_3 81 = 4$. In general:

$$y = a^x \implies \log_a y = x$$

Notice that since $3^4 = 81$, $3^{\log_3 81} = 81$. This is an example of a general result:

$$a^{\log_a x} = x$$

EXAMPLE 2.1

(i) Find the logarithm to the base 2 of each of these numbers.

(a) 64   (b) $\frac{1}{2}$ (c) 1   (d) $\sqrt{2}$

(ii) Show that $2^{\log_2 64} = 64$.

SOLUTION

(i) (a) $64 = 2^6$ and so $\log_2 64 = 6$

(b) $\frac{1}{2} = 2^{-1}$ and so $\log_2 \frac{1}{2} = -1$

(c) $1 = 2^0$ and so $\log_2 1 = 0$

(d) $\sqrt{2} = 2^{\frac{1}{2}}$ and so $\log_2 \sqrt{2} = \frac{1}{2}$

(ii) $2^{\log_2 64} = 2^6 = 64$ as required

Logarithms to the base 10

Any positive number can be expressed as a power of 10. Before the days of calculators, logarithms to the base 10 were used extensively as an aid to calculation. There is no need for that nowadays but the logarithm function remains an important part of mathematics, particularly the natural logarithm which you will meet later in this chapter. Base 10 logarithms continue to be a standard feature on calculators, and occur in some specialised contexts: the pH value of a liquid, for example, is a measure of its acidity or alkalinity and is given by $\log_{10}(1/\text{the concentration of H}^+ \text{ ions})$.

Since $1000 = 10^3$, $\log_{10} 1000 = 3$

Similarly $\log_{10} 100 = 2$

$\log_{10} 10 = 1$

$\log_{10} 1 = 0$

$\log_{10}\left(\frac{1}{10}\right) = \log_{10}(10^{-1}) = -1$

$\log_{10}\left(\frac{1}{100}\right) = \log_{10}(10^{-2}) = -2$

and so on.

INVESTIGATION

There are several everyday situations in which quantities are measured on logarithmic scales.

What are the relationships between the following?

(i) An earthquake of intensity 7 on the Richter Scale and one of intensity 8.
(ii) The frequency of the musical note middle C and that of the C above it.
(iii) The intensity of an 85 dB noise level and one of 86 dB.

The laws of logarithms

The laws of logarithms follow from those for indices.

Multiplication

Writing $xy = x \times y$ in the form of powers (or logarithms) to the base a and using the result that $x = a^{\log_a x}$ gives

$$a^{\log_a xy} = a^{\log_a x} \times a^{\log_a y}$$

and so $\quad a^{\log_a xy} = a^{\log_a x + \log_a y}$.

Consequently $\log_a xy = \log_a x + \log_a y$.

Division

Similarly $\log_a \left(\dfrac{x}{y}\right) = \log_a x - \log_a y$.

Power zero

Since $a^0 = 1$, $\log_a 1 = 0$.

However, it is more usual to state such laws without reference to the base of the logarithms except where necessary, and this convention is adopted in the key points at the end of this chapter. As well as the laws given above, others may be derived from them, as follows.

Indices

Since $\quad\quad\quad\quad x^n = x \times x \times x \times \ldots \times x \;\; (n \text{ times})$
it follows that $\quad \log x^n = \log x + \log x + \log x + \ldots + \log x \;\; (n \text{ times})$,
and so $\quad\quad\quad \log x^n = n \log x$.

This result is also true for non-integer values of n and is particularly useful because it allows you to solve equations in which the unknown quantity is the power, as in the next example.

EXAMPLE 2.2

Solve the equation $2^n = 1000$.

SOLUTION

$$2^n = 1000$$

Taking logarithms to the base 10 of both sides (since these can be found on a calculator),

$$\log_{10}(2^n) = \log_{10} 1000$$
$$n \log_{10} 2 = \log_{10} 1000$$
$$n = \dfrac{\log_{10} 1000}{\log_{10} 2} = 9.97 \text{ to 3 significant figures}$$

Note

Most calculators just have 'log' and not '$\log_{10}$' on their keys.

EXAMPLE 2.3

A geometric sequence begins 0.2, 1, 5, ...
The kth term is the first term in the sequence that is greater than 500 000.
Find the value of k.

SOLUTION

The kth term of a geometric sequence is given by $a_k = a \times r^{k-1}$.

In this case $a = 0.2$ and $r = 5$, so:

$$0.2 \times 5^{k-1} > 500\,000$$

$$5^{k-1} > \frac{500\,000}{0.2}$$

$$5^{k-1} > 2\,500\,000$$

Taking logarithms to the base 10 of both sides:

$$\log_{10} 5^{k-1} > \log_{10} 2\,500\,000$$
$$\Rightarrow (k-1)\log_{10} 5 > \log_{10} 2\,500\,000$$
$$\Rightarrow k - 1 > \frac{\log_{10} 2\,500\,000}{\log_{10} 5}$$
$$\Rightarrow k - 1 > 9.15$$
$$\Rightarrow k > 10.15$$

Since k is an integer, then $k = 11$.
So the 11th term is the first term greater than 500 000.

Check: 10th term $= 0.2 \times 5^{10-1} = 390\,625$ ($< 500\,000$) ✓
11th term $= 0.2 \times 5^{11-1} = 1\,953\,125$ ($> 500\,000$) ✓

Roots

A similar line of reasoning leads to the conclusion that:

$$\log \sqrt[n]{x} = \frac{1}{n} \log x$$

The logic runs as follows:

Since $\underbrace{\sqrt[n]{x} \times \sqrt[n]{x} \times \sqrt[n]{x} \times \ldots \times \sqrt[n]{x}}_{n \text{ times}} = x$

it follows that $n \log \sqrt[n]{x} = \log x$

and so $\log \sqrt[n]{x} = \frac{1}{n} \log x$

The logarithm of a number to its own base

Since $5^1 = 5$, it follows that $\log_5 5 = 1$.

Clearly the same is true for any number, and in general,

$$\log_a a = 1$$

Reciprocals

Another useful result is that, for any base,

$$\log\left(\frac{1}{y}\right) = -\log y$$

This is a direct consequence of the division law

$$\log_a\left(\frac{x}{y}\right) = \log_a x - \log_a y$$

with x set equal to 1:

$$\log\left(\frac{1}{y}\right) = \log 1 - \log y$$
$$= 0 - \log y$$
$$= -\log y$$

If the number y is greater than 1, it follows that $\frac{1}{y}$ lies between 0 and 1 and $\log\left(\frac{1}{y}\right)$ is negative. So for any base (>1), the logarithm of a number between 0 and 1 is negative. You saw an example of this on page 24: $\log_{10}\left(\frac{1}{10}\right) = -1$.

The result $\log\left(\frac{1}{y}\right) = -\log y$ is often useful in simplifying expressions involving logarithms.

ACTIVITY 2.1 Draw the graph of $y = \log_2 x$, taking values of x like $\frac{1}{8}, \frac{1}{4}, \frac{1}{2}, 1, 2, 4, 8, 16$. Use your graph to estimate the value of $\sqrt{2}$.

Graphs of logarithms

Whatever the value, a, of the base ($a > 1$), the graph of $y = \log_a x$ has the same general shape (shown in figure 2.1).

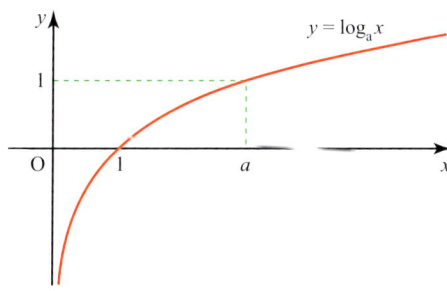

Figure 2.1

The graph has the following properties.

- The curve crosses the x axis at $(1, 0)$.
- The curve only exists for positive values of x.
- The line $x = 0$ is an asymptote and for values of x between 0 and 1 the curve lies below the x axis.
- There is no limit to the height of the curve for large values of x, but its gradient progressively decreases.
- The curve passes through the point $(a, 1)$.

❓ Each of the points above can be justified by work that you have already covered. How?

Exponential functions

The relationship $y = \log_a x$ may be rewritten as $x = a^y$, and so the graph of $x = a^y$ is exactly the same as that of $y = \log_a x$. Interchanging x and y has the effect of reflecting the graph in the line $y = x$, and changing the relationship into $y = a^x$, as shown in figure 2.2.

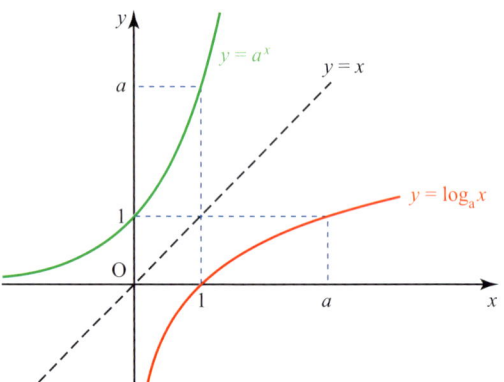

Figure 2.2

The function $y = a^x$, $x \in \mathbb{R}$ is called an *exponential function*. Notice that while the domain of $y = a^x$ is all real numbers ($x \in \mathbb{R}$), the range is strictly the positive real numbers. $y = a^x$ is the *inverse* of the logarithm function so the domain of the logarithm function is strictly the positive real numbers and its range is all real numbers. Remember the effect of applying a function followed by it inverse is to bring you back to where you started.

Thus $\log_a (a^x) = x$ and $a^{(\log_a x)} = x$.

EXERCISE 2A

1 $2^x = 32 \iff x = \log_2 32$

Write similar logarithmic equivalents of these equations. In each case find also the value of x, using your knowledge of indices and not using your calculator.

(i) $3^x = 9$
(ii) $4^x = 64$
(iii) $2^x = \frac{1}{4}$
(iv) $5^x = \frac{1}{5}$
(v) $7^x = 1$
(vi) $16^x = 2$

2 Write the equivalent of these equations in exponential form. Without using your calculator, find also the value of y in each case.

(i) $y = \log_3 9$
(ii) $y = \log_5 125$
(iii) $y = \log_2 16$
(iv) $y = \log_6 1$
(v) $y = \log_{64} 8$
(vi) $y = \log_5 \left(\frac{1}{25}\right)$

3 Write down the values of the following without using a calculator. Use your calculator to check your answers for those questions which use base 10.

(i) $\log_{10} 10\,000$
(ii) $\log_{10}\left(\frac{1}{10\,000}\right)$
(iii) $\log_{10} \sqrt{10}$
(iv) $\log_{10} 1$
(v) $\log_3 81$
(vi) $\log_3 \left(\frac{1}{81}\right)$
(vii) $\log_3 \sqrt{27}$
(viii) $\log_3 \sqrt[4]{3}$
(ix) $\log_4 2$
(x) $\log_5 \left(\frac{1}{125}\right)$

4 Write the following expressions in the form $\log x$ where x is a number.

(i) $\log 5 + \log 2$
(ii) $\log 6 - \log 3$
(iii) $2 \log 6$
(iv) $-\log 7$
(v) $\frac{1}{2} \log 9$
(vi) $\frac{1}{4} \log 16 + \log 2$
(vii) $\log 5 + 3 \log 2 - \log 10$
(viii) $\log 12 - 2 \log 2 - \log 9$
(ix) $\frac{1}{2} \log \sqrt{16} + 2 \log \left(\frac{1}{2}\right)$
(x) $2 \log 4 + \log 9 - \frac{1}{2} \log 144$

5 Express the following in terms of $\log x$.

(i) $\log x^2$
(ii) $\log x^5 - 2 \log x$
(iii) $\log \sqrt{x}$
(iv) $\log x^{\frac{3}{2}} + \log \sqrt[3]{x}$
(v) $3 \log x + \log x^3$
(vi) $\log \left(\sqrt{x}\right)^5$

6 Solve these inequalities.

(i) $2^x < 128$
(ii) $3^x + 5 \geqslant 32$
(iii) $4^x + 6 \geqslant 70$
(iv) $0.6^x < 0.8$
(v) $0.4^x - 0.1 \geqslant 0.3$
(vi) $0.5^x + 0.2 \leqslant 1$
(vii) $2 \leqslant 5^x < 8$
(viii) $1 \leqslant 7^x < 5$
(ix) $|2^x - 4| < 2$
(x) $|5^x - 7| < 4$

7 Express the following as a single logarithm.

$$2\log_{10} x - \log_{10} 7$$

Hence solve

$$2\log_{10} x - \log_{10} 7 = \log_{10} 63.$$

8 Use logarithms to the base 10 to solve the following equations.

(i) $2^x = 1\,000\,000$ (ii) $2^x = 0.001$
(iii) $1.08^x = 2$ (iv) $1.1^x = 100$
(v) $0.99^x = 0.000\,001$

9 A geometric sequence has first term 5 and common ratio 7. The kth term is $28\,824\,005$.

Use logarithms to find the value of k.

10 Find how many terms there are in these geometric sequences.

(i) $-1, 2, -4, 8, \ldots, -16\,777\,216$
(ii) $0.1, 0.3, 0.9, 2.7, \ldots, 4\,304\,672.1$

11 (i) Solve the inequality $|y - 5| < 1$.
   (ii) Hence solve the inequality $|3^x - 5| < 1$, giving 3 significant figures in your answer.

[Cambridge International AS & A Level Mathematics 9709, Paper 2 Q3 November 2007]

12 Given that $x = 4(3^{-y})$, express y in terms of x.

[Cambridge International AS & A Level Mathematics 9709, Paper 3 Q1 June 2006]

13 Using the substitution $u = 3^x$, or otherwise, solve, correct to 3 significant figures, the equation

$$3^x = 2 + 3^{-x}.$$

[Cambridge International AS & A Level Mathematics 9709, Paper 3 Q4 June 2007]

Modelling curves

When you obtain experimental data, you are often hoping to establish a mathematical relationship between the variables in question. Should the data fall on a straight line, you can do this easily because you know that a straight line with gradient m and intercept c has equation $y = mx + c$.

EXAMPLE 2.4

In an experiment the temperature θ (in °C) was measured at different times t (in seconds), in the early stages of a chemical reaction.
The results are shown in the table below.

t	20	40	60	80	100	120
θ	16.3	20.4	24.2	28.5	32.0	36.3

(i) Plot a graph of θ against t.

(ii) What is the relationship between θ and t?

SOLUTION

(i)

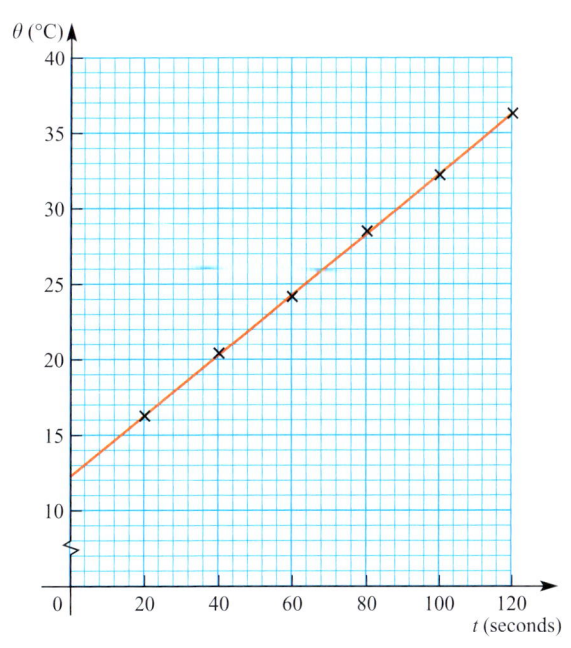

Figure 2.3

(ii) Figure 2.3 shows that the points lie reasonably close to a straight line and so it is possible to estimate its gradient and intercept.

Intercept: $c = 12.3$

Gradient: $m = \dfrac{36.3 - 16.3}{120 - 20} = 0.2$

In this case the equation is not $y = mx + c$ but $\theta = mt + c$, and so is given by

$\theta = 0.2t + 12.3$

It is often the case, however, that your results do not end up lying on a straight line but on a curve, so that this straightforward technique cannot be applied. The appropriate use of logarithms can convert some curved graphs into straight lines. This is the case if the relationship has one of two forms, $y = kx^n$ or $y = ka^x$.

The techniques used in these two cases are illustrated in the following examples. In theory, logarithms to any base may be used, but in practice you would only use those available on your calculator: logarithms to the base 10 and natural logarithms. The base of natural logarithms is a number, 2.718 28…, and is denoted by e. In the next section you will see how this apparently unnatural number arises naturally; for the moment what is important is that you can apply the techniques using base 10.

Relationships of the form $y = kx^n$

EXAMPLE 2.5

A water pipe is going to be laid between two points and an investigation is carried out as to how, for a given pressure difference, the rate of flow R litres per second varies with the diameter of the pipe d cm. The following data are collected.

d	1	2	3	5	10
R	0.02	0.32	1.62	12.53	199.80

It is suspected that the relationship between R and d may be of the form $R = kd^n$ where k is a constant.

(i) Explain how a graph of log d against log R tells you whether this is a good model for the relationship.
(ii) Make out a table of values of $\log_{10} d$ against $\log_{10} R$ and plot these on a graph.
(iii) If appropriate, use your graph to estimate the values of n and k.

SOLUTION

(i) If the relationship is of the form $R = kd^n$, then taking logarithms gives

$$\log R = \log k + \log d^n$$
or $\quad \log R = n \log d + \log k.$

This is in the form $y = mx + c$ as n and log k are constants (so can replace m and c) and log R and log d are variables (so can replace y and x).

$$\begin{array}{ccccc} \log R & = & n \log d & + & \log k \\ \updownarrow & & \updownarrow \; \updownarrow & & \updownarrow \\ y & = & m \; x & + & c \end{array}$$

So $\log R = n \log d + \log k$ is the equation of a straight line.

Consequently if the graph of log R against log d is a straight line, the model $R = kd^n$ is appropriate for the relationship and n is given by the gradient of the graph. The value of k is found from the intercept, log k, of the graph with the vertical axis.

$$\log_{10} k = \text{intercept} \Rightarrow k = 10^{\text{intercept}}$$

(ii) Working to 2 decimal places (you would find it hard to draw the graph to greater accuracy) the logarithmic data are as follows.

$\log_{10} d$	0	0.30	0.48	0.70	1.00
$\log_{10} R$	−1.70	−0.49	0.21	1.10	2.30

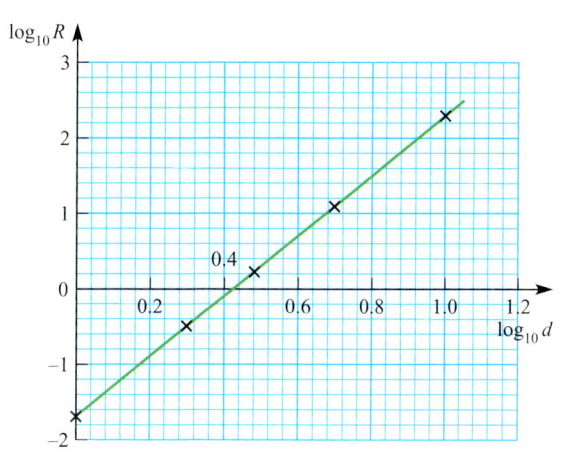

Figure 2.4

(iii) In this case the graph in figure 2.4 is indeed a straight line, with gradient 4 and intercept −1.70, so $n = 4$ and $k = 10^{-1.70} = 0.020$ (to 2 significant figures).

The proposed equation linking R and d is a good model for their relationship, and may be written as:

$$R = 0.02d^4$$

Exponential relationships

EXAMPLE 2.6 The temperature in °C, θ, of a cup of coffee at time t minutes after it is made is recorded as follows.

t	2	4	6	8	10	12
θ	81	70	61	52	45	38

(i) Plot the graph of θ against t.

(ii) Show how it is possible, by drawing a suitable graph, to test whether the relationship between θ and t is of the form $\theta = ka^t$, where k and a are constants.

(iii) Carry out the procedure.

SOLUTION

(i)

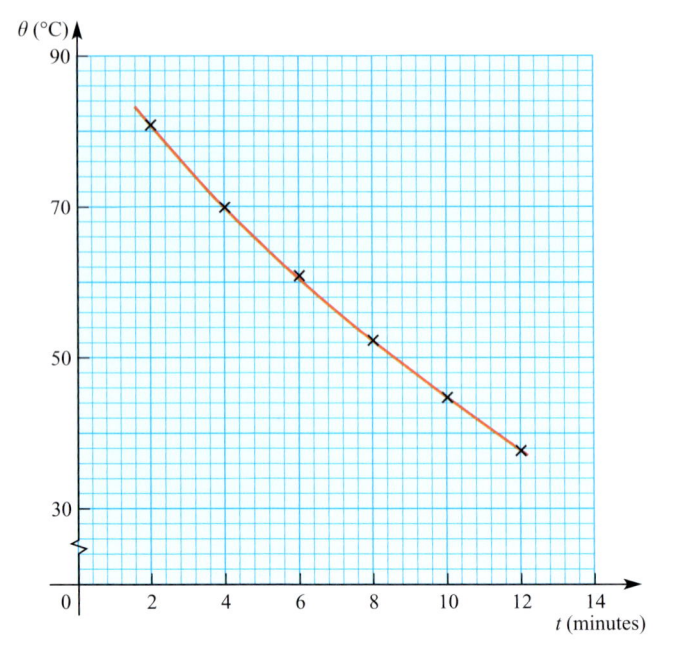

Figure 2.5

(ii) If the relationship is of the form $\theta = ka^t$, taking logarithms of both sides gives

$$\log \theta = \log k + \log a^t$$

or $\log \theta = t \log a + \log k$.

This is in the form $y = mx + c$ as $\log a$ and $\log k$ are constants (so can replace m and c) and $\log \theta$ and t are variable (so can replace y and x).

$$\begin{array}{ccccc} \log \theta & = & \log a \cdot t & + & \log k \\ \updownarrow & & \updownarrow \quad \updownarrow & & \updownarrow \\ y & = & m \quad x & + & c \end{array}$$

So $\log \theta = t \log a + \log k$ is the equation of a straight line.

Consequently if the graph of $\log \theta$ against t is a straight line, the model $\theta = ka^t$ is appropriate for the relationship, and $\log a$ is given by the gradient of the graph. The value of a is therefore found as $a = 10^{\text{gradient}}$. Similarly, the value of k is found from the intercept, $\log_{10} k$, of the line with the vertical axis: $k = 10^{\text{intercept}}$.

(iii) The table gives values of $\log_{10} \theta$ for the given values of t.

t	2	4	6	8	10	12
$\log_{10} \theta$	1.908	1.845	1.785	1.716	1.653	1.580

The graph of $\log_{10} \theta$ against t is as shown in figure 2.6.

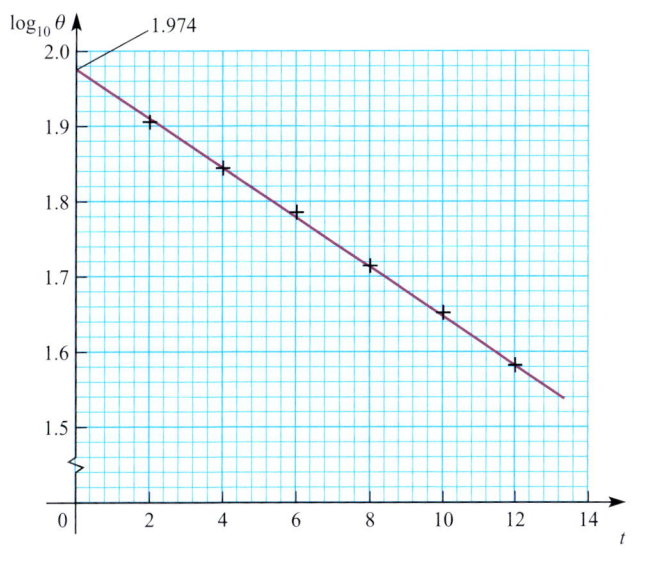

Figure 2.6

The graph is indeed a straight line so the proposed model is appropriate.
The gradient is -0.033 and so $a = 10^{-0.033} = 0.927$.
The intercept is 1.974 and so $k = 10^{1.974} = 94.2$.

The relationship between θ and t is given by:

$$\theta = 94.2 \times 0.927^t$$

Note

Because the base of the exponential function, 0.927, is less than 1, the function's value decreases rather than increases with t.

EXERCISE 2B

1 The planet Saturn has many moons. The table below gives the mean radius of orbit and the time taken to complete one orbit for five of the best-known of them.

Moon	Tethys	Dione	Rhea	Titan	Iapetus
Radius R ($\times 10^5$ km)	2.9	3.8	5.3	12.2	35.6
Period T (days)	1.9	2.7	4.5	15.9	79.3

It is believed that the relationship between R and T is of the form $R = kT^n$.

(i) How can this be tested by plotting log R against log T?
(ii) Make out a table of values of log R and log T and draw the graph.
(iii) Use your graph to estimate the values of k and n.

In 1980 a Voyager spacecraft photographed several previously unknown moons of Saturn. One of these, named 1980 S.27, has a mean orbital radius of 1.4×10^5 km.

(iv) Estimate how many days it takes this moon to orbit Saturn.

2 The table below shows the area, A cm^2, occupied by a patch of mould at time t days since measurements were started.

t	0	1	2	3	4	5
A	0.9	1.3	1.8	2.5	3.5	5.2

It is believed that A may be modelled by a relationship of the form $A = kb^t$.

(i) Show that the model may be written as $\log A = t \log b + \log k$.
(ii) What graph must be plotted to test this model?
(iii) Plot the graph and use it to estimate the values of b and k.
(iv) (a) Estimate the time when the area of mould was 2 cm^2.
  (b) Estimate the area of the mould after 3.5 days.
(v) How is this sort of growth pattern described?

3 The inhabitants of an island are worried about the rate of deforestation taking place. A research worker uses records over the last 200 years to estimate the number of trees at different dates.

It is suggested that the number of trees N has been decreasing exponentially with the number of years, t, since 1930, so that N may be modelled by the equation

$$N = ka^t$$

where k and a are constants.

(i) Show that the model may be written as $\log N = t \log a + \log k$.

The diagram shows the graph of $\log N$ against t.

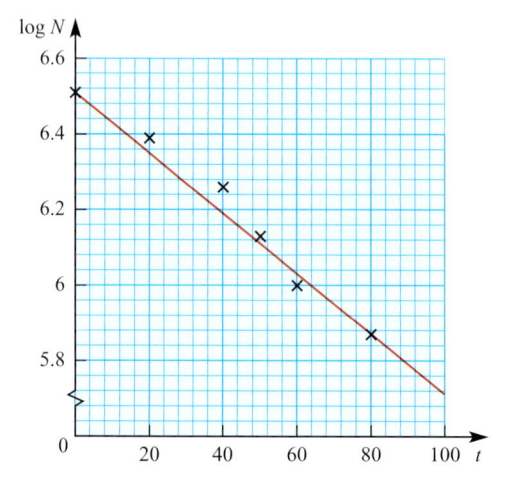

(ii) Estimate the values of k and a.
What is the significance of k?

4 The time after a train leaves a station is recorded in minutes as t and the distance that it has travelled in metres as s. It is suggested that the relationship between s and t is of the form $s = kt^n$ where k and n are constants.

  (i) Show that the graph of $\log s$ against $\log t$ produces a straight line.

  The diagram shows the graph of $\log s$ against $\log t$.

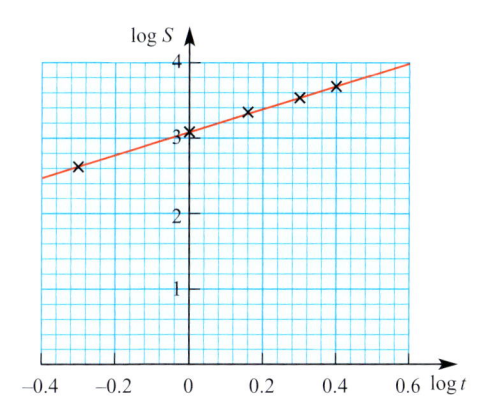

  (ii) Estimate the values of k and n.
  (iii) Estimate how far the train travelled in its first 100 seconds.
  (iv) Explain why you would be wrong to use your results to estimate the distance the train has travelled after 10 minutes.

5 The variables t and A satisfy the equation $A = kb^t$, where b and k are constants.

  (i) Show that the graph of $\log A$ against t produces a straight line.

  The graph of $\log A$ against t passes through the points $(0, 0.2)$ and $(4, 0.75)$.

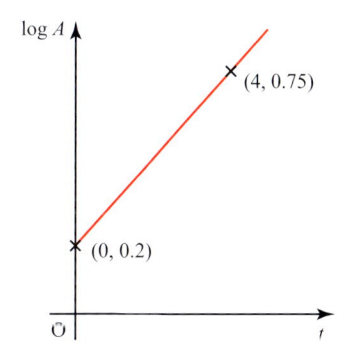

  (ii) Find the values of b and k.

6 All but one of the following pairs of readings satisfy, to 3 significant figures, a formula of the type $y = A \times x^B$.

x	1.51	2.13	3.50	4.62	5.07	7.21
y	2.09	2.75	4.09	5.10	6.21	7.28

Find the values of A and B, explaining your method. If the values of x are correct, state which value of y appears to be wrong and estimate what the value should be.

[MEI]

7 An experimenter takes observations of a quantity *y* for various values of a variable *x*. He wishes to test whether these observations conform to a formula $y = A \times x^B$ and, if so, to find the values of the constants *A* and *B*.

Take logarithms of both sides of the formula. Use the result to explain what he should do, what will happen if there is no relationship, and if there is one, how to find *A* and *B*.

Carry this out accurately on graph paper for the observations in the table, and record clearly the resulting formula if there is one.

x	4	7	10	13	20
y	3	3.97	4.74	5.41	6.71

8 It is believed that the relationship between the variables *x* and *y* is of the form $y = Ax^n$. In an experiment the data in the table are obtained.

x	3	6	10	15	20
y	10.4	29.4	63.2	116.2	178.19

In order to estimate the constants *A* and *n*, $\log_{10} y$ is plotted against $\log_{10} x$.

(i) Draw the graph of $\log_{10} y$ against $\log_{10} x$.
(ii) Explain and justify how the shape of your graph enables you to decide whether the relationship is indeed of the form $y = Ax^n$.
(iii) Estimate the values of *A* and *n*.

[MEI]

9 In a spectacular experiment on cell growth the following data were obtained, where *N* is the number of cells at a time *t* minutes after the start of the growth.

t	1.5	2.7	3.4	8.1	10
N	9	19	32	820	3100

At *t* = 10 a chemical was introduced which killed off the culture.

The relationship between *N* and *t* was thought to be modelled by $N = ab^t$, where *a* and *b* are constants.

(i) Show that the relationship is equivalent to $\log N = t \log b + \log a$.
(ii) Plot the values of log *N* against *t* and say how they confirm the supposition that the relationship is of the form $N = ab^t$.
(iii) Find the values of *a* and *b*.
(iv) If the growth had not been stopped at *t* = 10 and had continued according to your model, how many cells would there have been after 20 minutes?

[MEI]

10 It is believed that two quantities, z and d, are connected by a relationship of the form $z = kd^n$, where k and n are constants, provided that d does not exceed some fixed (but unknown) value, D.

An experiment produced the following data.

d	780	810	870	930	990	1050	1110	1170
z	2.1	2.6	3.2	4.0	4.8	5.6	5.9	6.1

(i) Explain why, if $z = kd^n$, then plotting $\log_{10} z$ against $\log_{10} d$ should produce a straight-line graph.

(ii) Draw up a table and plot the values of $\log_{10} z$ against $\log_{10} d$.

(iii) Use these points to suggest a value for D.

(iv) It is known that, for $d < D$, n is a whole number.
Use your graph to find the value of n.
Show also that $k \approx 5 \times 10^{-9}$.

(v) Use your value for n and the estimate $k = 5 \times 10^{-9}$ to find the value of d for which $z = 3.0$.

[MEI]

11 The variables x and y satisfy the relation $3^y = 4^{x+2}$.

(i) By taking logarithms, show that the graph of y against x is a straight line. Find the exact value of the gradient of this line.

(ii) Calculate the x co-ordinate of the point of intersection of this line with the line $y = 2x$, giving your answer correct to 2 decimal places.

[Cambridge International AS & A Level Mathematics 9709, Paper 2 Q2 June 2007]

The natural logarithm function

The shaded region in figure 2.7 is bounded by the x axis, the lines $x = 1$ and $x = 3$, and the curve $y = \dfrac{1}{x}$. The area of this region may be represented by $\int_1^3 \dfrac{1}{x}\, dx$.

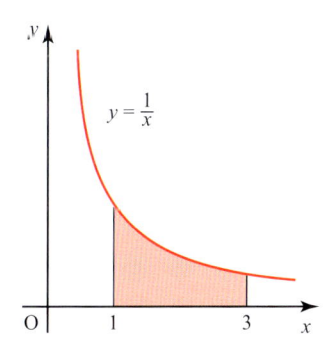

Figure 2.7

? Explain why you cannot apply the rule

$$\int kx^n \, dx = \frac{kx^{n+1}}{n+1} + c$$

to this integral.

However, the area in the diagram clearly has a definite value, and so we need to find ways to express and calculate it.

INVESTIGATION

Estimate, using numerical integration (for example by dividing the area up into a number of strips), the areas represented by these integrals.

(i) $\int_1^3 \frac{1}{x} \, dx$ (ii) $\int_1^2 \frac{1}{x} \, dx$ (iii) $\int_1^6 \frac{1}{x} \, dx$

What relationship can you see between your answers?

The area under the curve $y = \frac{1}{x}$ between $x = 1$ and $x = a$, that is $\int_1^a \frac{1}{x} \, dx$, depends on the value a. For every value of a (greater than 1) there is a definite value of the area. Consequently, the area is a function of a.

To investigate this function you need to give it a name, say L, so that L(a) is the area from 1 to a and L(x) is the area from 1 to x. Then look at the properties of L(x) to see if its behaviour is like that of any other function with which you are familiar.

The investigation you have just done should have suggested to you that

$$\int_1^3 \frac{1}{x} \, dx + \int_1^2 \frac{1}{x} \, dx = \int_1^6 \frac{1}{x} \, dx.$$

This can now be written as

L(3) + L(2) = L(6).

This suggests a possible law, that

L(a) + L(b) = L(ab).

At this stage this is just a conjecture, based on one particular example. To prove it, you need to take the general case and this is done in the activity below. (At first reading you may prefer to leave the activity, accepting that the result can be proved.)

ACTIVITY 2.2 Prove that $L(a) + L(b) = L(ab)$, by following the steps below.

(i) Explain, with the aid of a diagram, why

$$L(a) + \int_a^{ab} \frac{1}{x}\,dx = L(ab).$$

(ii) Now call $x = az$, so that dx can be replaced by $a\,dz$. Show that

$$\int_a^{ab} \frac{1}{x}\,dx = \int_1^b \frac{1}{z}\,dz.$$

> Notice that the limits of the left-hand integral, ab and a, are values for x but those for the right-hand integral, b and 1, are values for z. So, to find the new limits for the right-hand integral, you should find z when $x = a$ (the lower limit) and when $x = ab$ (the upper limit). Remember $az = x$.

Explain why $\int_1^b \frac{1}{z}\,dz = L(b)$.

(iii) Use the results from parts (i) and (ii) to show that

$$L(a) + L(b) = L(ab).$$

What function has this property? For all logarithms

$$\log(a) + \log(b) = \log(ab).$$

Could it be that this is a logarithmic function?

ACTIVITY 2.3 Satisfy yourself that the function has the following properties of logarithms.

(i) $L(1) = 0$

(ii) $L(a) - L(b) = L\left(\dfrac{a}{b}\right)$

(iii) $L(a^n) = nL(a)$

The base of the logarithm function L(x)

Having accepted that $L(x)$ is indeed a logarithmic function (for $x > 0$), the remaining problem is to find the base of the logarithm. By convention this is denoted by the letter e. A further property of logarithms is that for any base p

$$\log_p p = 1 \quad (p > 1).$$

So to find the base e, you need to find the point such that the area $L(e)$ under the graph is 1. See figure 2.8.

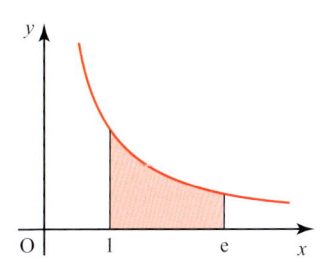

Figure 2.8

You have already estimated the value of $L(2)$ to be about 0.7 and that of $L(3)$ to be about 1.1 so the value of e is between 2 and 3.

ACTIVITY 2.4

You will need a calculator with an area-finding facility, or other suitable technology, to do this. If you do not have this, read on.

Use the fact that $\int_1^e \frac{1}{x} dx = 1$ to find the value of e, knowing that it lies between 2 and 3, to 2 decimal places.

The value of e is given to 9 decimal places in the key points on page 50. Like π, e is a number which occurs naturally within mathematics. It is irrational: when written as a decimal, it never terminates and has no recurring pattern.

The function L(x) is thus the logarithm of x to the base e, $\log_e x$. This is often called the natural logarithm of x, and written as ln x.

Values of x between 0 and 1

So far it has been assumed that the domain of the function ln x is the real numbers greater than 1 ($x \in \mathbb{R}, x > 1$). However, the domain of ln x also includes values of x between 0 and 1. As an example of a value of x between 0 and 1, look at $\ln \frac{1}{2}$.

Since $\quad \ln\left(\frac{a}{b}\right) = \ln a - \ln b$

$\Rightarrow \quad \ln\left(\frac{1}{2}\right) = \ln 1 - \ln 2 = -\ln 2$ (since ln 1 = 0)

In the same way, you can show that for any value of x between 0 and 1, the value of ln x is negative.

When the value of x is very close to zero, the value of ln x is a large negative number.

$$\ln\left(\frac{1}{1000}\right) = -\ln 1000 = -6.9$$

$$\ln\left(\frac{1}{1000000}\right) = -\ln 1\,000\,000 = -13.8$$

So as $x \to 0$, $\ln x \to -\infty$ (for positive values of x).

The graph of the natural logarithm function

The graph of the natural logarithm function (shown in figure 2.9) has the characteristic shape of all logarithmic functions and like other such functions it is only defined for $x > 0$. The value of ln x increases without limit, but ever more slowly: it has been described as 'the slowest way to get to infinity'.

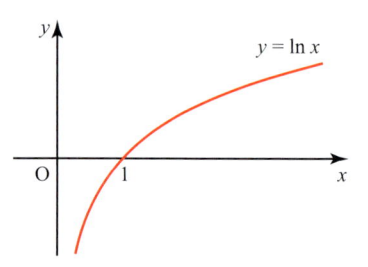

Figure 2.9

Historical note | Logarithms were discovered independently by John Napier (1550–1617), who lived at Merchiston Castle in Edinburgh, and Jolst Bürgi (1552–1632) from Switzerland. It is generally believed that Napier had the idea first, and so he is credited with their discovery. Natural logarithms are also called Naperian logarithms but there is no basis for this since Napier's logarithms were definitely not the same as natural logarithms. Napier was deeply involved in the political and religious events of his day and mathematics and science were little more than hobbies for him. He was a man of remarkable ingenuity and imagination and also drew plans for war chariots that look very like modern tanks, and for submarines.

The exponential function

Making x the subject of $y = \ln x$, using the theory of logarithms you obtain $x = e^y$.

Interchanging x and y, which has the effect of reflecting the graph in the line $y = x$, gives the *exponential function* $y = e^x$.

The graphs of the natural logarithm function and its inverse are shown in figure 2.10.

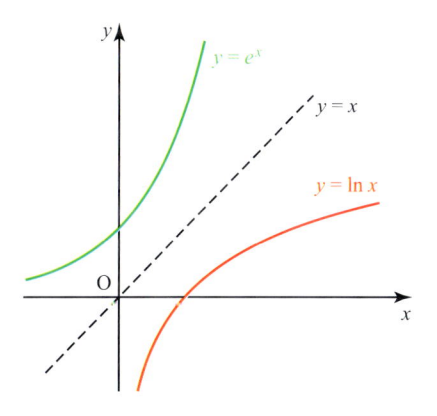

Figure 2.10

You saw in *Pure Mathematics 1* Chapter 4 that reflecting in the line $y = x$ gives an *inverse* function, so it follows that e^x and $\ln x$ are each the inverse of the other.

Notice that $e^{\ln x} = x$, using the definition of logarithms, and $\ln(e^x) = x \ln e = x$.

Although the function e^x is called *the* exponential function, in fact any function of the form a^x is exponential. Figure 2.11 shows several exponential curves.

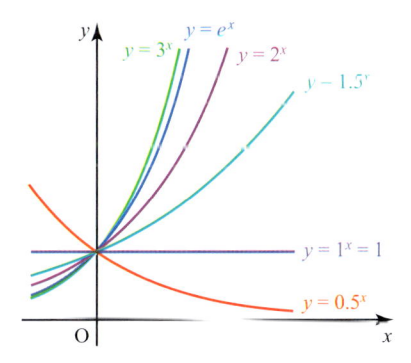

Figure 2.11

The exponential function $y = e^x$ increases at an ever-increasing rate. This is described as *exponential growth*.

By contrast, the graph of $y = e^{-x}$, shown in figure 2.12, approaches the x axis ever more slowly as x increases. This is called *exponential decay*.

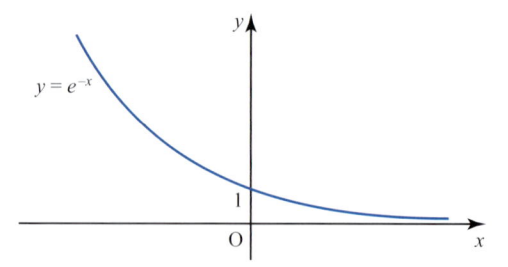

Figure 2.12

You will meet e^x and $\ln x$ again later in this book. In Chapter 4 you learn how to differentiate these functions and in Chapter 5 you learn how to integrate them. In this secion you focus on practical applications which require you to use the [ln] key on your calculator.

EXAMPLE 2.7

The number, N, of insects in a colony is given by $N = 2000\, e^{0.1t}$ where t is the number of days after observations have begun.

(i) Sketch the graph of N against t.
(ii) What is the population of the colony after 20 days?
(iii) How long does it take the colony to reach a population of 10 000?

SOLUTION

(i)

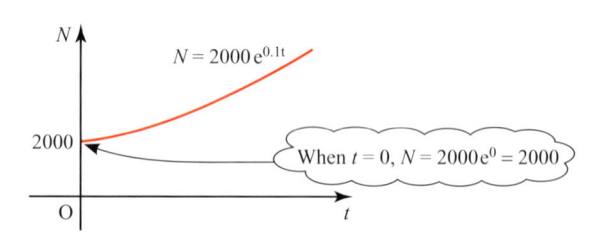

When $t = 0$, $N = 2000 e^0 = 2000$

Figure 2.13

(ii) When $t = 20$, $\quad N = 2000\, e^{0.1 \times 20} = 14\,778$

The population is 14 778 insects.

(iii) When $N = 10\,000$, $\quad 10\,000 = 2000\, e^{0.1t}$
$$5 = e^{0.1t}$$

Taking natural logarithms of both sides,

$$\ln 5 = \ln(e^{0.1t})$$
$$\ln 5 = 0.1t$$

Remember $\ln(e^x) = x$.

and so
$$t = 10 \ln 5$$
$$t = 16.09\ldots$$

It takes just over 16 days for the population to reach 10 000.

EXAMPLE 2.8

The radioactive mass, M grams in a lump of material is given by $M = 25e^{-0.0012t}$ where t is the time in seconds since the first observation.

(i) Sketch the graph of M against t.
(ii) What is the initial size of the mass?
(iii) What is the mass after 1 hour?
(iv) The half-life of a radioactive substance is the time it takes to decay to half of its mass. What is the half-life of this material?

SOLUTION

(i)

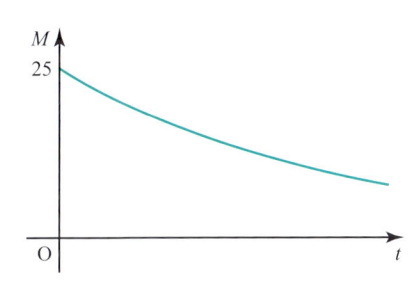

Figure 2.14

(ii) When $t = 0$, $\quad M = 25e^0$
$\qquad\qquad\qquad M = 25$

The initial mass is 25 g.

(iii) After 1 hour, $\quad t = 3600$
$\qquad\qquad\qquad M = 25e^{-0.0012 \times 3600}$
$\qquad\qquad\qquad M = 0.3324...$

The mass after 1 hour is 0.33 g (to 2 decimal places).

(iv) The initial mass is 25 g, so after one half-life,

$$M = \tfrac{1}{2} \times 25 = 12.5 \,\text{g}$$

At this point the value of t is given by

$\qquad\qquad 12.5 = 25e^{-0.0012t}$
$\Rightarrow \qquad 0.5 = e^{-0.0012t}$

Taking logarithms of both sides:

$\qquad\qquad \ln 0.5 = \ln e^{-0.0012t}$
$\qquad\qquad \ln 0.5 = -0.0012t$
$\Rightarrow \qquad t = \dfrac{\ln 0.5}{-0.0012}$
$\qquad\qquad t = 577.6$ (to 1 decimal place).

The half-life is 577.6 seconds. (This is just under 10 minutes, so the substance is highly radioactive.)

EXAMPLE 2.9 Make p the subject of $\ln(p) - \ln(1-p) = t$.

SOLUTION

$$\ln\left(\frac{p}{1-p}\right) = t$$

Using $\log a - \log b = \log\left(\frac{a}{b}\right)$

Writing both sides as powers of e gives

$$e^{\ln\left(\frac{p}{1-p}\right)} = e^t$$

Remember $e^{\ln x} = x$

$$\Rightarrow \quad \frac{p}{1-p} = e^t$$

$$p = e^t(1-p)$$
$$p = e^t - pe^t$$
$$p + pe^t = e^t$$
$$p(1 + e^t) = e^t$$
$$p = \frac{e^t}{1 + e^t}$$

EXAMPLE 2.10 Solve these equations.

(i) $\ln(x-4) = \ln x - 4$

(ii) $e^{2x} + e^x = 6$

SOLUTION

(i) $\quad \ln(x-4) = \ln x - 4$

$\Rightarrow \quad x - 4 = e^{\ln x - 4}$

$\quad x - 4 = e^{\ln x} e^{-4}$

$\quad x - 4 = xe^{-4}$

Rearrange to get all the x terms on one side:

$$x - xe^{-4} = 4$$
$$x(1 - e^{-4}) = 4$$
$$x = \frac{4}{1 - e^{-4}}$$

So $\quad x = 4.07$

(ii) $e^{2x} + e^x = 6$ is a quadratic equation in e^x.

Substituting $u = e^x$:
$$u^2 + u = 6$$

So $\quad u^2 + u - 6 = 0$

Factorising: $(u-2)(u+3) = 0$

So $u = 2$ or $u = -3$.

Since $u = e^x$ then $e^x = 2$ or $e^x = -3$.

$e^x = -3$ has no solution.

$e^x = 2 \Rightarrow x = \ln 2$

So $\quad x = 0.693$

EXERCISE 2C

1 Make x the subject of $\ln x - \ln x_0 = kt$.

2 Make t the subject of $s = s_0 e^{-kt}$.

3 Make p the subject of $\ln p = -0.02t$.

4 Make x the subject of $y - 5 = (y_0 - 5)e^x$.

5 Solve these equations.

(i) $\ln(3-x) = 4 + \ln x$
(ii) $\ln(x+5) = 5 + \ln x$
(iii) $\ln(2-x) = 2 + \ln x$
(iv) $e^x = \dfrac{4}{e^x}$
(v) $e^{2x} - 8e^x + 16 = 0$
(vi) $e^{2x} + e^x = 12$

6 A colony of humans settles on a previously uninhabited planet. After t years, their population, P, is given by $P = 100e^{0.05t}$.

(i) Sketch the graph of P against t.
(ii) How many settlers land on the planet initially?
(iii) What is the population after 50 years?
(iv) How long does it take the population to reach 1 million?

7 The height h metres of a species of pine tree t years after planting is modelled by the equation $h = 20 - 19 \times 0.9^t$.

   (i) What is the height of the trees when they are planted?
   (ii) Calculate the height of the trees after 2 years, and the time taken for the height to reach 10 metres.

The relationship between the market value $\$y$ of the timber from the tree and the height h metres of the tree is modelled by the equation $y = ah^b$, where a and b are constants.

The diagram shows the graph of $\ln y$ plotted against $\ln h$.

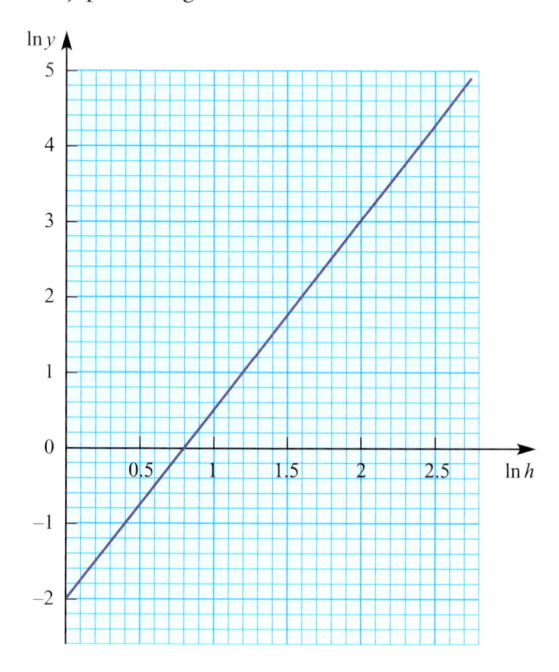

   (iii) Use the graph to calculate the values of a and b.
   (iv) Calculate how long it takes to grow trees worth $100.

[MEI, *adapted*]

8 It is given that $\ln(y + 5) - \ln y = 2 \ln x$. Express y in terms of x, in a form not involving logarithms.

[Cambridge International AS & A Level Mathematics 9709, Paper 2 Q2 November 2009]

9 Given that $(1.25)^x = (2.5)^y$, use logarithms to find the value of $\dfrac{x}{y}$ correct to 3 significant figures.

[Cambridge International AS & A Level Mathematics 9709, Paper 2 Q1 June 2009]

10 Solve, correct to 3 significant figures, the equation

$$e^x + e^{2x} = e^{3x}.$$

[Cambridge International AS & A Level Mathematics 9709, Paper 3 Q2 June 2008]

11 The variables x and y satisfy the equation $y = A(b^{-x})$, where A and b are constants. The graph of $\ln y$ against x is a straight line passing through the points $(0, 1.3)$ and $(1.6, 0.9)$, as shown in the diagram. Find the values of A and b, correct to 2 decimal places.

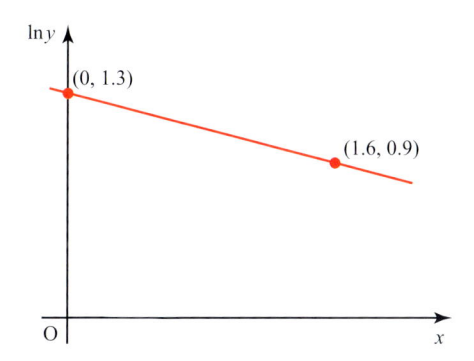

[Cambridge International AS & A Level Mathematics 9709, Paper 2 Q3 November 2008]

12 Solve the equation $\ln(2 + e^{-x}) = 2$, giving your answer correct to 2 decimal places.

[Cambridge International AS & A Level Mathematics 9709, Paper 3 Q1 June 2009]

13 Two variable quantities x and y are related by the equation $y = Ax^n$, where A and n are constants. The diagram shows the result of plotting $\ln y$ against $\ln x$ for four pairs of values of x and y. Use the diagram to estimate the values of A and n.

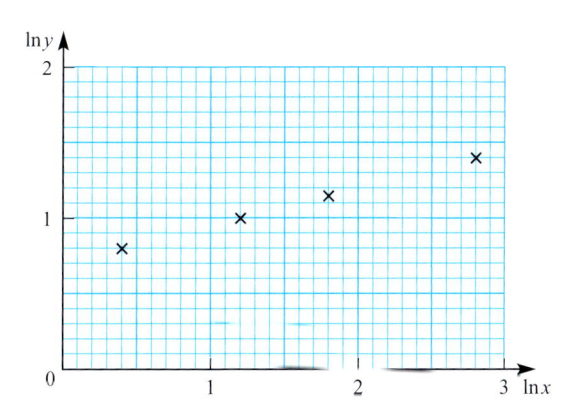

[Cambridge International AS & A Level Mathematics 9709, Paper 3 Q2 November 2005]

KEY POINTS

1. A function of the form a^x is described as exponential.

2. $y = \log_a x \Leftrightarrow a^y = x$.

3. Logarithms to any base

   Multiplication: $\log xy = \log x + \log y$

   Division: $\log\left(\dfrac{x}{y}\right) = \log x - \log y$

   Logarithm of 1: $\log 1 = 0$

   Powers: $\log x^n = n \log x$

   Reciprocals: $\log\left(\dfrac{1}{y}\right) = -\log y$

   Roots: $\log \sqrt[n]{x} = \dfrac{1}{n}\log x$

   Logarithm to its own base: $\log_a a = 1$

4. Logarithms may be used to discover the relationship between the variables in two types of situation.

   $y = kx^n \Leftrightarrow \log y = \log k + n\log x$

   Plot $\log y$ against $\log x$: this relationship gives a straight line where n is the gradient and $\log k$ is the intercept.

   $y = ka^x \Leftrightarrow \log y = \log k + x\log a$

   Plot $\log y$ against x: this relationship gives a straight line where $\log a$ is the gradient and $\log k$ is the intercept.

5. $\int \dfrac{1}{x}\,dx = \log_e |x| + c$.

6. $\log_e x$ is called the natural logarithm of x and denoted by $\ln x$.

7. $e = 2.718\,281\,828\,4\ldots$ is the base of natural logarithms.

8. e^x and $\ln x$ are inverse functions: $e^{\ln x} = x$ and $\ln(e^x) = x$.

3 Trigonometry

**Music, when soft voices die,
Vibrates in the memory –**

P.B. Shelley

? Both of these photographs show forms of waves. In each case, estimate the wavelength and the amplitude in metres (see figure 3.1).

Use your measurements to suggest, for each curve, values of a and b which would make $y = a \sin bx$ a suitable model for the curve.

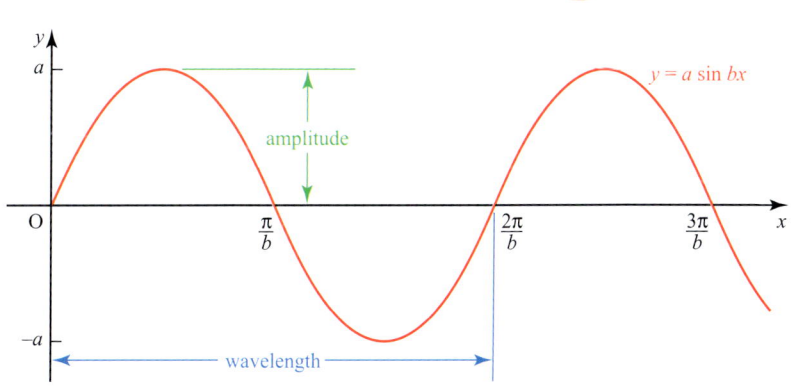

Figure 3.1

Reciprocal trigonometrical functions

As well as the three main trigonometrical functions, there are three more which are commonly used. These are their reciprocals – cosecant (cosec), secant (sec) and cotangent (cot), defined by

$$\operatorname{cosec}\theta = \frac{1}{\sin\theta}; \qquad \sec\theta = \frac{1}{\cos\theta}; \qquad \cot\theta = \frac{1}{\tan\theta}\left(=\frac{\cos\theta}{\sin\theta}\right).$$

Each of these is undefined for certain values of θ. For example, $\operatorname{cosec}\theta$ is undefined for $\theta = 0°, 180°, 360°, \ldots$ since $\sin\theta$ is zero for these values of θ.

Figure 3.2 shows the graphs of these functions. Notice how all three of the functions have asymptotes at intervals of 180°. Each of the graphs shows one of the main trigonometrical functions as a red line and the related reciprocal function as a blue line.

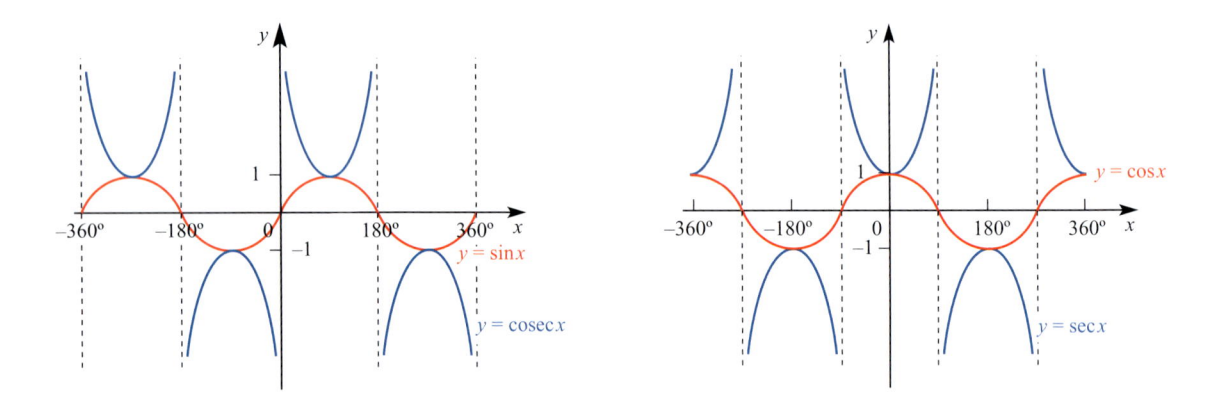

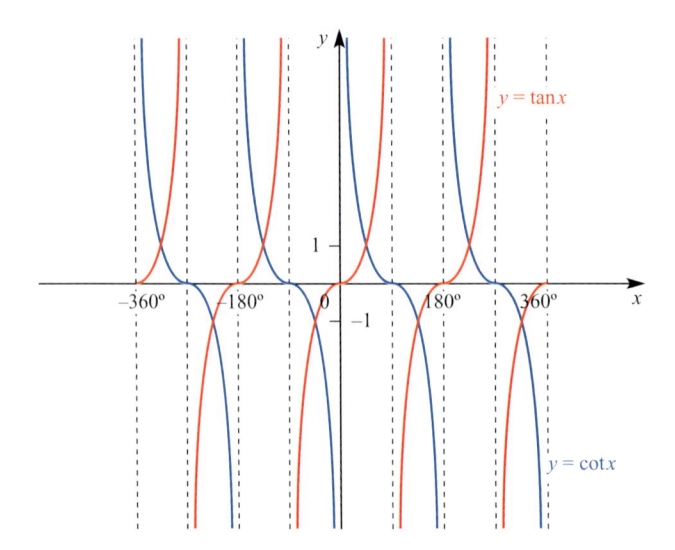

Figure 3.2

Using the definitions of the reciprocal functions two alternative trigonometrical forms of Pythagoras' theorem can be obtained.

(i) $\sin^2\theta + \cos^2\theta \equiv 1$

Dividing both sides by $\cos^2\theta$: $\dfrac{\sin^2\theta}{\cos^2\theta} + \dfrac{\cos^2\theta}{\cos^2\theta} \equiv \dfrac{1}{\cos^2\theta}$

$\Rightarrow \tan^2\theta + 1 \equiv \sec^2\theta.$

This identity is sometimes used in mechanics.

(ii) $\sin^2\theta + \cos^2\theta \equiv 1$

Dividing both sides by $\sin^2\theta$: $\dfrac{\sin^2\theta}{\sin^2\theta} + \dfrac{\cos^2\theta}{\sin^2\theta} \equiv \dfrac{1}{\sin^2\theta}$

$\Rightarrow 1 + \cot^2\theta \equiv \text{cosec}^2\theta.$

Questions concerning reciprocal functions are usually most easily solved by considering the related function, as in the following examples.

EXAMPLE 3.1   Find cosec 120° leaving your answer in surd form.

SOLUTION

$\text{cosec}\,120° = \dfrac{1}{\sin 120°}$

$= 1 \div \dfrac{\sqrt{3}}{2}$

$= \dfrac{2}{\sqrt{3}}$

EXAMPLE 3.2   Find values of θ in the interval $0° \leq \theta \leq 360°$ for which $\sec^2\theta = 4 + 2\tan\theta$.

SOLUTION

First you need to obtain an equation containing only one trigonometrical function.

$\sec^2\theta = 4 + 2\tan\theta$

$\Rightarrow \tan^2\theta + 1 = 4 + 2\tan\theta$

$\Rightarrow \tan^2\theta - 2\tan\theta - 3 = 0$

$\Rightarrow (\tan\theta - 3)(\tan\theta + 1) = 0$

$\Rightarrow \tan\theta = 3 \text{ or } \tan\theta = -1$

$\tan\theta = 3 \quad \Rightarrow \quad \theta = 71.6°$ (calculator)

or $\qquad \theta = 71.6° + 180° = 251.6°$ (see figure 3.3, overleaf)

$\tan\theta = -1 \Rightarrow \theta = -45°$ (not in the required range)

or $\theta = -45° + 180° = 135°$ (see figure 3.3)

or $\theta = 135° + 180° = 315°$

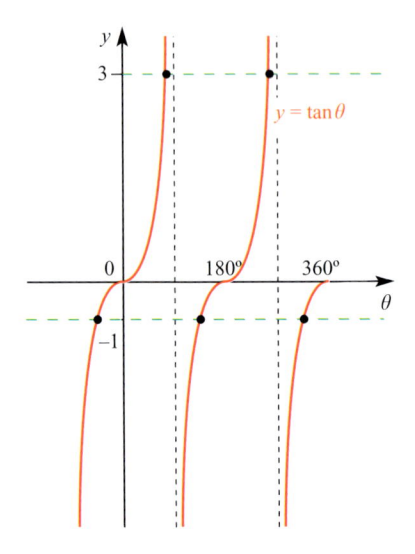

Figure 3.3

The values of θ are $71.6°, 135°, 251.6°, 315°$.

EXERCISE 3A

1 Solve the following equations for $0° \leq x \leq 360°$.

(i) $\operatorname{cosec} x = 1$ (ii) $\sec x = 2$ (iii) $\cot x = 4$

(iv) $\sec x = -3$ (v) $\cot x = -1$ (vi) $\operatorname{cosec} x = -2$

2 Find the following giving your answers as fractions or in surd form. You should not need your calculator.

(i) $\cot 135°$ (ii) $\sec 150°$ (iii) $\operatorname{cosec} 240°$

(iv) $\sec 210°$ (v) $\cot 270°$ (vi) $\operatorname{cosec} 225°$

3 In triangle ABC, angle $A = 90°$ and $\sec B = 2$.

(i) Find the angles B and C.

(ii) Find $\tan B$.

(iii) Show that $1 + \tan^2 B = \sec^2 B$.

4 In triangle LMN, angle $M = 90°$ and $\cot N = 1$.

(i) Find the angles L and N.

(ii) Find $\sec L$, $\operatorname{cosec} L$, and $\tan L$.

(iii) Show that $1 + \tan^2 L = \sec^2 L$.

5   Malini is 1.5 m tall.
    At 8 pm one evening her shadow is 6 m long.
    Given that the angle of elevation of the sun at that moment is α

    (i) show that $\cot \alpha = 4$
    (ii) find α.

6   (i)   For what values of α, where $0° \leq \alpha \leq 360°$, are $\sec \alpha$, $\csc \alpha$ and $\cot \alpha$ all positive?
    (ii) Are there any values of α for which $\sec \alpha$, $\csc \alpha$ and $\cot \alpha$ are all negative? Explain your answer.
    (iii) Are there any values of α for which $\sec \alpha$, $\csc \alpha$ and $\cot \alpha$ are all equal? Explain your answer.

7   Solve the following equations for $0° \leq x \leq 360°$.

    (i) $\cos x = \sec x$
    (ii) $\csc x = \sec x$
    (iii) $2 \sin x = 3 \cot x$
    (iv) $\csc^2 x + \cot^2 x = 2$
    (v) $3 \sec^2 x - 10 \tan x = 0$
    (vi) $1 + \cot^2 x = 2 \tan^2 x$

Compound-angle formulae

The photographs at the start of this chapter show just two of the countless examples of waves and oscillations that are part of the world around us.

Because such phenomena are modelled by trigonometrical (and especially sine and cosine) functions, trigonometry has an importance in mathematics far beyond its origins in right-angled triangles.

ACTIVITY 3.1

Find an acute angle θ so that $\sin(\theta + 60°) = \cos(\theta - 60°)$.

Hint: Try drawing graphs and searching for a numerical solution.

You should be able to find the solution using either of these methods, but replacing 60° by, for example, 35° would make both of these methods rather tedious. In this chapter you will meet some formulae which help you to solve such equations more efficiently.

⚠ It is tempting to think that $\sin(\theta + 60°)$ should equal $\sin \theta + \sin 60°$, but this is not so, as you can see by substituting a numerical value of θ. For example, putting $\theta = 30°$ gives $\sin(\theta + 60°) = 1$, but $\sin \theta + \sin 60° \approx 1.366$.

To find an expression for sin(θ + 60°), you would use the *compound-angle formula*

$$\sin(\theta + \phi) = \sin\theta\cos\phi + \cos\theta\sin\phi.$$

This is proved below in the case when θ and ϕ are acute angles. It is, however, true for all values of the angles. It is an *identity*.

P As you work through this proof make a list of all the results you are assuming.

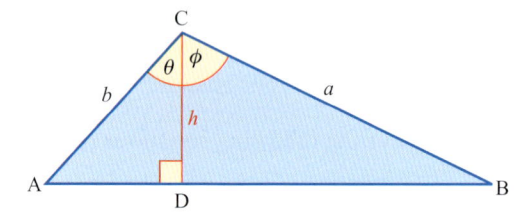

Figure 3.4

Using the trigonometrical formula for the area of a triangle (see figure 3.4):

$$\text{area ABC} = \text{area ADC} + \text{area DBC}$$

$$\tfrac{1}{2}ab\sin(\theta + \phi) = \tfrac{1}{2}bh\sin\theta + \tfrac{1}{2}ah\sin\phi$$

($h = a\cos\phi$ from $\triangle$DBC)
($h = b\cos\theta$ from $\triangle$ADC)

$$\Rightarrow \quad ab\sin(\theta + \phi) = ab\sin\theta\cos\phi + ab\cos\theta\sin\phi$$

which gives

$$\sin(\theta + \phi) = \sin\theta\cos\phi + \cos\theta\sin\phi \qquad \text{①}$$

This is the first of the compound-angle formulae (or expansions), and it can be used to prove several more. These are true for all values of θ and ϕ.

Replacing ϕ by $-\phi$ in ① gives

$$\sin(\theta - \phi) = \sin\theta\cos(-\phi) + \cos\theta\sin(-\phi)$$

($\cos(-\phi) = \cos\phi$) ($\sin(-\phi) = -\sin\phi$)

$$\Rightarrow \quad \sin(\theta - \phi) = \sin\theta\cos\phi - \cos\theta\sin\phi \qquad \text{②}$$

ACTIVITY 3.2 Derive the rest of these formulae.

(i) To find an expansion for $\cos(\theta - \phi)$ replace θ by $(90° - \theta)$ in the expansion of $\sin(\theta + \phi)$.

Hint: $\sin(90° - \theta) = \cos\theta$ and $\cos(90° - \theta) = \sin\theta$

(ii) To find an expansion for $\cos(\theta + \phi)$ replace ϕ by $(-\phi)$ in the expansion of $\cos(\theta - \phi)$.

(iii) To find an expansion for $\tan(\theta + \phi)$, write $\tan(\theta + \phi) = \dfrac{\sin(\theta + \phi)}{\cos(\theta + \phi)}$.

Hint: After using the expansions of $\sin(\theta + \phi)$ and $\cos(\theta + \phi)$, divide the numerator and the denominator of the resulting fraction by $\cos\theta\cos\phi$ to give an expansion in terms of $\tan\theta$ and $\tan\phi$.

(iv) To find an expansion for $\tan(\theta - \phi)$ in terms of $\tan\theta$ and $\tan\phi$, replace ϕ by $(-\phi)$ in the expansion of $\tan(\theta + \phi)$.

? Are your results valid for all values of θ and ϕ?

Test your results with $\theta = 60°$, $\phi = 30°$.

The four results obtained in Activity 3.2, together with the two previous results, form the set of compound-angle formulae.

$$\sin(\theta + \phi) = \sin\theta\cos\phi + \cos\theta\sin\phi$$

$$\sin(\theta - \phi) = \sin\theta\cos\phi - \cos\theta\sin\phi$$

$$\cos(\theta + \phi) = \cos\theta\cos\phi - \sin\theta\sin\phi$$

$$\cos(\theta - \phi) = \cos\theta\cos\phi + \sin\theta\sin\phi$$

$$\tan(\theta + \phi) = \frac{\tan\theta + \tan\phi}{1 - \tan\theta\tan\phi} \quad (\theta + \phi) \neq 90°, 270°, \ldots$$

$$\tan(\theta - \phi) = \frac{\tan\theta - \tan\phi}{1 + \tan\theta\tan\phi} \quad (\theta - \phi) \neq 90°, 270°, \ldots$$

You are now in a position to solve the earlier problem more easily. To find an acute angle θ such that $\sin(\theta + 60°) = \cos(\theta - 60°)$, you expand each side using the compound-angle formulae.

$$\sin(\theta + 60°) = \sin\theta\cos 60° + \cos\theta\sin 60°$$

$$= \tfrac{1}{2}\sin\theta + \tfrac{\sqrt{3}}{2}\cos\theta \qquad ①$$

$$\cos(\theta - 60°) = \cos\theta\cos 60° + \sin\theta\sin 60°$$

$$= \tfrac{1}{2}\cos\theta + \tfrac{\sqrt{3}}{2}\sin\theta \qquad ②$$

From ① and ②

$$\tfrac{1}{2}\sin\theta + \tfrac{\sqrt{3}}{2}\cos\theta = \tfrac{1}{2}\cos\theta + \tfrac{\sqrt{3}}{2}\sin\theta$$

$$\sin\theta + \sqrt{3}\cos\theta = \cos\theta + \sqrt{3}\sin\theta$$

Collect like terms:

$$\Rightarrow (\sqrt{3} - 1)\cos\theta = (\sqrt{3} - 1)\sin\theta$$

$$\cos\theta = \sin\theta$$

Divide by $\cos\theta$: $1 = \tan\theta$

$\theta = 45°$

This gives an equation in one trigonometrical ratio.

Since an acute angle was required, this is the only root.

Uses of the compound-angle formulae

You have already seen compound-angle formulae used in solving a trigonometrical equation and this is quite a common application of them. However, their significance goes well beyond that since they form the basis for a number of important techniques. Those covered in this book are as follows.

- **The derivation of double-angle formulae**
  The derivation and uses of these are covered on pages 61 to 63.

- **The addition of different sine and cosine functions**
  This is covered on pages 66 to 70. It is included here because the basic wave form is a sine curve. It has many applications, for example in applied mathematics, physics and chemistry.

- **Calculus of trigonometrical functions**
  This is covered in Chapters 4 and 5 and also in Chapter 8 if you are studying *Pure Mathematics 3*. Proofs of the results depend on using either the compound-angle formulae or the factor formulae which are derived from them.

You will see from this that the compound-angle formulae are important in the development of the subject. Some people learn them by heart, others think it is safer to look them up when they are needed. Whichever policy you adopt, you should understand these formulae and recognise their form. Without that you will be unable to do the next example, which uses one of them in reverse.

EXAMPLE 3.3 Simplify $\cos\theta\cos 3\theta - \sin\theta\sin 3\theta$.

SOLUTION

The formula which has the same pattern of $\cos\cos - \sin\sin$ is

$$\cos(\theta + \phi) = \cos\theta\cos\phi - \sin\theta\sin\phi$$

Using this, and replacing ϕ by 3θ, gives

$$\cos\theta\cos 3\theta - \sin\theta\sin 3\theta = \cos(\theta + 3\theta)$$
$$= \cos 4\theta$$

EXERCISE 3B

1 Use the compound-angle formulae to write the following as surds.

  (i) $\sin 75° = \sin(45° + 30°)$
  (ii) $\cos 135° = \cos(90° + 45°)$
  (iii) $\tan 15° = \tan(45° - 30°)$
  (iv) $\tan 75° = \tan(45° + 30°)$

2 Expand each of the following expressions.

  (i) $\sin(\theta + 45°)$
  (ii) $\cos(\theta - 30°)$
  (iii) $\sin(60° - \theta)$
  (iv) $\cos(2\theta + 45°)$
  (v) $\tan(\theta + 45°)$
  (vi) $\tan(\theta - 45°)$

3 Simplify each of the following expressions.

  (i) $\sin 2\theta \cos\phi - \cos 2\theta \sin\theta$
  (ii) $\cos\phi \cos 7\phi - \sin\phi \sin 7\phi$
  (iii) $\sin 120° \cos 60° + \cos 120° \sin 60°$
  (iv) $\cos\theta\cos\theta - \sin\theta\sin\theta$

4 Solve the following equations for values of θ in the range $0° \leqslant \theta \leqslant 180°$.

  (i) $\cos(60° + \theta) = \sin\theta$
  (ii) $\sin(45° - \theta) = \cos\theta$
  (iii) $\tan(45° + \theta) = \tan(45° - \theta)$
  (iv) $2\sin\theta - 3\cos(\theta - 60°)$
  (v) $\sin\theta = \cos(\theta + 120°)$

5 Solve the following equations for values of θ in the range $0 \leqslant \theta \leqslant \pi$.
   (When the range is given in radians, the solutions should be in radians, using multiples of π where appropriate.)

  (i) $\sin\left(\theta + \dfrac{\pi}{4}\right) = \cos\theta$

  (ii) $2\cos\left(\theta - \dfrac{\pi}{3}\right) = \cos\left(\theta + \dfrac{\pi}{2}\right)$

6 Calculators are not to be used in this question.

The diagram shows three points L(–2, 1), M(0, 2) and N(3, –2) joined to form a triangle. The angles α and β and the point P are shown in the diagram.

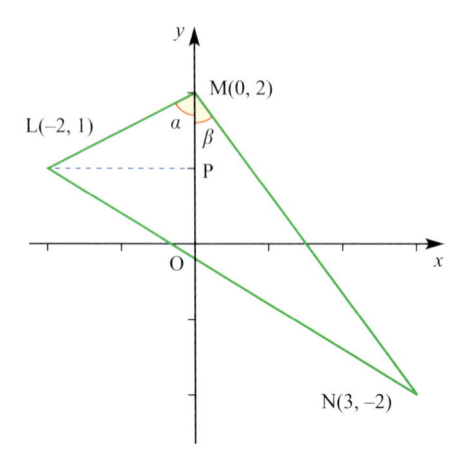

(i) Show that $\sin\alpha = \dfrac{2}{\sqrt{5}}$ and write down the value of $\cos\alpha$.

(ii) Find the values of $\sin\beta$ and $\cos\beta$.

(iii) Show that $\sin\angle LMN = \dfrac{11}{5\sqrt{5}}$.

(iv) Show that $\tan\angle LNM = \dfrac{11}{27}$.

[MEI]

7 (i) Show that the equation

$$\sin(x+30°) = 2\cos(x+60°)$$

can be written in the form

$$(3\sqrt{3})\sin x = \cos x.$$

(ii) Hence solve the equation

$$\sin(x+30°) = 2\cos(x+60°),$$

for $-180° \leqslant x \leqslant 180°$.

[Cambridge International AS & A Level Mathematics 9709, Paper 2 Q4 November 2008]

8 (i) Show that the equation

$$\tan(45° + x) - \tan x = 2$$

can be written in the form

$$\tan^2 x + 2\tan x - 1 = 0.$$

(ii) Hence solve the equation

$$\tan(45° + x) - \tan x = 2,$$

giving all solutions in the interval $0° \leqslant x \leqslant 180°$.

[Cambridge International AS & A Level Mathematics 9709, Paper 3 Q5 November 2007]

9 The angles α and β lie in the interval $0° < x < 180°$, and are such that

$$\tan \alpha = 2\tan \beta \quad \text{and} \quad \tan(\alpha + \beta) = 3.$$

Find the possible values of α and β.

[Cambridge International AS & A Level Mathematics 9709, Paper 32 Q4 November 2009]

10 (i) Show that the equation $\tan(30° + \theta) = 2\tan(60° - \theta)$ can be written in the form

$$\tan^2 \theta + (6\sqrt{3})\tan \theta - 5 = 0.$$

(ii) Hence, or otherwise, solve the equation

$$\tan(30° + \theta) = 2\tan(60° - \theta),$$

for $0° \leqslant \theta \leqslant 180°$.

[Cambridge International AS & A Level Mathematics 9709, Paper 3 Q4 June 2008]

Double-angle formulae

p As you work through these proofs, think how you can check the results.

Is a check the same as a proof?

Substituting $\phi = \theta$ in the relevant compound-angle formulae leads immediately to expressions for $\sin 2\theta$, $\cos 2\theta$ and $\tan 2\theta$, as follows.

(i) $\sin(\theta + \phi) = \sin\theta\cos\phi + \cos\theta\sin\phi$

When $\phi = \theta$, this becomes

$\sin(\theta + \theta) = \sin\theta\cos\theta + \cos\theta\sin\theta$

giving $\quad \sin 2\theta = 2\sin\theta\cos\theta.$

(ii) $\quad \cos(\theta + \phi) = \cos\theta\cos\phi - \sin\theta\sin\phi$

When $\phi = \theta$, this becomes

$$\cos(\theta + \theta) = \cos\theta\cos\theta - \sin\theta\sin\theta$$

giving $\cos 2\theta = \cos^2\theta - \sin^2\theta$.

Using the Pythagorean identity $\cos^2\theta + \sin^2\theta = 1$, two other forms for $\cos 2\theta$ can be obtained.

$$\cos 2\theta = (1 - \sin^2\theta) - \sin^2\theta \quad \Rightarrow \quad \cos 2\theta = 1 - 2\sin^2\theta$$

$$\cos 2\theta = \cos^2\theta - (1 - \cos^2\theta) \quad \Rightarrow \quad \cos 2\theta = 2\cos^2\theta - 1$$

These alternative forms are often more useful since they contain only one trigonometrical function.

(iii) $\quad \tan(\theta + \phi) = \dfrac{\tan\theta + \tan\phi}{1 - \tan\theta\tan\phi} \quad (\theta + \phi) \neq 90°, 270°, \ldots$

When $\phi = \theta$, this becomes

$$\tan(\theta + \theta) = \frac{\tan\theta + \tan\theta}{1 - \tan\theta\tan\theta}$$

giving $\quad \tan 2\theta = \dfrac{2\tan\theta}{1 - \tan^2\theta} \qquad \theta \neq 45°, 135°, \ldots$.

Uses of the double-angle formulae

e *In modelling situations*

You will meet situations, such as that below, where using a double-angle formula not only allows you to write an expression more neatly but also thereby allows you to interpret its meaning more clearly.

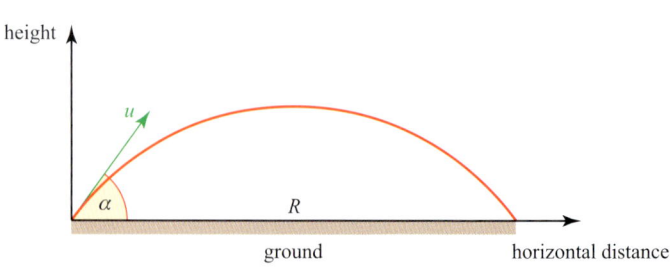

Figure 3.5

When an object is projected, such as a golf ball being hit as in figure 3.5, with speed u at an angle α to the horizontal over level ground, the horizontal distance it travels before striking the ground, called its range, R, is given by the product of the horizontal component of the velocity $u\cos\alpha$ and its time of flight $\frac{2u\sin\alpha}{g}$.

$$R = \frac{2u^2\sin\alpha\cos\alpha}{g}$$

Using the double-angle formula, $\sin 2\alpha = 2\sin\alpha\cos\alpha$ allows this to be written as

$$R = \frac{u^2\sin 2\alpha}{g}.$$

Since the maximum value of $\sin 2\alpha$ is 1, it follows that the greatest value of the range R is $\frac{u^2}{g}$ and that this occurs when $2\alpha = 90°$ and so $\alpha = 45°$. Thus an angle of projection of 45° will give the maximum range of the projectile over level ground. (This assumes that air resistance may be ignored.)

In this example, the double-angle formula enabled the expression for R to be written tidily. However, it did more than that because it made it possible to find the maximum value of R by inspection and without using calculus.

In calculus

The double-angle formulae allow a number of functions to be integrated and you will meet some of these later (see page 125).

The formulae for $\cos 2\theta$ are particularly useful in this respect since

$$\cos 2\theta = 1 - 2\sin^2\theta \quad \Rightarrow \quad \sin^2\theta = \tfrac{1}{2}(1 - \cos 2\theta)$$

and

$$\cos 2\theta = 2\cos^2\theta - 1 \quad \Rightarrow \quad \cos^2\theta = \tfrac{1}{2}(1 + \cos 2\theta)$$

and these identities allow you to integrate $\sin^2\theta$ and $\cos^2\theta$.

In solving equations

You will sometimes need to solve equations involving both single and double angles as shown by the next two examples.

EXAMPLE 3.4 Solve the equation $\sin 2\theta = \sin\theta$ for $0° \leq \theta \leq 360°$.

SOLUTION

$\sin 2\theta = \sin\theta$

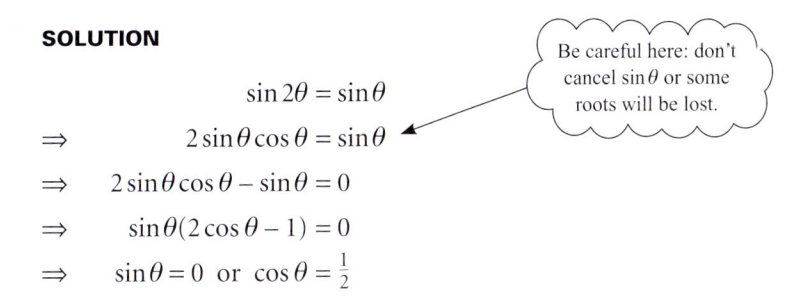

Be careful here: don't cancel $\sin\theta$ or some roots will be lost.

$\Rightarrow \quad 2\sin\theta\cos\theta = \sin\theta$

$\Rightarrow \quad 2\sin\theta\cos\theta - \sin\theta = 0$

$\Rightarrow \quad \sin\theta(2\cos\theta - 1) = 0$

$\Rightarrow \quad \sin\theta = 0 \text{ or } \cos\theta = \tfrac{1}{2}$

$\sin\theta = 0 \Rightarrow \theta = 0°$ (principal value) or $180°$ or $360°$ (see figure 3.6)

> The principal value is the one which comes from your calculator.

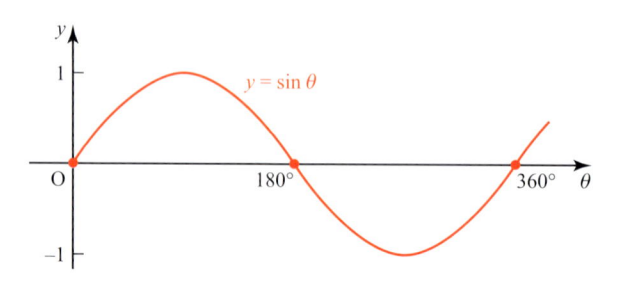

Figure 3.6

$\cos\theta = \frac{1}{2} \Rightarrow \theta = 60°$ (principal value) or $300°$ (see figure 3.7)

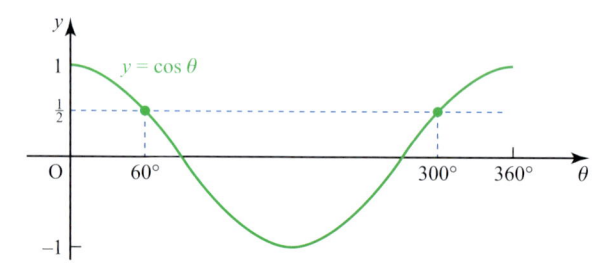

Figure 3.7

The full set of roots for $0° \leq \theta \leq 360°$ is $\theta = 0°, 60°, 180°, 300°, 360°$.

When an equation contains $\cos 2\theta$, you will save time if you take care to choose the most suitable expansion.

EXAMPLE 3.5

Solve $2 + \cos 2\theta = \sin\theta$ for $0 \leq \theta \leq 2\pi$. (Notice that the request for $0 \leq \theta \leq 2\pi$, i.e. in radians, is an invitation to give the answer in radians.)

SOLUTION

Using $\cos 2\theta = 1 - 2\sin^2\theta$ gives

$$2 + (1 - 2\sin^2\theta) = \sin\theta$$

$\Rightarrow \quad 2\sin^2\theta + \sin\theta - 3 = 0$

$\Rightarrow \quad (2\sin\theta + 3)(\sin\theta - 1) = 0$

$\Rightarrow \quad \sin\theta = -\frac{3}{2}$ (not valid since $-1 \leq \sin\theta \leq 1$)

or $\quad \sin\theta = 1$

> This is the most suitable expansion since the right-hand side contains $\sin\theta$.

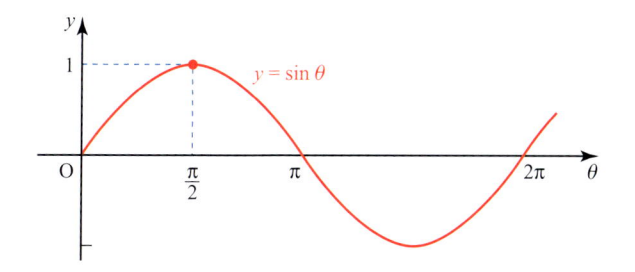

Figure 3.8 shows that the principal value $\theta = \frac{\pi}{2}$ is the only root for $0 \leq \theta \leq 2\pi$.

Figure 3.8

EXERCISE 3C

1. Solve the following equations for $0° \leq \theta \leq 360°$.
   (i) $2\sin 2\theta = \cos\theta$
   (ii) $\tan 2\theta = 4\tan\theta$
   (iii) $\cos 2\theta + \sin\theta = 0$
   (iv) $\tan\theta \tan 2\theta = 1$
   (v) $2\cos 2\theta = 1 + \cos\theta$

2. Solve the following equations for $-\pi \leq \theta \leq \pi$.
   (i) $\sin 2\theta = 2\sin\theta$
   (ii) $\tan 2\theta = 2\tan\theta$
   (iii) $\cos 2\theta - \cos\theta = 0$
   (iv) $1 + \cos 2\theta = 2\sin^2\theta$
   (v) $\sin 4\theta = \cos 2\theta$

   Hint: Write the expression in part (v) as an equation in 2θ.

3. By first writing $\sin 3\theta$ as $\sin(2\theta + \theta)$, express $\sin 3\theta$ in terms of $\sin\theta$. Hence solve the equation $\sin 3\theta = \sin\theta$ for $0 \leq \theta \leq 2\pi$.

4. Solve $\cos 3\theta = 1 - 3\cos\theta$ for $0° \leq \theta \leq 360°$.

5. Simplify $\dfrac{1 + \cos 2\theta}{\sin 2\theta}$.

6. Express $\tan 3\theta$ in terms of $\tan\theta$.

7. Show that $\dfrac{1 - \tan^2\theta}{1 + \tan^2\theta} = \cos 2\theta$.

8. (i) Show that $\tan\left(\dfrac{\pi}{4} + \theta\right)\tan\left(\dfrac{\pi}{4} - \theta\right) = 1$.

   (ii) Given that $\tan 26.6° = 0.5$, solve $\tan\theta = 2$ without using your calculator. Give θ to 1 decimal place, where $0° < \theta < 90°$.

9. (i) Sketch on the same axes the graphs of
   $$y = \cos 2x \quad \text{and} \quad y = 3\sin x - 1 \quad \text{for} \quad 0 \leq x \leq 2\pi.$$

   (ii) Show that these curves meet at points whose x co-ordinates are solutions of the equation $2\sin^2 x + 3\sin x - 2 = 0$.

   (iii) Solve this equation to find the values of x in terms of π for $0 \leq x \leq 2\pi$.

   [MEI]

10 (i) Prove the identity

$$\cos 4\theta + 4\cos 2\theta \equiv 8\cos^4\theta - 3.$$

(ii) Hence solve the equation

$$\cos 4\theta + 4\cos 2\theta = 2,$$

for $0° \leq \theta \leq 360°$.

[Cambridge International AS & A Level Mathematics 9709, Paper 3 Q6 June 2005]

11 (i) Prove the identity $\operatorname{cosec} 2\theta + \cot 2\theta \equiv \cot\theta$.

(ii) Hence solve the equation $\operatorname{cosec} 2\theta + \cot 2\theta = 2$, for $0° \leq \theta \leq 360°$.

[Cambridge International AS & A Level Mathematics 9709, Paper 3 Q3 June 2009]

12 It is given that $\cos a = \frac{3}{5}$, where $0° \leq a \leq 90°$. Showing your working and without using a calculator to evaluate a,

(i) find the exact value of $\sin(a - 30)°$,

(ii) find the exact value of $\tan 2a$, and hence find the exact value of $\tan 3a$.

[Cambridge International AS & A Level Mathematics 9709, Paper 32 Q3 June 2010]

The forms $r\cos(\theta \pm \alpha)$, $r\sin(\theta \pm \alpha)$

Another modification of the compound-angle formulae allows you to simplify expressions such as $4\sin\theta + 3\cos\theta$ and hence solve equations of the form

$$a\sin\theta + b\cos\theta = c.$$

To find a single expression for $4\sin\theta + 3\cos\theta$, you match it to the expression

$$r\sin(\theta + \alpha) = r(\sin\theta\cos\alpha + \cos\theta\sin\alpha).$$

This is because the expansion of $r\sin(\theta + \alpha)$ has $\sin\theta$ in the first term, $\cos\theta$ in the second term and a plus sign in between them. It is then possible to choose appropriate values of r and α.

$$4\sin\theta + 3\cos\theta \equiv r(\sin\theta\cos\alpha + \cos\theta\sin\alpha)$$

Coefficients of $\sin\theta$: $\quad 4 = r\cos\alpha$

Coefficients of $\cos\theta$: $\quad 3 = r\sin\alpha$.

Looking at the right-angled triangle in figure 3.9 gives the values for r and α.

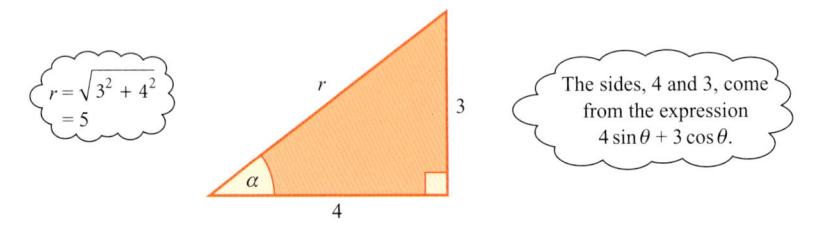

Figure 3.9

In this triangle, the hypotenuse is $\sqrt{4^2 + 3^2} = 5$, which corresponds to r in the expression above.

The angle α is given by

$$\sin\alpha = \tfrac{3}{5} \quad \text{and} \quad \cos\alpha = \tfrac{4}{5} \quad \Rightarrow \quad \alpha = 36.9°.$$

So the expression becomes

$$4\sin\theta + 3\cos\theta = 5\sin(\theta + 36.9°).$$

The steps involved in this procedure can be generalised to write

$$a\sin\theta + b\cos\theta = r\sin(\theta + \alpha)$$

where

$$r = \sqrt{a^2 + b^2} \qquad \sin\alpha = \frac{b}{\sqrt{a^2+b^2}} = \frac{b}{r} \qquad \cos\alpha = \frac{a}{\sqrt{a^2+b^2}} = \frac{a}{r}$$

The same expression may also be written as a cosine function. In this case, rewrite $4\sin\theta + 3\cos\theta$ as $3\cos\theta + 4\sin\theta$ and notice that:

(i) The expansion of $\cos(\theta - \beta)$ starts with $\cos\theta$ … just like the expression $3\cos\theta + 4\sin\theta$.
(ii) The expansion of $\cos(\theta - \beta)$ has $+$ in the middle, just like the expression $3\cos\theta + 4\sin\theta$.

The expansion of $r\cos(\theta - \beta)$ is given by

$$r\cos(\theta - \beta) = r(\cos\theta\cos\beta + \sin\theta\sin\beta).$$

To compare this with $3\cos\theta + 4\sin\theta$, look at the triangle in figure 3.10 in which

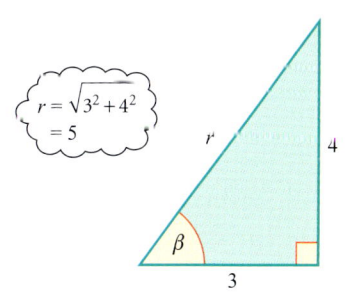

$$r = \sqrt{3^2 + 4^2} = 5 \qquad \cos\beta = \tfrac{3}{5} \qquad \sin\beta = \tfrac{4}{5} \quad \Rightarrow \quad \beta = 53.1°.$$

Figure 3.10

This means that you can write $3\cos\theta + 4\sin\theta$ in the form

$$r\cos(\theta - \beta) = 5\cos(\theta - 53.1°).$$

The procedure used here can be generalised to give the result

$$a\cos\theta + b\sin\theta = r\cos(\theta - \alpha)$$

where

$$r = \sqrt{a^2 + b^2} \qquad \cos\alpha = \frac{a}{r} \qquad \sin\alpha = \frac{b}{r}.$$

Note

The value of *r* will always be positive, but $\cos\alpha$ and $\sin\alpha$ may be positive or negative, depending on the values of *a* and *b*. In all cases, it is possible to find an angle α for which $-180° < \alpha < 180°$.

You can derive alternative expressions of this type based on other compound-angle formulae if you wish α to be an acute angle, as is done in the next example.

EXAMPLE 3.6

(i) Express $\sqrt{3}\sin\theta - \cos\theta$ in the form $r\sin(\theta - \alpha)$, where $r > 0$ and $0 < \alpha < \frac{\pi}{2}$.

(ii) State the maximum and minimum values of $\sqrt{3}\sin\theta - \cos\theta$.

(iii) Sketch the graph of $y = \sqrt{3}\sin\theta - \cos\theta$ for $0 \leq \theta \leq 2\pi$.

(iv) Solve the equation $\sqrt{3}\sin\theta - \cos\theta = 1$ for $0 \leq \theta \leq 2\pi$.

SOLUTION

(i) $r\sin(\theta - \alpha) = r(\sin\theta\cos\alpha - \cos\theta\sin\alpha)$
$\qquad\qquad\quad = (r\cos\alpha)\sin\theta - (r\sin\alpha)\cos\theta$

Comparing this with $\sqrt{3}\sin\theta - \cos\theta$, the two expressions are identical if

$$r\cos\alpha = \sqrt{3} \qquad \text{and} \qquad r\sin\alpha = 1.$$

From the triangle in figure 3.11

$$r = \sqrt{1+3} = 2 \quad \text{and} \quad \tan\alpha = \frac{1}{\sqrt{3}} \Rightarrow \alpha = \frac{\pi}{6}$$

so $\qquad \sqrt{3}\sin\theta - \cos\theta = 2\sin\left(\theta - \frac{\pi}{6}\right).$

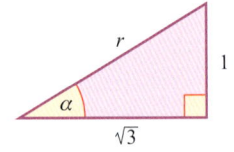

Figure 3.11

(ii) The sine function oscillates between 1 and −1, so $2\sin\left(\theta - \frac{\pi}{6}\right)$ oscillates between 2 and −2.

Maximum value = 2
Minimum value = −2.

(iii) To sketch the curve $y = 2\sin\left(\theta - \frac{\pi}{6}\right)$, notice that

- it is a sine curve
- its y values go from -2 to 2
- it crosses the horizontal axis where $\theta = \frac{\pi}{6}, \frac{7\pi}{6}, \frac{13\pi}{6}$

The curve is shown in Figure 3.12.

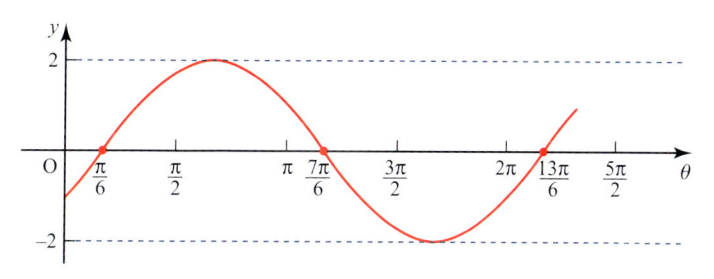

Figure 3.12

(iv) The equation $\sqrt{3}\sin\theta - \cos\theta = 1$ is equivalent to

$$2\sin\left(\theta - \frac{\pi}{6}\right) = 1$$

$$\Rightarrow \sin\left(\theta - \frac{\pi}{6}\right) = \frac{1}{2}$$

Let $x = \left(\theta - \frac{\pi}{6}\right)$ and solve $\sin x = \frac{1}{2}$.

Solving $\sin x = \frac{1}{2}$ gives $x = \frac{\pi}{6}$ (principal value)

or $\qquad x = \pi - \frac{\pi}{6} = \frac{5\pi}{6}$ (from the graph in figure 3.13)

giving $\qquad \theta = \frac{\pi}{6} + \frac{\pi}{6} = \frac{\pi}{3}$ or $\theta = \frac{5\pi}{6} + \frac{\pi}{6} = \pi$.

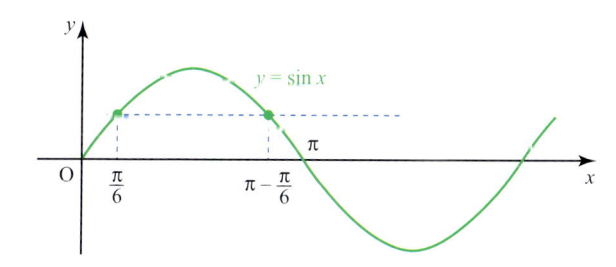

Figure 3.13

The roots in $0 \leq \theta \leq 2\pi$ are $\theta = \frac{\pi}{3}$ and π.

⚠ Always check (for example by reference to a sketch graph) that the number of roots you have found is consistent with the number you are expecting. When solving equations of the form $\sin(\theta - \alpha) = c$ by considering $\sin x = c$, it is sometimes necessary to go outside the range specified for θ since, for example, $0 \leq \theta \leq 2\pi$ is the same as $-\alpha \leq x \leq 2\pi - \alpha$.

Using these forms

There are many situations which produce expressions which can be tidied up using these forms. They are also particularly useful for solving equations involving both the sine and cosine of the same angle.

The fact that $a\cos\theta + b\sin\theta$ can be written as $r\cos(\theta - \alpha)$ is an illustration of the fact that any two waves of the same frequency, whatever their amplitudes, can be added together to give a single combined wave, also of the same frequency.

EXERCISE 3D

1 Express each of the following in the form $r\cos(\theta - \alpha)$, where $r > 0$ and $0° < \alpha < 90°$.

 (i) $\cos\theta + \sin\theta$
 (ii) $20\cos\theta + 21\sin\theta$
 (iii) $\cos\theta + \sqrt{3}\sin\theta$
 (iv) $\sqrt{5}\cos\theta + 2\sin\theta$

2 Express each of the following in the form $r\cos(\theta + \alpha)$, where $r > 0$ and $0 < \alpha < \dfrac{\pi}{2}$.

 (i) $\cos\theta - \sin\theta$
 (ii) $\sqrt{3}\cos\theta - \sin\theta$

3 Express each of the following in the form $r\sin(\theta + \alpha)$, where $r > 0$ and $0° < \alpha < 90°$.

 (i) $\sin\theta + 2\cos\theta$
 (ii) $2\sin\theta + \sqrt{5}\cos\theta$

4 Express each of the following in the form $r\sin(\theta - \alpha)$, where $r > 0$ and $0 < \alpha < \dfrac{\pi}{2}$.

 (i) $\sin\theta - \cos\theta$
 (ii) $\sqrt{7}\sin\theta - \sqrt{2}\cos\theta$

5 Express each of the following in the form $r\cos(\theta - \alpha)$, where $r > 0$ and $-180° < \alpha < 180°$.

 (i) $\cos\theta - \sqrt{3}\sin\theta$
 (ii) $2\sqrt{2}\cos\theta - 2\sqrt{2}\sin\theta$
 (iii) $\sin\theta + \sqrt{3}\cos\theta$
 (iv) $5\sin\theta + 12\cos\theta$
 (v) $\sin\theta - \sqrt{3}\cos\theta$
 (vi) $\sqrt{2}\sin\theta - \sqrt{2}\cos\theta$

6 (i) Express $5\cos\theta - 12\sin\theta$ in the form $r\cos(\theta + \alpha)$, where $r > 0$ and $0° < \alpha < 90°$.

(ii) State the maximum and minimum values of $5\cos\theta - 12\sin\theta$.

(iii) Sketch the graph of $y = 5\cos\theta - 12\sin\theta$ for $0° \leq \theta \leq 360°$.

(iv) Solve the equation $5\cos\theta - 12\sin\theta = 4$ for $0° \leq \theta \leq 360°$.

7 (i) Express $3\sin\theta - \sqrt{3}\cos\theta$ in the form $r\sin(\theta - \alpha)$, where $r > 0$ and $0 < \alpha < \dfrac{\pi}{2}$.

(ii) State the maximum and minimum values of $3\sin\theta - \sqrt{3}\cos\theta$ and the smallest positive values of θ for which they occur.

(iii) Sketch the graph of $y = 3\sin\theta - \sqrt{3}\cos\theta$ for $0 \leq \theta \leq 2\pi$.

(iv) Solve the equation $3\sin\theta - \sqrt{3}\cos\theta = \sqrt{3}$ for $0 \leq \theta \leq 2\pi$.

8 (i) Express $2\sin 2\theta + 3\cos 2\theta$ in the form $r\sin(2\theta + \alpha)$, where $r > 0$ and $0° < \alpha < 90°$.

(ii) State the maximum and minimum values of $2\sin 2\theta + 3\cos 2\theta$ and the smallest positive values of θ for which they occur.

(iii) Sketch the graph of $y = 2\sin 2\theta + 3\cos 2\theta$ for $0° \leq \theta \leq 360°$.

(iv) Solve the equation $2\sin 2\theta + 3\cos 2\theta = 1$ for $0° \leq \theta \leq 360°$.

9 (i) Express $\cos\theta + \sqrt{2}\sin\theta$ in the form $r\cos(\theta - \alpha)$, where $r > 0$ and $0° < \alpha < 90°$.

(ii) State the maximum and minimum values of $\cos\theta + \sqrt{2}\sin\theta$ and the smallest positive values of θ for which they occur.

(iii) Sketch the graph of $y = \cos\theta + \sqrt{2}\sin\theta$ for $0° \leq \theta \leq 360°$.

(iv) State the maximum and minimum values of

$$\dfrac{1}{3 + \cos\theta + \sqrt{2}\sin\theta}$$

and the smallest positive values of θ for which they occur.

10 The diagram shows a table jammed in a corridor. The table is 120 cm long and 80 cm wide, and the width of the corridor is 130 cm.

(i) Show that $12\sin\theta + 8\cos\theta = 13$.

(ii) Hence find the angle θ. (There are two answers.)

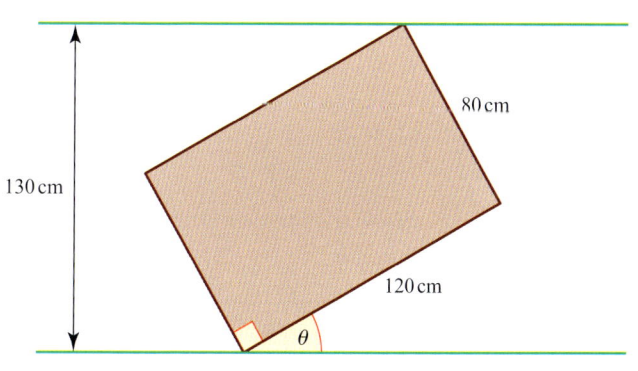

11 (i) Use a trigonometrical formula to expand $\cos(x+\alpha)$.

(ii) Express $y = 2\cos x - 5\sin x$ in the form $r\cos(x+\alpha)$, giving the positive value of r and the smallest positive value of α.

(iii) State the maximum and minimum values of y and the corresponding values of x for $0° \leqslant x \leqslant 360°$.

(iv) Solve the equation
$$2\cos x - 5\sin x = 3, \quad \text{for } 0° \leqslant x \leqslant 360°.$$
[MEI]

12 (i) Find the value of the acute angle α for which
$$5\cos x - 3\sin x = \sqrt{34}\cos(x+\alpha)$$
for all x.

Giving your answers correct to 1 decimal place,

(ii) solve the equation $5\cos x - 3\sin x = 4$ for $0° \leqslant x \leqslant 360°$

(iii) solve the equation $5\cos 2x - 3\sin 2x = 4$ for $0° \leqslant x \leqslant 360°$.
[MEI]

13 (i) Find the positive value of R and the acute angle α for which
$$6\cos x + 8\sin x = R\cos(x-\alpha).$$

(ii) Sketch the curve with equation
$$y = 6\cos x + 8\sin x, \quad \text{for } 0° \leqslant x \leqslant 360°.$$

Mark your axes carefully and indicate the angle α on the x axis.

(iii) Solve the equation
$$6\cos x + 8\sin x = 4, \quad \text{for } 0° \leqslant x \leqslant 360°.$$

(iv) Solve the equation
$$8\cos\theta + 6\sin\theta = 4, \quad \text{for } 0° \leqslant \theta \leqslant 360°.$$
[MEI]

14 In the diagram below, angle QPT = angle SQR = θ, angle QPR = α, PQ = a, QR = b, PR = c, angle QSR = angle QTP = $90°$, SR = TU.

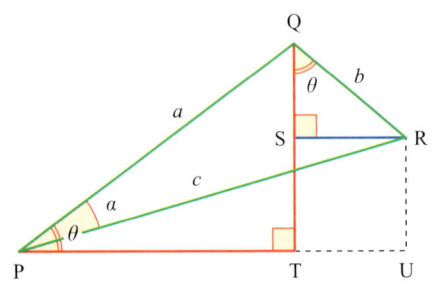

(i) Show that angle PQR = $90°$, and write down the length of c in terms of a and b.

(ii) Show that PU may be written as $a\cos\theta + b\sin\theta$ and as $c\cos(\theta - \alpha)$. Write down the value of $\tan\alpha$ in terms of a and b.

(iii) In the case when $a = 4$, $b = 3$, find the acute angle α.

(iv) Solve the equation

$$4\cos\theta + 3\sin\theta = 2 \quad \text{for} \quad 0° \leq \theta \leq 360°.$$

[MEI]

15 (i) Express $3\cos x + 4\sin x$ in the form $R\cos(x - \alpha)$, where $R > 0$ and $0° < \alpha < 90°$, stating the exact value of R and giving the value of α correct to 2 decimal places.

(ii) Hence solve the equation

$$3\cos x + 4\sin x = 4.5,$$

giving all solutions in the interval $0° < x < 360°$.

[Cambridge International AS & A Level Mathematics 9709, Paper 22 Q6 November 2009]

16 (i) Express $5\cos\theta - \sin\theta$ in the form $R\cos(\theta + \alpha)$, where $R > 0$ and $0° < \alpha < 90°$, giving the exact value of R and the value of α correct to 2 decimal places.

(ii) Hence solve the equation

$$5\cos\theta - \sin\theta = 4,$$

giving all solutions in the interval $0° \leq \theta \leq 360°$.

[Cambridge International AS & A Level Mathematics 9709, Paper 2 Q5 June 2008]

17 (i) Express $7\cos\theta + 24\sin\theta$ in the form $R\cos(\theta - \alpha)$, where $R > 0$ and $0° < \alpha < 90°$, giving the exact value of R and the value of α correct to 2 decimal places.

(ii) Hence solve the equation

$$7\cos\theta + 24\sin\theta = 15,$$

giving all solutions in the interval $0° \leq \theta \leq 360°$.

[Cambridge International AS & A Level Mathematics 9709, Paper 3 Q4 June 2006]

18 By expressing $8\sin\theta - 6\cos\theta$ in the form $R\sin(\theta - \alpha)$, solve the equation

$$8\sin\theta - 6\cos\theta = 7,$$

for $0° \leq \theta \leq 360°$.

[Cambridge International AS & A Level Mathematics 9709, Paper 3 Q5 November 2005]

INVESTIGATION

The simplest alternating current is one which varies with time t according to

$$I = A \sin 2\pi ft,$$

where f is the frequency and A is the maximum value. The frequency of the public AC supply is 50 hertz (cycles per second).

Investigate what happens when two alternating currents $A_1 \sin 2\pi ft$ and $A_2 \sin(2\pi ft + \alpha)$ with the same frequency f but a phase difference of α are added together.

The previous exercises have each concentrated on just one of the many trigonometrical techniques which you will need to apply confidently. The following exercise requires you to identify which technique is the correct one.

EXERCISE 3E

1 Simplify the following.

   (i) $2 \sin 3\theta \cos 3\theta$
   (ii) $\cos^2 3\theta - \sin^2 3\theta$
   (iii) $\cos^2 3\theta + \sin^2 3\theta$
   (iv) $1 - 2\sin^2\left(\dfrac{\theta}{2}\right)$
   (v) $\sin(\theta - \alpha)\cos\alpha + \cos(\theta - \alpha)\sin\alpha$
   (vi) $3 \sin\theta \cos\theta$
   (vii) $\dfrac{\sin 2\theta}{2 \sin\theta}$
   (viii) $\cos 2\theta - 2\cos^2\theta$

2 Express

   (i) $(\cos x - \sin x)^2$ in terms of $\sin 2x$
   (ii) $\cos^4 x - \sin^4 x$ in terms of $\cos 2x$
   (iii) $2\cos^2 x - 3\sin^2 x$ in terms of $\cos 2x$.

3 Prove that

   (i) $\dfrac{1 - \cos 2\theta}{1 + \cos 2\theta} \equiv \tan^2\theta$
   (ii) $\operatorname{cosec} 2\theta + \cot 2\theta \equiv \cot\theta$
   (iii) $\tan 4\theta \equiv \dfrac{4t(1 - t^2)}{1 - 6t^2 + t^4}$ where $t = \tan\theta$.

4 Solve the following equations.

(i) $\sin(\theta + 40°) = 0.7$ $0° \leq \theta \leq 360°$

(ii) $3\cos^2\theta + 5\sin\theta - 1 = 0$ $0° \leq \theta \leq 360°$

(iii) $2\cos\left(\theta - \dfrac{\pi}{6}\right) = 1$ $-\pi \leq \theta \leq \pi$

(iv) $\cos(45° - \theta) = 2\sin(30° + \theta)$ $-180° \leq \theta \leq 180°$

(v) $\cos 2\theta + 3\sin\theta = 2$ $0 \leq \theta \leq 2\pi$

(vi) $\cos\theta + 3\sin\theta = 2$ $0° \leq \theta \leq 360°$

(vii) $\tan^2\theta - 3\tan\theta - 4 = 0$ $0° \leq \theta \leq 180°$

e The general solutions of trigonometrical equations

The equation $\tan\theta = 1$ has infinitely many roots:

$$\ldots, -315°, -135°, 45°, 225°, 405°, \ldots \text{ (in degrees)}$$

$$\ldots, -\dfrac{7\pi}{4}, -\dfrac{3\pi}{4}, -\dfrac{\pi}{4}, \dfrac{5\pi}{4}, \dfrac{9\pi}{4}, \ldots \text{ (in radians)}.$$

Only one of these roots, namely 45° or $\dfrac{\pi}{4}$, is denoted by the function $\tan^{-1} 1$. This is the value which your calculator will give you. It is called the *principal value*.

The principal value for any inverse trigonometrical function is unique and lies within a specified range:

$$-\dfrac{\pi}{2} < \tan^{-1} x < \dfrac{\pi}{2}$$

$$-\dfrac{\pi}{2} \leq \sin^{-1} x \leq \dfrac{\pi}{2}$$

$$0 \leq \cos^{-1} x \leq \pi.$$

It is possible to deduce all other roots from the principal value and this is shown below.

To solve the equation $\tan\theta = c$, notice how all possible values of θ occur at intervals of 180° or π radians (see figure 3.14). So the general solution is

$$\theta = \tan^{-1} c + n\pi \quad n \in \mathbb{Z} \quad \text{(in radians)}.$$

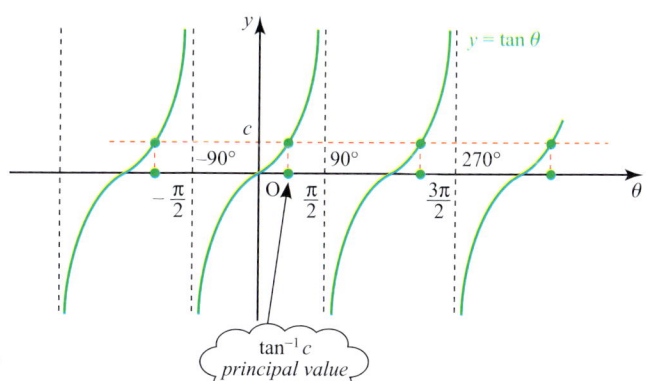

Figure 3.14

The cosine graph (see figure 3.15) has the y axis as a line of symmetry. Notice how the values $\pm\cos^{-1} c$ generate all the other roots at intervals of 360° or 2π. So the general solution is

$$\theta = \pm\cos^{-1} c + 2n\pi \qquad n \in \mathbb{Z} \quad \text{(in radians)}.$$

Figure 3.15

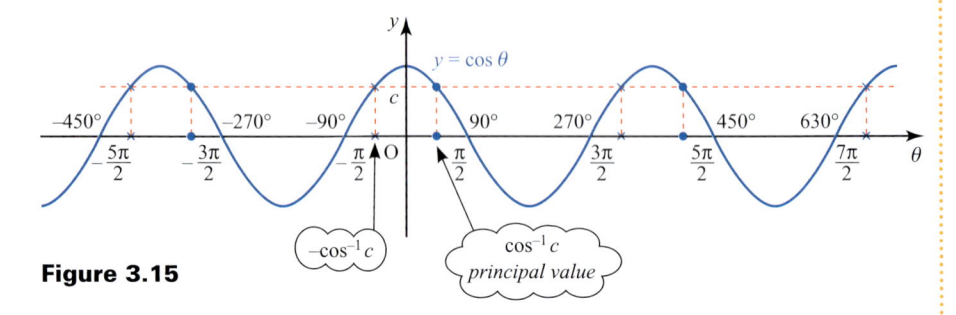

Now look at the sine graph (see figure 3.16). As for the cosine graph, there are two roots located symmetrically. The line of symmetry for the sine graph is $\theta = \dfrac{\pi}{2}$, which generates all the other possible roots. This gives rise to the slightly more complicated expressions

$$\theta = \frac{\pi}{2} \pm \left(\frac{\pi}{2} - \sin^{-1} c\right) + 2n\pi$$

or $\quad \theta = \left(2n + \dfrac{1}{2}\right)\pi \pm \left(\dfrac{\pi}{2} - \sin^{-1} c\right) \quad n \in \mathbb{Z}.$

You may, however, find it easier to remember these as two separate formulae:

$$\theta = 2n\pi + \sin^{-1} c \qquad \text{or} \qquad \theta = (2n+1)\pi - \sin^{-1} c.$$

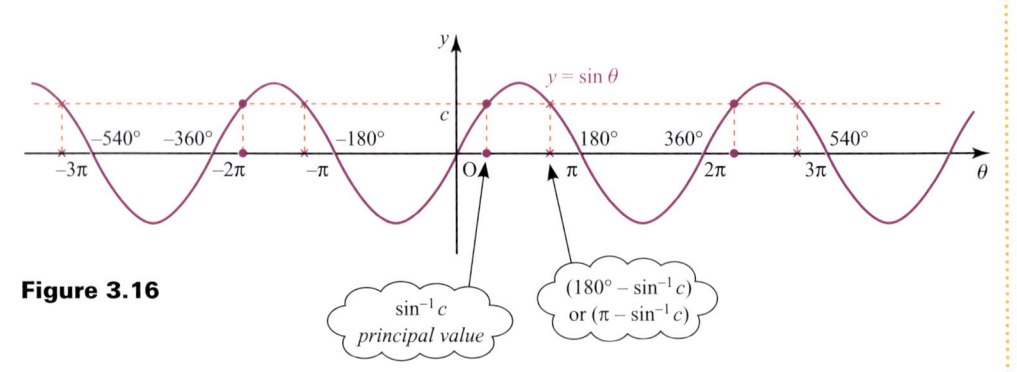

Figure 3.16

ACTIVITY 3.3 Show that the general solution of the equation $\sin\theta = c$ may also be written

$$\theta = n\pi + (-1)^n \sin^{-1} c.$$

KEY POINTS

1. $\sec\theta = \dfrac{1}{\cos\theta}$; $\operatorname{cosec}\theta = \dfrac{1}{\sin\theta}$; $\cot\theta = \dfrac{1}{\tan\theta}$

2. $\tan^2\theta + 1 = \sec^2\theta$; $1 + \cot^2\theta = \operatorname{cosec}^2\theta$

3. **Compound-angle formulae**
   - $\sin(\theta + \phi) = \sin\theta\cos\phi + \cos\theta\sin\phi$
   - $\sin(\theta - \phi) = \sin\theta\cos\phi - \cos\theta\sin\phi$
   - $\cos(\theta + \phi) = \cos\theta\cos\phi - \sin\theta\sin\phi$
   - $\cos(\theta - \phi) = \cos\theta\cos\phi + \sin\theta\sin\phi$
   - $\tan(\theta + \phi) = \dfrac{\tan\theta + \tan\phi}{1 - \tan\theta\tan\phi}$ $(\theta + \phi) \neq 90°, 270°, \ldots$
   - $\tan(\theta - \phi) = \dfrac{\tan\theta - \tan\phi}{1 + \tan\theta\tan\phi}$ $(\theta - \phi) \neq 90°, 270°, \ldots$

4. **Double-angle and related formulae**
   - $\sin 2\theta = 2\sin\theta\cos\theta$
   - $\cos 2\theta = \cos^2\theta - \sin^2\theta = 1 - 2\sin^2\theta = 2\cos^2\theta - 1$
   - $\tan 2\theta = \dfrac{2\tan\theta}{1 - \tan^2\theta}$ $\theta \neq 45°, 135°, \ldots$
   - $\sin^2\theta = \tfrac{1}{2}(1 - \cos 2\theta)$
   - $\cos^2\theta = \tfrac{1}{2}(1 + \cos 2\theta)$

5. **The r, α formulae**
   - $a\sin\theta + b\cos\theta = r\sin(\theta + \alpha)$
   - $a\sin\theta - b\cos\theta = r\sin(\theta - \alpha)$
   - $a\cos\theta + b\sin\theta = r\cos(\theta - \alpha)$
   - $a\cos\theta - b\sin\theta = r\cos(\theta + \alpha)$

   where $r = \sqrt{a^2 + b^2}$
   $\cos\alpha = \dfrac{a}{r}$
   $\sin\alpha = \dfrac{b}{r}$

Differentiation

A mathematician, like a painter or poet, is a maker of patterns. If his patterns are more permanent than theirs it is because they are made with ideas.

G.H. Hardy

The product rule

Figure 4.1 shows a sketch of the curve of $y = 20x(x-1)^6$.

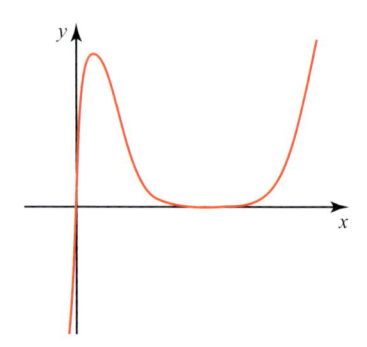

Figure 4.1

If you wanted to find the gradient function, $\dfrac{dy}{dx}$, for the curve, you could expand the right-hand side then differentiate it term by term – a long and cumbersome process!

There are other functions like this, made up of the product of two or more simpler functions, which are not just time-consuming to expand – they are *impossible* to expand. One such function is

$$y = (x-1)^{\frac{1}{2}}(x+1)^6 \quad \text{(for } x > 1\text{)}.$$

Clearly you need a technique for differentiating functions that are products of simpler ones, and a suitable notation with which to express it.

The most commonly used notation involves writing

$$y = uv,$$

where the variables u and v are both functions of x. Using this notation, $\dfrac{dy}{dx}$ is given by

$$\frac{dy}{dx} = u\frac{dv}{dx} + v\frac{du}{dx}.$$

This is called the *product rule* and it is derived from first principles in the next section.

The product rule from first principles

A small increase δx in x leads to corresponding small increases δu, δv and δy in u, v and y. And so

$$y + \delta y = (u + \delta u)(v + \delta v)$$
$$= uv + v\delta u + u\delta v + \delta u \delta v.$$

Since $y = uv$, the increase in y is given by

$$\delta y = v\delta u + u\delta v + \delta u \delta v.$$

Dividing both sides by δx,

$$\frac{\delta y}{\delta x} = v\frac{\delta u}{\delta x} + u\frac{\delta v}{\delta x} + \delta u\frac{\delta v}{\delta x}.$$

In the limit, as $\delta x \to 0$, so do δu, δv and δy, and

$$\frac{\delta u}{\delta x} \to \frac{du}{dx}, \quad \frac{\delta v}{\delta x} \to \frac{dv}{dx} \quad \text{and} \quad \frac{\delta y}{\delta x} \to \frac{dy}{dx}.$$

The expression becomes

$$\frac{dy}{dx} = v\frac{du}{dx} + u\frac{dv}{dx}.$$

Notice that since $\delta u \to 0$ the last term on the right-hand side has disappeared.

EXAMPLE 4.1 Given that $y = (2x + 3)(x^2 - 5)$, find $\frac{dy}{dx}$ using the product rule.

SOLUTION

$y = (2x + 3)(x^2 - 5)$

Let $u = 2x + 3$ and $v = x^2 - 5$.

Then $\frac{du}{dx} = 2$ and $\frac{dv}{dx} = 2x$.

Using the product rule, $\frac{dy}{dx} = v\frac{du}{dx} + u\frac{dv}{dx}$
$$= (x^2 - 5) \times 2 + (2x + 3) \times 2x$$
$$= 2(x^2 - 5 + 2x^2 + 3x)$$
$$= 2(3x^2 + 3x - 5)$$

| Note

In this case you could have multiplied out the expression for y.

$$y = 2x^3 + 3x^2 - 10x - 15$$
$$\frac{dy}{dx} = 6x^2 + 6x - 10$$
$$= 2(3x^2 + 3x - 5)$$

EXAMPLE 4.2 Differentiate $y = 20x(x-1)^6$.

SOLUTION

Let $u = 20x$ and $v = (x-1)^6$.

Then $\dfrac{du}{dx} = 20$ and $\dfrac{dv}{dx} = 6(x-1)^5$ (using the chain rule).

Using the product rule, $\dfrac{dy}{dx} = v\dfrac{du}{dx} + u\dfrac{dv}{dx}$

$= (x-1)^6 \times 20 + 20x \times 6(x-1)^5$
$= 20(x-1)^5 \times (x-1) + 20(x-1)^5 \times 6x$
$= 20(x-1)^5[(x-1) + 6x]$
$= 20(x-1)^5(7x-1)$

$20(x-1)^5$ is a common factor.

The factorised result is the most useful form for the solution, as it allows you to find stationary points easily. You should always try to factorise your answer as much as possible. Once you have used the product rule, look for factors straight away and do not be tempted to multiply out.

The quotient rule

In the last section, you met a technique for differentiating the product of two functions. In this section you will see how to differentiate a function which is the quotient of two simpler functions.

As before, you start by identifying the simpler functions. For example, the function

$$y = \dfrac{3x+1}{x-2} \quad \text{(for } x \neq 2\text{)}$$

can be written as $y = \dfrac{u}{v}$ where $u = 3x + 1$ and $v = x - 2$. Using this notation, $\dfrac{dy}{dx}$ is given by

$$\dfrac{dy}{dx} = \dfrac{v\dfrac{du}{dx} - u\dfrac{dv}{dx}}{v^2}$$

This is called the *quotient rule* and it is derived from first principles in the next section.

The quotient rule from first principles

A small increase, δx in x results in corresponding small increases δu, δv and δy in u, v and y. The new value of y is given by

$$y + \delta y = \frac{u + \delta u}{v + \delta v}$$

and since $y = \frac{u}{v}$, you can rearrange this to obtain an expression for δy in terms of u and v.

$$\delta y = \frac{u + \delta u}{v + \delta v} - \frac{u}{v}$$

$$= \frac{v(u + \delta u) - u(v + \delta v)}{v(v + \delta v)}$$

$$= \frac{uv + v\delta u - uv - u\delta v}{v(v + \delta v)}$$

$$= \frac{v\delta u - u\delta v}{v(v + \delta v)}$$

Dividing both sides by δx gives

$$\frac{\delta y}{\delta x} = \frac{v\frac{\delta u}{\delta x} - u\frac{\delta v}{\delta x}}{v(v + \delta v)}$$

> To divide the right-hand side by δx you only divide the numerator by δx.

In the limit as $\delta x \to 0$, this is written in the form you met on the previous page.

$$\frac{dy}{dx} = \frac{v\frac{du}{dx} - u\frac{dv}{dx}}{v^2}$$

ACTIVITY 4.1 Verify that the quotient rule gives $\frac{dy}{dx}$ correctly when $u = x^{10}$ and $v = x^7$.

EXAMPLE 4.3 Given that $y = \frac{3x + 1}{x - 2}$, find $\frac{dy}{dx}$ using the quotient rule.

SOLUTION

Letting $u = 3x + 1$ and $v = x - 2$ gives

$$\frac{du}{dx} = 3 \quad \text{and} \quad \frac{dv}{dx} = 1.$$

Using the quotient rule, $\frac{dy}{dx} = \frac{v\frac{du}{dx} - u\frac{dv}{dx}}{v^2}$

$$= \frac{(x - 2) \times 3 - (3x + 1) \times 1}{(x - 2)^2}$$

$$= \frac{3x - 6 - 3x - 1}{(x - 2)^2}$$

$$= \frac{-7}{(x - 2)^2}$$

EXAMPLE 4.4 Given that $y = \dfrac{x^2 + 1}{3x - 1}$, find $\dfrac{dy}{dx}$ using the quotient rule.

SOLUTION

Letting $u = x^2 + 1$ and $v = 3x - 1$ gives

$$\frac{du}{dx} = 2x \quad \text{and} \quad \frac{dv}{dx} = 3.$$

Using the quotient rule, $\dfrac{dy}{dx} = \dfrac{v\dfrac{du}{dx} - u\dfrac{dv}{dx}}{v^2}$

$$= \frac{(3x - 1) \times 2x - (x^2 + 1) \times 3}{(3x - 1)^2}$$

$$= \frac{6x^2 - 2x - 3x^2 - 3}{(3x - 1)^2}$$

$$= \frac{3x^2 - 2x - 3}{(3x - 1)^2}$$

EXERCISE 4A

1 Differentiate the following using the product rule or the quotient rule.

(i) $y = (x^2 - 1)(x^3 + 3)$
(ii) $y = x^5(3x^2 + 4x - 7)$
(iii) $y = x^2(2x + 1)^4$
(iv) $y = \dfrac{2x}{3x - 1}$
(v) $y = \dfrac{x^3}{x^2 + 1}$
(vi) $y = (2x + 1)^2(3x^2 - 4)$
(vii) $y = \dfrac{2x - 3}{2x^2 + 1}$
(viii) $y = \dfrac{x - 2}{(x + 3)^2}$
(ix) $y = (x + 1)\sqrt{x - 1}$

2 The diagram shows the graph of $y = \dfrac{x}{x - 1}$.

(i)   Find $\dfrac{dy}{dx}$.
(ii) Find the gradient of the curve at $(0, 0)$, and the equation of the tangent at $(0, 0)$.
(iii) Find the gradient of the curve at $(2, 2)$, and the equation of the tangent at $(2, 2)$.
(iv) What can you deduce about the two tangents?

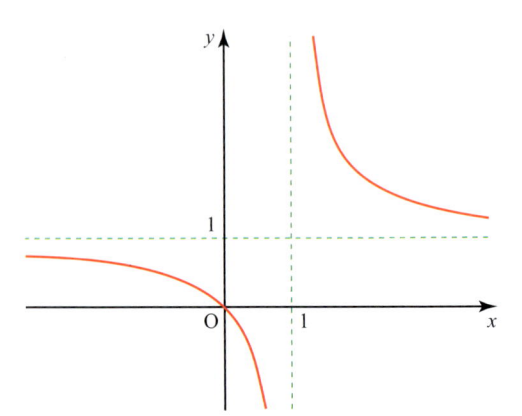

3 Given that $y = (x+1)(x-2)^2$
  - (i) find $\dfrac{dy}{dx}$
  - (ii) find any stationary points and determine their nature
  - (iii) sketch the curve.

4 Given that $y = \dfrac{x-3}{x-4}$
  - (i) find $\dfrac{dy}{dx}$
  - (ii) find the equation of the tangent to the curve at the point (6, 1.5)
  - (iii) find the equation of the normal to the curve at the point (5, 2)
  - (iv) use your answer from part (i) to deduce that the curve has no stationary points, and sketch the graph.

5 The diagram shows the graph of $y = \dfrac{2x}{\sqrt{x-1}}$, which is undefined for $x < 0$ and $x = 1$. P is a minimum point.

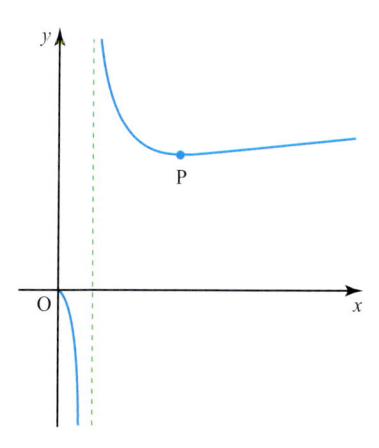

  - (i) Find $\dfrac{dy}{dx}$.
  - (ii) Find the gradient of the curve at (9, 9), and show that the equation of the normal at (9, 9) is $y = -4x + 45$.
  - (iii) Find the co-ordinates of P and verify that it is a minimum point.
  - (iv) Write down the equation of the tangent and the normal to the curve at P.
  - (v) Write down the point of intersection of the normal found in part (ii) and
    - (a) the tangent found in part (iv), call it Q
    - (b) the normal found in part (iv), call it R.
  - (vi) Show that the area of the triangle PQR is $\dfrac{441}{8}$.

6 The diagram shows the graph of $y = \dfrac{x^2 - 2x - 5}{2x + 3}$.

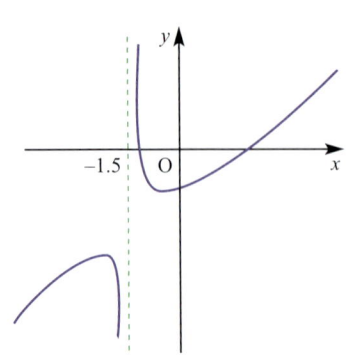

(i) Find $\dfrac{dy}{dx}$.

(ii) Use your answer from part (i) to find any stationary points of the curve.

(iii) Classify each of the stationary points and use calculus to justify your answer.

7 A curve has the equation $y = \dfrac{x^2}{2x + 1}$.

(i) Find $\dfrac{dy}{dx}$.

Hence find the co-ordinates of the stationary points on the curve.

(ii) You are given that $\dfrac{d^2y}{dx^2} = \dfrac{2}{(2x + 1)^3}$.

Use this information to determine the nature of the stationary points in part (i).

[MEI]

8 The diagram shows part of the graph with the equation $y = x\sqrt{9 - 2x^2}$.
It crosses the x axis at $(a, 0)$.

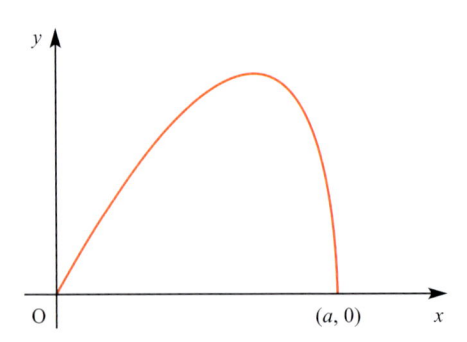

(i) Find the value of a, giving your answer as a multiple of $\sqrt{2}$.

(ii) Show that the result of differentiating $\sqrt{9-2x^2}$ is $\dfrac{-2x}{\sqrt{9-2x^2}}$.

Hence show that if $y = x\sqrt{9-2x^2}$ then

$$\frac{dy}{dx} = \frac{9-4x^2}{\sqrt{9-2x^2}}.$$

(iii) Find the x co-ordinate of the maximum point on the graph of $y = x\sqrt{9-2x^2}$.

Write down the gradient of the curve at the origin.

What can you say about the gradient at the point $(a, 0)$?

Differentiating natural logarithms and exponentials

In Chapter 2 you learnt that the integral of $\dfrac{1}{x}$ is $\ln x$. It follows, therefore, that the differential of $\ln x$ is $\dfrac{1}{x}$.

So $\quad y = \ln x \;\Rightarrow\; \dfrac{dy}{dx} = \dfrac{1}{x}$

The differential of the inverse function, $y = e^x$, may be found by interchanging y and x.

$$x = \ln y \;\Rightarrow\; \frac{dx}{dy} = \frac{1}{y}$$

$$\Rightarrow\; \frac{dy}{dx} = \frac{1}{\frac{dx}{dy}} = y = e^x.$$

Therefore $\dfrac{d}{dx}e^x = e^x$.

The differential of e^x is itself e^x. This may at first seem rather surprising.

P The function $f(x)$ $(x \in \mathbb{R})$ is a polynomial in x of order n.

So

$$f(x) = a_n x^n + a_{n-1} x^{n-1} + \ldots + a_1 x + a_0$$

where $a_n, a_{n-1}, \ldots, a_0$ are all constants and at least a_n is not zero.

How can you prove that $\dfrac{d}{dx} f(x)$ cannot equal $f(x)$?

Since the differential of e^x is e^x, it follows that the integral of e^x is also e^x.

$$\int e^x \, dx = e^x + c.$$

This may be summarised as in the following table.

Differentiation	Integration
$y \longrightarrow \dfrac{dy}{dx}$	$y \longrightarrow \int y\,dx$
$\ln x \longrightarrow \dfrac{1}{x}$	$\dfrac{1}{x} \longrightarrow \ln x + c$
$e^x \longrightarrow e^x$	$e^x \longrightarrow e^x + c$

These results allow you to extend very considerably the range of functions which you are able to differentiate and integrate.

EXAMPLE 4.5 Differentiate $y = e^{5x}$.

SOLUTION

Make the substitution $u = 5x$ to give $y = e^u$.

Now $\dfrac{dy}{du} = e^u = e^{5x}$ and $\dfrac{du}{dx} = 5$.

By the chain rule,

$$\dfrac{dy}{dx} = \dfrac{dy}{du} \times \dfrac{du}{dx}$$

$$= e^{5x} \times 5$$

$$= 5e^{5x}$$

This result can be generalised as follows.

$$y = e^{ax} \implies \dfrac{dy}{dx} = ae^{ax} \quad \text{where } a \text{ is any constant.}$$

This is an important standard result, and you would normally use it automatically, without recourse to the chain rule.

EXAMPLE 4.6 Differentiate $y = \dfrac{4}{e^{2x}}$.

SOLUTION

$$y = \dfrac{4}{e^{2x}} = 4e^{-2x}$$

$$\implies \dfrac{dy}{dx} = 4 \times \left(-2e^{-2x}\right)$$

$$= -8e^{-2x}$$

EXAMPLE 4.7 Differentiate $y = 3e^{(x^2+1)}$.

SOLUTION

Let $u = x^2 + 1$, then $y = 3e^u$.

$$\Rightarrow \quad \frac{dy}{du} = 3e^u = 3e^{(x^2+1)} \quad \text{and} \quad \frac{du}{dx} = 2x$$

By the chain rule,

$$\frac{dy}{dx} = \frac{dy}{du} \times \frac{du}{dx}$$

$$= 3e^{(x^2+1)} \times 2x$$

$$= 6xe^{(x^2+1)}$$

EXAMPLE 4.8 Differentiate the following.

(i) $y = 2 \ln x$ (ii) $y = \ln(3x)$

SOLUTION

(i) $\frac{dy}{dx} = 2 \times \frac{1}{x}$

$$= \frac{2}{x}$$

(ii) Let $u = 3x$, then $y = \ln u$

$$\Rightarrow \quad \frac{dy}{du} = \frac{1}{u} = \frac{1}{3x} \quad \text{and} \quad \frac{du}{dx} = 3$$

By the chain rule,

$$\frac{dy}{dx} = \frac{dy}{du} \times \frac{du}{dx}$$

$$= \frac{1}{3x} \times 3$$

$$= \frac{1}{x}$$

Note

An alternative solution to part (ii) is

$$y = \ln(3x) = \ln 3 + \ln x \quad \Rightarrow \quad \frac{dy}{dx} = 0 + \frac{1}{x} = \frac{1}{x}.$$

❓ The gradient function found in part (ii) above for $y = \ln(3x)$ is the same as that for $y = \ln(x)$. What does this tell you about the shapes of the two curves? For what values of x is it valid?

EXAMPLE 4.9 Differentiate the following.

(i) $y = \ln(x^4)$ (ii) $y = \ln(x^2 + 1)$

SOLUTION

(i) By the properties of logarithms

$$y = \ln(x^4)$$
$$= 4\ln(x)$$
$$\Rightarrow \frac{dy}{dx} = \frac{4}{x}$$

(ii) Let $u = x^2 + 1$, then $y = \ln u$

$$\Rightarrow \frac{dy}{du} = \frac{1}{u} = \frac{1}{x^2 + 1} \quad \text{and} \quad \frac{du}{dx} = 2x$$

By the chain rule,

$$\frac{dy}{dx} = \frac{dy}{du} \times \frac{du}{dx}$$
$$= \frac{1}{x^2 + 1} \times 2x$$
$$= \frac{2x}{x^2 + 1}$$

If you need to differentiate expressions similar to those in the examples above, follow exactly the same steps. The results can be generalised as follows.

$y = a\ln x \Rightarrow \frac{dy}{dx} = \frac{a}{x}$	$y = ae^x \Rightarrow \frac{dy}{dx} = ae^x$
$y = \ln(ax) \Rightarrow \frac{dy}{dx} = \frac{1}{x}$	$y = e^{ax} \Rightarrow \frac{dy}{dx} = ae^{ax}$
$y = \ln(f(x)) \Rightarrow \frac{dy}{dx} = \frac{f'(x)}{f(x)}$	$y = e^{f(x)} \Rightarrow \frac{dy}{dx} = f'(x)e^{f(x)}$

EXAMPLE 4.10 Differentiate $y = \frac{\ln x}{x}$.

SOLUTION

Here y is of the form $\frac{u}{v}$ where $u = \ln x$ and $v = x$

$$\Rightarrow \frac{du}{dx} = \frac{1}{x} \quad \text{and} \quad \frac{dv}{dx} = 1.$$

By the quotient rule,

$$\frac{dy}{dx} = \frac{v\frac{du}{dx} - u\frac{dv}{dx}}{v^2}$$

$$= \frac{x \times \frac{1}{x} - 1 \times \ln x}{x^2}$$

$$= \frac{1 - \ln x}{x^2}$$

EXERCISE 4B

1 Differentiate the following.

(i) $y = 3\ln x$
(ii) $y = \ln(4x)$
(iii) $y = \ln(x^2)$
(iv) $y = \ln(x^2 + 1)$
(v) $y = \ln\left(\frac{1}{x}\right)$
(vi) $y = x \ln x$
(vii) $y = x^2 \ln(4x)$
(viii) $y = \ln\left(\frac{x+1}{x}\right)$
(ix) $y = \ln\sqrt{x^2 - 1}$
(x) $y = \frac{\ln x}{x^2}$

2 Differentiate the following.

(i) $y = 3e^x$
(ii) $y = e^{2x}$
(iii) $y = e^{x^2}$
(iv) $y = e^{(x+1)^2}$
(v) $y = xe^{4x}$
(vi) $y = 2x^3 e^{-x}$
(vii) $y = \frac{x}{e^x}$
(viii) $y = (e^{2x} + 1)^3$

3 Knowing how much rain has fallen in a river basin, hydrologists are often able to give forecasts of what will happen to a river level over the next few hours. In one case it is predicted that the height h, in metres, of a river above its normal level during the next 3 hours will be $0.12e^{0.9t}$, where t is the time elapsed, in hours, after the prediction.

(i) Find $\frac{dh}{dt}$, the rate at which the river is rising.
(ii) At what rate will the river be rising after 0, 1, 2 and 3 hours?

4 The graph of $y = xe^x$ is shown below.

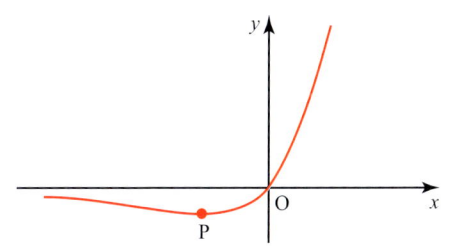

(i) Find $\frac{dy}{dx}$ and $\frac{d^2y}{dx^2}$.
(ii) Find the co-ordinates of the minimum point P.

5 The graph of $f(x) = x\ln(x^2)$ is shown below.

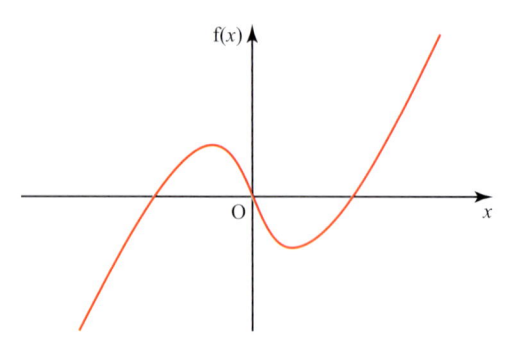

(i) Describe, giving a reason, any symmetries of the graph.
(ii) Find $f'(x)$ and $f''(x)$.
(iii) Find the co-ordinates of any stationary points.

6 Given that $y = \dfrac{e^x}{x}$

(i) find $\dfrac{dy}{dx}$

(ii) find the co-ordinates of any stationary points on the curve
(iii) sketch the curve.

7 (i) Differentiate $\ln x$ and $x \ln x$ with respect to x.

The sketch shows the graph of $y = x \ln x$ for $0 \leq x \leq 3$.

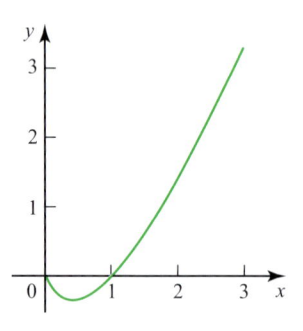

(ii) Show that the curve has a stationary point $\left(\dfrac{1}{e}, -\dfrac{1}{e}\right)$.

[MEI]

8 The diagram shows the graph of $y = xe^{-x}$.

(i) Differentiate xe^{-x}.
(ii) Find the co-ordinates of the point A, the maximum point on the curve.

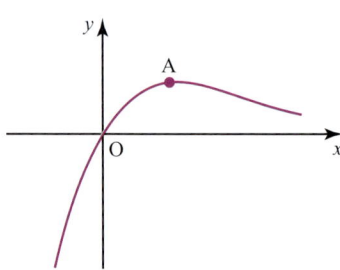

[MEI]

9 The diagram shows a sketch of the graph of $y = f(x)$, where

$$f(x) = \frac{\ln x}{x} \quad (x > 0).$$

The graph crosses the x axis at the point P and has a turning point at Q.

(i) Write down the x co-ordinate of P.

(ii) Find the first and second derivatives, $f'(x)$ and $f''(x)$, simplifying your answers as far as possible.

(iii) Hence show that the x co-ordinate of Q is e.
Find the y co-ordinate of Q in terms of e.
Find $f''(e)$, and use this result to verify that Q is a maximum point.

[MEI, *part*]

10 Find the exact co-ordinates of the point on the curve $y = xe^{-\frac{1}{2}x}$ at which $\dfrac{d^2y}{dx^2} = 0$.

[Cambridge International AS & A Level Mathematics 9709, Paper 2 Q6 November 2008]

11 It is given that the curve $y = (x-2)e^x$ has one stationary point.

(i) Find the exact co-ordinates of this point.

(ii) Determine whether this point is a maximum or a minimum point.

[Cambridge International AS & A Level Mathematics 9709, Paper 2 Q6 June 2008]

12 The equation of a curve is $y = x^3 e^{-x}$.

(i) Show that the curve has a stationary point where $x = 3$.

(ii) Find the equation of the tangent to the curve at the point where $x = 1$.

[Cambridge International AS & A Level Mathematics 9709, Paper 22 Q5 June 2010]

Differentiating trigonometrical functions

ACTIVITY 4.2 Figure 4.2 shows the graph of $y = \sin x$, with x measured in radians, together with the graph of $y = x$. You are going to sketch the graph of the gradient function for the graph of $y = \sin x$.

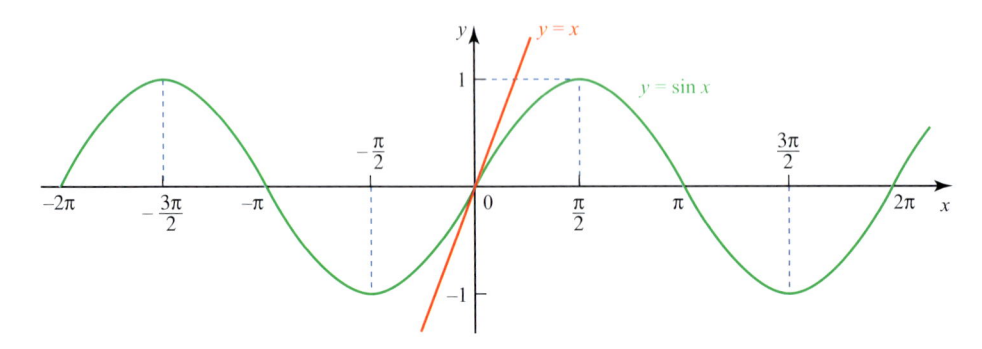

Figure 4.2

Draw a horizontal axis for the angles, marked from -2π to 2π, and a vertical axis for the gradient, marked from -1 to 1, as shown in Figure 4.3.

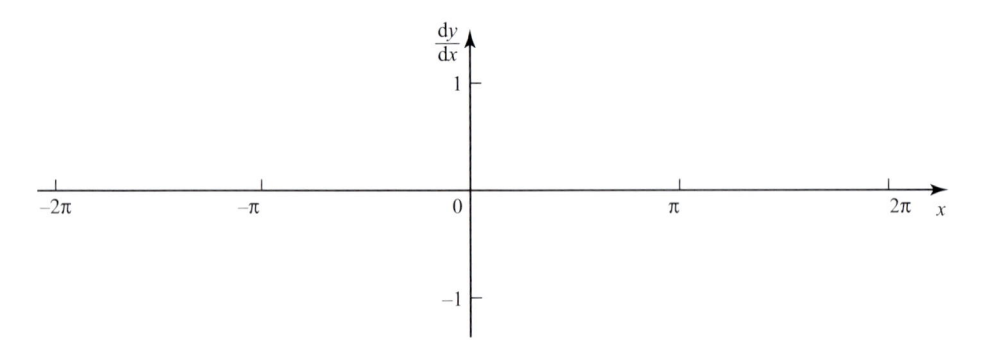

Figure 4.3

First, look for the angles for which the gradient of $y = \sin x$ is zero. Mark zeros at these angles on your gradient graph.

Decide which parts of $y = \sin x$ have a positive gradient and which have a negative gradient. This will tell you whether your gradient graph should be above or below the x axis at any point.

Look at the part of the graph of $y = \sin x$ near $x = 0$ and compare it with the graph of $y = x$. What do you think the gradient of $y = \sin x$ is at this point? Mark this point on your gradient graph. Also mark on any other points with plus or minus the same gradient.

Now, by considering whether the gradient of $y = \sin x$ is increasing or decreasing at any particular point, sketch in the rest of the gradient graph.

The gradient graph that you have drawn should look like a familiar graph. What graph do you think it is?

Sketch the graph of $y = \cos x$, with x measured in radians, and use it as above to obtain a sketch of the graph of the gradient function of $y = \cos x$.

❓ Is $y = x$ still a tangent of $y = \sin x$ if x is measured in degrees?

Activity 4.2 showed you that the graph of the gradient function of $y = \sin x$ resembled the graph of $y = \cos x$. You will also have found that the graph of the gradient function of $y = \cos x$ looks like the graph of $y = \sin x$ reflected in the x axis to become $y = -\sin x$.

🅟 Both of these results are in fact true but the work above does not amount to a proof. Explain why.

Summary of results

$$\frac{d}{dx}(\sin x) = \cos x \qquad \frac{d}{dx}(\cos x) = -\sin x$$

⚠ Remember that these results are only valid when the angle is measured in radians, so when you are using any of the derivatives of trigonometrical functions you need to work in radians.

ACTIVITY 4.3 By writing $\tan x = \dfrac{\sin x}{\cos x}$, use the quotient rule to show that

$$\frac{d}{dx}(\tan x) = \sec^2 x \text{ where } x \text{ is measured in radians.}$$

You can use the three results met so far to differentiate a variety of functions involving trigonometrical functions, by using the chain rule, product rule or quotient rule, as in the following examples.

EXAMPLE 4.11 Differentiate $y = \cos 2x$.

SOLUTION

As $\cos 2x$ is a function of a function, you may use the chain rule.

Let $u = 2x \Rightarrow \dfrac{du}{dx} = 2$

$y = \cos u \Rightarrow \dfrac{dy}{du} = -\sin u$

$\dfrac{dy}{dx} = \dfrac{dy}{du} \times \dfrac{du}{dx}$

$= -\sin u \times 2$

$= -2 \sin 2x$

With practice it should be possible to do this in your head, without needing to write down the substitution.

This result may be generalised.

$$y = \cos kx \Rightarrow \dfrac{dy}{dx} = -k \sin kx.$$

Similarly

$$y = \sin kx \Rightarrow \dfrac{dy}{dx} = k \cos kx$$

and

$$y = \tan kx \Rightarrow \dfrac{dy}{dx} = k \sec^2 kx.$$

EXAMPLE 4.12 Differentiate $y = x^2 \sin x$.

SOLUTION

$x^2 \sin x$ is of the form uv, so the product rule can be used with $u = x^2$ and $v = \sin x$.

$\dfrac{du}{dx} = 2x \qquad \dfrac{dv}{dx} = \cos x$

Using the product rule

$\dfrac{dy}{dx} = v\dfrac{du}{dx} + u\dfrac{dv}{dx}$

$\Rightarrow \dfrac{dy}{dx} = 2x \sin x + x^2 \cos x$

EXAMPLE 4.13 Differentiate $y = e^{\tan x}$.

SOLUTION

$e^{\tan x}$ is a function of a function, so the chain rule may be used.

Let $\quad u = \tan x \quad \Rightarrow \quad \dfrac{du}{dx} = \sec^2 x$

$\qquad y = e^u \quad \Rightarrow \quad \dfrac{dy}{du} = e^u$

Using the chain rule

$$\dfrac{dy}{dx} = \dfrac{dy}{du} \times \dfrac{du}{dx}$$

$$= e^u \sec^2 x$$

$$= e^{\tan x} \sec^2 x$$

EXAMPLE 4.14 Differentiate $y = \dfrac{1 + \sin x}{\cos x}$.

SOLUTION

$\dfrac{1 + \sin x}{\cos x}$ is of the form $\dfrac{u}{v}$ so the quotient rule can be used, with

$\qquad u = 1 + \sin x \qquad$ and $\qquad v = \cos x$

$\Rightarrow \quad \dfrac{du}{dx} = \cos x \qquad$ and $\qquad \dfrac{dv}{dx} = -\sin x$

The quotient rule is

$$\dfrac{dy}{dx} = \dfrac{v \dfrac{du}{dx} - u \dfrac{dv}{dx}}{v^2}$$

Substituting for u and v and their derivatives gives

$$\dfrac{dy}{dx} = \dfrac{(\cos x)(\cos x) - (1 + \sin x)(-\sin x)}{(\cos x)^2}$$

$$= \dfrac{\cos^2 x + \sin x + \sin^2 x}{\cos^2 x}$$

$$= \dfrac{1 + \sin x}{\cos^2 x} \qquad (\text{using } \sin^2 x + \cos^2 x = 1)$$

$$= (\sec^2 x)(1 + \sin x)$$

EXERCISE 4C

1. Differentiate each of the following.
   - (i) $2\cos x + \sin x$
   - (ii) $\tan x + 5$
   - (iii) $\sin x - \cos x$

2. Use the product rule to differentiate each of the following.
   - (i) $x \tan x$
   - (ii) $\sin x \cos x$
   - (iii) $e^x \sin x$

3. Use the quotient rule to differentiate each of the following.
   - (i) $\dfrac{\sin x}{x}$
   - (ii) $\dfrac{e^x}{\cos x}$
   - (iii) $\dfrac{x + \cos x}{\sin x}$

4. Use the chain rule to differentiate each of the following.
   - (i) $\tan(x^2 + 1)$
   - (ii) $\sin 2x$
   - (iii) $\ln(\sin x)$

5. Use an appropriate method to differentiate each of the following.
   - (i) $\sqrt{\cos x}$
   - (ii) $e^x \tan x$
   - (iii) $\sin 4x^2$
   - (iv) $e^{\cos 2x}$
   - (v) $\dfrac{\sin x}{1 + \cos x}$
   - (vi) $\ln(\tan x)$

6. (i) Differentiate $y = x \cos x$.
   (ii) Find the gradient of the curve $y = x \cos x$ at the point where $x = \pi$.
   (iii) Find the equation of the tangent to the curve $y = x \cos x$ at the point where $x = \pi$.
   (iv) Find the equation of the normal to the curve $y = x \cos x$ at the point where $x = \pi$.

7. If $y = e^x \cos 3x$, find $\dfrac{dy}{dx}$ and $\dfrac{d^2y}{dx^2}$ and hence show that
$$\dfrac{d^2y}{dx^2} - 2\dfrac{dy}{dx} + 10y = 0.$$
[MEI]

8. Consider the function $y = e^{-x} \sin x$, where $-\pi \leqslant x \leqslant \pi$.
   (i) Find $\dfrac{dy}{dx}$.
   (ii) Show that, at stationary points, $\tan x = 1$.
   (iii) Determine the co-ordinates of the stationary points, correct to 2 significant figures.
   (iv) Explain how you could determine whether your stationary points are maxima or minima. You are not required to do any calculations.
[MEI]

9. The equation of a curve is $y = x + 2\cos x$. Find the x co-ordinates of the stationary points of the curve for $0 \leqslant x \leqslant 2\pi$, and determine the nature of each of these stationary points.

[Cambridge International AS & A Level Mathematics 9709, Paper 2 Q3 June 2006]

10 The equation of a curve is $y = x + \cos 2x$. Find the x co-ordinates of the stationary points of the curve for which $0 \leqslant x \leqslant \pi$, and determine the nature of each of these stationary points.

[Cambridge International AS & A Level Mathematics 9709, Paper 3 Q3 November 2005]

11 The curve with equation $y = e^{-x} \sin x$ has one stationary point for which $0 \leqslant x \leqslant \pi$.

(i) Find the x co-ordinate of this point.

(ii) Determine whether this point is a maximum or a minimum point.

[Cambridge International AS & A Level Mathematics 9709, Paper 3 Q4 November 2007]

12 The curve $y = \dfrac{e^x}{\cos x}$, for $-\tfrac{1}{2}\pi < x < \tfrac{1}{2}\pi$, has one stationary point. Find the x co-ordinate of this point.

[Cambridge International AS & A Level Mathematics 9709, Paper 3 Q3 November 2008]

Differentiating functions defined implicitly

All the functions you have differentiated so far have been of the form $y = f(x)$. However, many functions cannot be arranged in this way at all, for example $x^3 + y^3 = xy$, and others can look clumsy when you try to make y the subject.

An example of this is the semi-circle $x^2 + y^2 = 4$, $y \geqslant 0$, illustrated in figure 4.4.

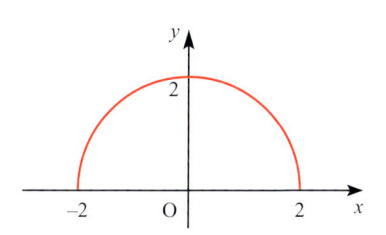

Figure 4.4

Because of Pythagoras' theorem, the curve is much more easily recognised in this form than in the equivalent $y = \sqrt{4 - x^2}$.

When a function is specified by an equation connecting x and y which does not have y as the subject it is called an *implicit function*.

The chain rule $\dfrac{dy}{dx} = \dfrac{dy}{du} \times \dfrac{du}{dx}$ and the product rule $\dfrac{d}{dx}(uv) = u\dfrac{dv}{dx} + v\dfrac{du}{dx}$ are used extensively to help in the differentiation of implicit functions.

EXAMPLE 4.15 Differentiate each of the following with respect to x.

(i) y^2 (ii) xy (iii) $3x^2y^3$ (iv) $\sin y$

SOLUTION

(i) $\dfrac{d}{dx}(y^2) = \dfrac{d}{dy}(y^2) \times \dfrac{dy}{dx}$ (chain rule)

$\qquad = 2y\dfrac{dy}{dx}$

(ii) $\dfrac{d}{dx}(xy) = x\dfrac{dy}{dx} + y$ (product rule)

(iii) $\dfrac{d}{dx}(3x^2y^3) = 3\left(x^2\dfrac{d}{dx}(y^3) + y^3\dfrac{d}{dx}(x^2)\right)$ (product rule)

$\qquad = 3\left(x^2 \times 3y^2\dfrac{dy}{dx} + y^3 \times 2x\right)$ (chain rule)

$\qquad = 3xy^2\left(3x\dfrac{dy}{dx} + 2y\right)$

(iv) $\dfrac{d}{dx}(\sin y) = \dfrac{d}{dy}(\sin y) \times \dfrac{dy}{dx}$ (chain rule)

$\qquad = (\cos y)\dfrac{dy}{dx}$

EXAMPLE 4.16 The equation of a curve is given by $y^3 + xy = 2$.

(i) Find an expression for $\dfrac{dy}{dx}$ in terms of x and y.

(ii) Hence find the gradient of the curve at $(1, 1)$ and the equation of the tangent to the curve at that point.

SOLUTION

(i) $y^3 + xy = 2$

$\Rightarrow \quad 3y^2\dfrac{dy}{dx} + \left(x\dfrac{dy}{dx} + y\right) = 0$

$\Rightarrow \quad (3y^2 + x)\dfrac{dy}{dx} = -y$

$\Rightarrow \quad \dfrac{dy}{dx} = \dfrac{-y}{3y^2 + x}$

(ii) At $(1, 1)$, $\dfrac{dy}{dx} = -\dfrac{1}{4}$

Substitute $x = 1$, $y = 1$ into the expression for $\dfrac{dy}{dx}$.

$\Rightarrow$ Using $y - y_1 = m(x - x_1)$ the equation of the tangent is $(y - 1) = -\dfrac{1}{4}(x - 1)$

$\Rightarrow \quad x + 4y - 5 = 0$

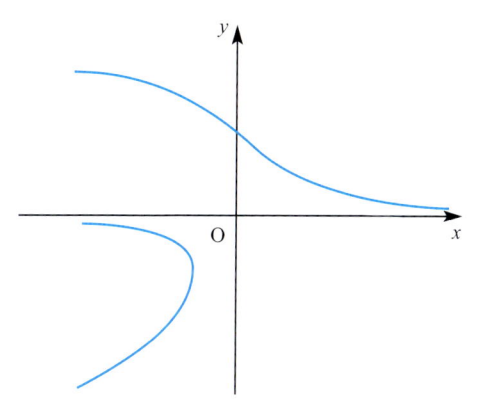

❓ Figure 4.5 shows the graph of the curve with the equation $y^3 + xy = 2$.

Figure 4.5

Why is this not a function?

Stationary points

As before these occur where $\dfrac{dy}{dx} = 0$.

Putting $\dfrac{dy}{dx} = 0$ will not usually give values of x directly, but will give a relationship between x and y. This needs to be solved simultaneously with the equation of the curve to find the co-ordinates.

EXAMPLE 4.17

(i) Differentiate $x^3 + y^3 = 3xy$ with respect to x.

(ii) Hence find the co-ordinates of any stationary points.

SOLUTION

(i) $\dfrac{d}{dx}(x^3) + \dfrac{d}{dx}(y^3) = \dfrac{d}{dx}(3xy)$

$\Rightarrow 3x^2 + 3y^2 \dfrac{dy}{dx} = 3\left(x\dfrac{dy}{dx} + y\right)$

(ii) At stationary points, $\dfrac{dy}{dx} = 0$

$\Rightarrow 3x^2 = 3y$

$\Rightarrow x^2 = y$

> Notice how it is not necessary to find an expression for $\dfrac{dy}{dx}$ unless you are told to.

To find the co-ordinates of the stationary points, solve

$$\left. \begin{array}{r} x^2 = y \\ x^3 + y^3 = 3xy \end{array} \right\} \text{simultaneously}$$

Substituting for y gives

$$x^3 + (x^2)^3 = 3x(x^2)$$
$$\Rightarrow \quad x^3 + x^6 = 3x^3$$
$$\Rightarrow \quad x^6 = 2x^3$$
$$\Rightarrow \quad x^3(x^3 - 2) = 0$$
$$\Rightarrow \quad x = 0 \quad \text{or} \quad x = \sqrt[3]{2}$$

$y = x^2$ so the stationary points are $(0, 0)$ and $\left(\sqrt[3]{2}, \sqrt[3]{4}\right)$.

The stationary points are A and B in figure 4.6.

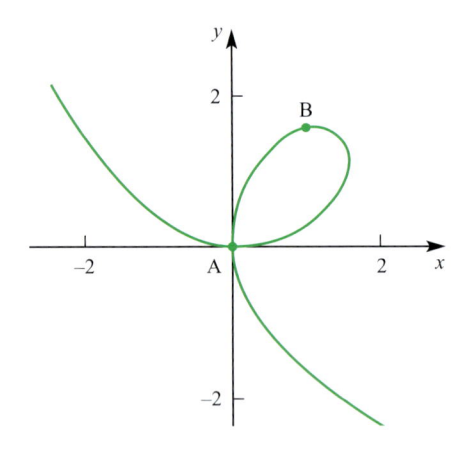

Figure 4.6

Types of stationary points

As with explicit functions, the nature of a stationary point can be determined by considering the sign of $\dfrac{d^2y}{dx^2}$ either side of the stationary point.

EXAMPLE 4.18

The curve with equation $\sin x + \sin y = 1$ for $0 \leqslant x \leqslant \pi$, $0 \leqslant y \leqslant \pi$ is shown in figure 4.7.

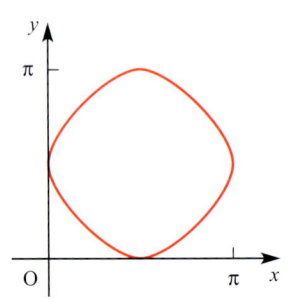

Figure 4.7

(i) Differentiate the equation of the curve with respect to x and hence find the co-ordinates of any stationary points.

(ii) Show that the points $\left(\frac{\pi}{6}, \frac{\pi}{6}\right)$, $\left(\frac{\pi}{6}, \frac{5\pi}{6}\right)$, $\left(\frac{5\pi}{6}, \frac{\pi}{6}\right)$ and $\left(\frac{5\pi}{6}, \frac{5\pi}{6}\right)$ all lie on the curve.
Find the gradient at each of these points.
What can you conclude about the natures of the stationary points?

SOLUTION

(i) $$\sin x + \sin y = 1$$
$$\Rightarrow \cos x + (\cos y)\frac{dy}{dx} = 0 \quad \text{①}$$
$$\Rightarrow \frac{dy}{dx} = -\frac{\cos x}{\cos y}$$

At any stationary point $\frac{dy}{dx} = 0 \Rightarrow \cos x = 0$
$$\Rightarrow x = \frac{\pi}{2} \text{ (only solution in range)}$$

Substitute in $\sin x + \sin y = 1$.

When $x = \frac{\pi}{2}$, $\sin x = 1 \Rightarrow \sin y = 0$
$$\Rightarrow y = 0 \text{ or } y = \pi$$

$\Rightarrow$ stationary points at $\left(\frac{\pi}{2}, 0\right)$ and $\left(\frac{\pi}{2}, \pi\right)$.

(ii) $\sin\frac{\pi}{6} = \frac{1}{2}$, $\sin\frac{5\pi}{6} = \frac{1}{2}$

So, for each of the four given points, $\sin x + \sin y = \frac{1}{2} + \frac{1}{2} = 1$.

Therefore they all lie on the curve.

The gradient of the curve is given by
$$\frac{dy}{dx} = -\frac{\cos x}{\cos y}$$

$\cos\frac{\pi}{6} = \frac{\sqrt{3}}{2}$, $\cos\frac{5\pi}{6} = -\frac{\sqrt{3}}{2}$

At $\left(\frac{\pi}{6}, \frac{\pi}{6}\right)$, $\frac{dy}{dx} = -\frac{\frac{\sqrt{3}}{2}}{\frac{\sqrt{3}}{2}} = -1$

At $\left(\frac{\pi}{6}, \frac{5\pi}{6}\right)$, $\frac{dy}{dx} = -\frac{\frac{\sqrt{3}}{2}}{-\frac{\sqrt{3}}{2}} = 1$

At $\left(\frac{5\pi}{6}, \frac{\pi}{6}\right)$, $\frac{dy}{dx} = -\frac{-\frac{\sqrt{3}}{2}}{\frac{\sqrt{3}}{2}} = 1$

At $\left(\frac{5\pi}{6}, \frac{5\pi}{6}\right)$, $\frac{dy}{dx} = -\frac{-\frac{\sqrt{3}}{2}}{-\frac{\sqrt{3}}{2}} = -1$

These results show that

$\left(\dfrac{\pi}{2}, 0\right)$ is a minimum $\left(\dfrac{\pi}{2}, \pi\right)$ is a maximum

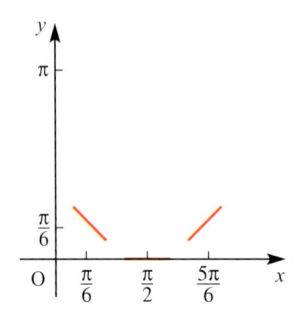

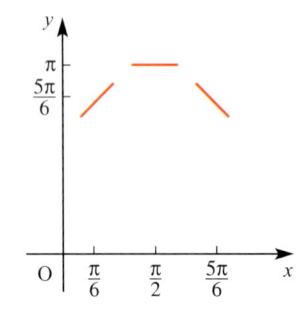

These points are confirmed by considering the sketch in figure 4.7 on page 100.

EXERCISE 4D

1 Differentiate each of the following with respect to x.

(i) y^4
(ii) $x^2 + y^3 - 5$
(iii) $xy + x + y$
(iv) $\cos y$
(v) $e^{(y+2)}$
(vi) xy^3
(vii) $2x^2 y^5$
(viii) $x + \ln y - 3$
(ix) $xe^y - \cos y$
(x) $x^2 \ln y$
(xi) $xe^{\sin y}$
(xii) $x \tan y - y \tan x$

2 Find the gradient of the curve $xy^3 = 5 \ln y$ at the point $(0, 1)$.

3 Find the gradient of the curve $e^{\sin x} + e^{\cos y} = e + 1$ at the point $\left(\dfrac{\pi}{2}, \dfrac{\pi}{2}\right)$.

4 (i) Find the gradient of the curve $x^2 + 3xy + y^2 = x + 3y$ at the point $(2, -1)$.
  (ii) Hence find the equation of the tangent to the curve at this point.

5 Find the co-ordinates of all the stationary points on the curve $x^2 + y^2 + xy = 3$.

6 A curve has the equation $(x - 6)(y + 4) = 2$.

  (i) Find an expression for $\dfrac{dy}{dx}$ in terms of x and y.
  (ii) Find the equation of the normal to the curve at the point $(7, -2)$.
  (iii) Find the co-ordinates of the point where the normal meets the curve again.
  (iv) By rewriting the equation in the form $y - a = \dfrac{b}{x - c}$ identify any asymptotes and sketch the curve.

7 A curve has the equation $y = x^x$ for $x > 0$.

   (i) Take logarithms to base e of both sides of the equation.

   (ii) Differentiate the resulting equation with respect to x.

   (iii) Find the co-ordinates of the stationary point, giving your answer to 3 decimal places.

   (iv) Sketch the curve for $x > 0$.

8 The equation of a curve is $3x^2 + 2xy + y^2 = 6$. It is given that there are two points on the curve where the tangent is parallel to the x axis.

   (i) Show by differentiation that, at these points, $y = -3x$.

   (ii) Hence find the co-ordinates of the two points.

   [Cambridge International AS & A Level Mathematics 9709, Paper 2 Q5 June 2006]

9 The equation of a curve is $x^3 + y^3 = 9xy$.

   (i) Show that $\dfrac{dy}{dx} = \dfrac{3y - x^2}{y^2 - 3x}$.

   (ii) Find the equation of the tangent to the curve at the point $(2, 4)$, giving your answer in the form $ax + by = c$.

   [Cambridge International AS & A Level Mathematics 9709, Paper 2 Q4 November 2005]

10 The equation of a curve is $x^2 + y^2 - 4xy + 3 = 0$.

   (i) Show that $\dfrac{dy}{dx} = \dfrac{2y - x}{y - 2x}$.

   (ii) Find the co-ordinates of each of the points on the curve where the tangent is parallel to the x axis.

   [Cambridge International AS & A Level Mathematics 9709, Paper 2 Q7 June 2008]

11 The equation of a curve is $x^3 - x^2y - y^3 = 3$.

   (i) Find $\dfrac{dy}{dx}$ in terms of x and y.

   (ii) Find the equation of the tangent to the curve at the point $(2, 1)$, giving your answer in the form $ax + by + c = 0$.

   [Cambridge International AS & A Level Mathematics 9709, Paper 32 Q3 November 2009]

12 The equation of a curve is $xy(x + y) = 2a^3$, where a is a non-zero constant. Show that there is only one point on the curve at which the tangent is parallel to the x axis, and find the co-ordinates of this point.

   [Cambridge International AS & A Level Mathematics 9709, Paper 3 Q6 June 2008]

Parametric equations

When you go on a ride like the one in the picture, your body follows a very unnatural path and this gives rise to sensations which you may find exhilarating or frightening.

You are accustomed to expressing curves as mathematical equations. How would you do so in a case like this?

Figure 4.8 shows a simplified version of such a ride.

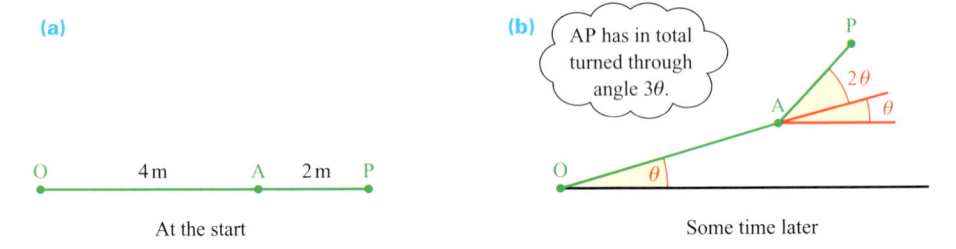

(a) At the start

(b) Some time later — AP has in total turned through angle 3θ.

Figure 4.8

The passenger's chair is on the end of a rod AP of length 2 m which is rotating about A. The rod OA is 4 m long and is itself rotating about O. The gearing of the mechanism ensures that the rod AP rotates twice as fast relative to OA as the rod OA does. This is illustrated by the angles marked on figure 4.8(b), at a time when OA has rotated through an angle θ.

At this time, the co-ordinates of the point P, taking O as the origin, are given by

$$x = 4\cos\theta + 2\cos 3\theta$$
$$y = 4\sin\theta + 2\sin 3\theta$$

(see figure 4.9).

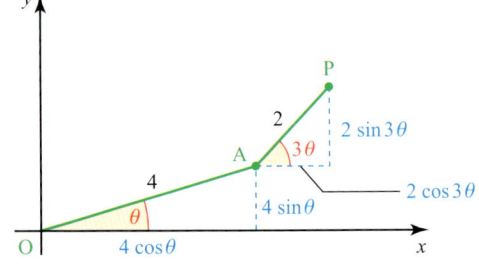

Figure 4.9

These two equations are called *parametric equations* of the curve. They do not give the relationship between *x* and *y* directly in the form $y = f(x)$ but use a third variable, θ, to do so. This third variable is called the *parameter*.

To plot the curve, you need to substitute values of θ and find the corresponding values of *x* and *y*.

Thus $\theta = 0°$ $\Rightarrow$ $x = 4 + 2 = 6$
$y = 0 + 0 = 0$ Point (6, 0)

$\theta = 30°$ $\Rightarrow$ $x = 4 \times 0.866 + 0 = 3.464$
$y = 4 \times 0.5 + 2 \times 1 = 4$ Point (3.46, 4)

and so on.

Joining points found in this way reveals the curve to have the shape shown in figure 4.10.

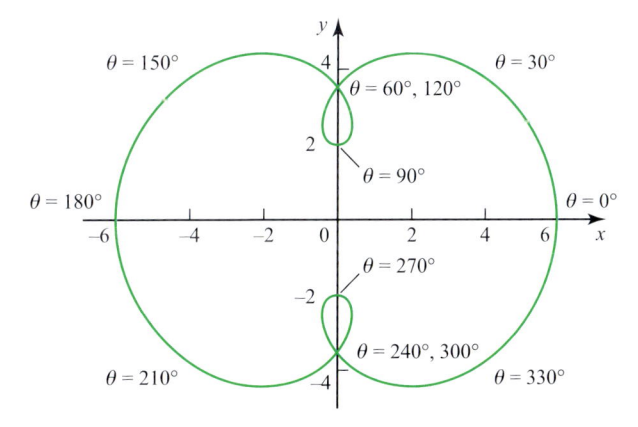

Figure 4.10

? At what points of the curve would you feel the greatest sensations?

ⓑ Graphs from parametric equations

Parametric equations are very useful in situations such as this, where an otherwise complicated equation may be expressed reasonably simply in terms of a parameter. Indeed, there are some curves which can be given by parametric equations but cannot be written as cartesian equations (in terms of *x* and *y* only).

The next example is based on a simpler curve. Make sure that you can follow the solution completely before going on to the rest of the chapter.

EXAMPLE 4.19 A curve has the parametric equations $x = 2t$, $y = \dfrac{36}{t^2}$.

(i) Find the co-ordinates of the points corresponding to $t = 1, 2, 3, -1, -2$ and -3.
(ii) Plot the points you have found and join them to give the curve.
(iii) Explain what happens as $t \to 0$.

SOLUTION

(i)

t	−3	−2	−1	1	2	3
x	−6	−4	−2	2	4	6
y	4	9	36	36	9	4

The points required are (−6, 4), (−4, 9), (−2, 36), (2, 36), (4, 9) and (6, 4).

(ii) The curve is shown in figure 4.11.

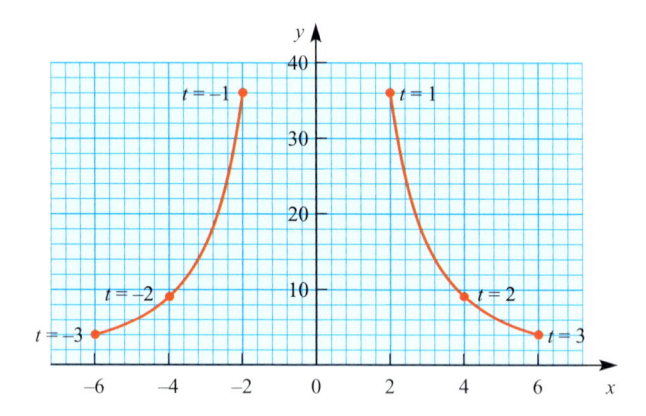

Figure 4.11

(iii) As $t \to 0$, $x \to 0$ and $y \to \infty$. The y axis is an asymptote for the curve.

EXAMPLE 4.20 A curve has the parametric equations $x = t^2$, $y = t^3 - t$.

(i) Find the co-ordinates of the points corresponding to values of t from -2 to $+2$ at half-unit intervals.
(ii) Sketch the curve for $-2 \leqslant t \leqslant 2$.
(iii) Are there any values of x for which the curve is undefined?

SOLUTION

(i)

t	−2	−1.5	−1	−0.5	0	0.5	1	1.5	2
x	4	2.25	1	0.25	0	0.25	1	2.25	4
y	−6	−1.875	0	0.375	0	−0.375	0	1.875	6

(ii)

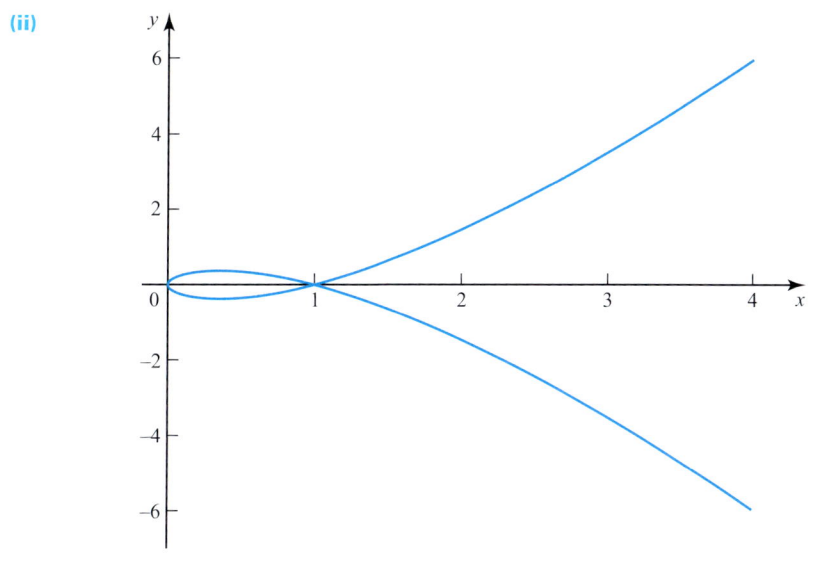

Figure 4.12

(iii) The curve in figure 4.12 is undefined for $x < 0$.

Graphic calculators can be used to sketch parametric curves but, as with cartesian curves, you need to be careful when choosing the range.

b Finding the equation by eliminating the parameter

For some pairs of parametric equations, it is possible to eliminate the parameter and obtain the cartesian equation for the curve. This is usually done by making the parameter the subject of one of the equations, and substituting this expression into the other.

EXAMPLE 4.21 Eliminate t from the equations $x = t^3 - 2t^2$, $y = \dfrac{t}{2}$.

SOLUTION

$y = \dfrac{t}{2} \quad \Rightarrow \quad t = 2y.$

Substituting this in the equation $x = t^3 - 2t^2$ gives

$$x = (2y)^3 - 2(2y)^2 \quad \text{or} \quad x = 8y^3 - 8y^2.$$

Parametric differentiation

To differentiate a function which is defined in terms of a parameter t, you need to use the chain rule:

$$\frac{dy}{dx} = \frac{dy}{dt} \times \frac{dt}{dx}.$$

Since

$$\frac{dt}{dx} = \frac{1}{\frac{dx}{dt}}$$

it follows that

$$\frac{dy}{dx} = \frac{\frac{dy}{dt}}{\frac{dx}{dt}}$$

provided that $\frac{dx}{dt} \neq 0$.

EXAMPLE 4.22 A curve has the parametric equations $x = t^2$, $y = 2t$.

(i) Find $\frac{dy}{dx}$ in terms of the parameter t.

(ii) Find the equation of the tangent to the curve at the general point $(t^2, 2t)$.

(iii) Find the equation of the tangent at the point where $t = 3$.

(iv) Eliminate the parameter, and hence sketch the curve and the tangent at the point where $t = 3$.

SOLUTION

(i) $x = t^2 \quad \Rightarrow \quad \frac{dx}{dt} = 2t$

$y = 2t \quad \Rightarrow \quad \frac{dy}{dt} = 2$

$$\frac{dy}{dx} = \frac{\frac{dy}{dt}}{\frac{dx}{dt}} = \frac{2}{2t} = \frac{1}{t}$$

(ii) Using $y - y_1 = m(x - x_1)$ and taking the point (x_1, y_1) as $(t^2, 2t)$, the equation of the tangent at the point $(t^2, 2t)$ is

$$y - 2t = \frac{1}{t}(x - t^2)$$

$\Rightarrow \quad ty - 2t^2 = x - t^2$

$\Rightarrow \quad x - ty + t^2 = 0$

This equation still contains the parameter, and is called the equation of the tangent at the general point.

(iii) Substituting $t = 3$ into this equation gives the equation of the tangent at the point where $t = 3$.

The tangent is $x - 3y + 9 = 0$.

(iv) Eliminating t from $x = t^2$, $y = 2t$ gives

$$x = \left(\frac{y}{2}\right)^2 \quad \text{or} \quad y^2 = 4x.$$

This is a parabola with the x axis as its line of symmetry.

The point where $t = 3$ has co-ordinates $(9, 6)$.

The tangent $x - 3y + 9 = 0$ crosses the axes at $(0, 3)$ and $(-9, 0)$.

The curve is shown in figure 4.13.

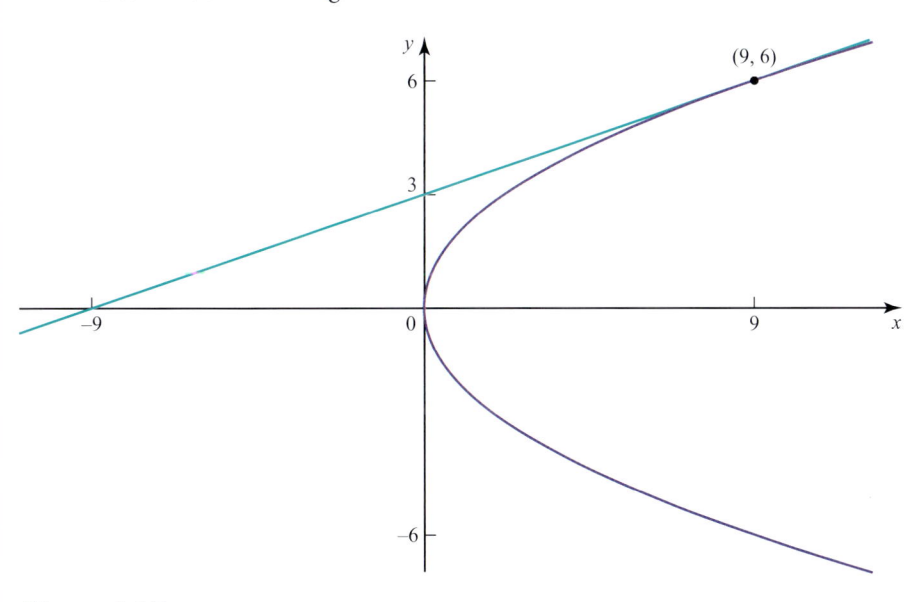

Figure 4.13

EXAMPLE 4.23 A curve has parametric equations $x = 4\cos\theta$, $y = 3\sin\theta$.

(i) Find $\dfrac{dy}{dx}$ at the point with parameter θ.

(ii) Find the equation of the normal at the general point $(4\cos\theta, 3\sin\theta)$.

(iii) Find the equation of the normal at the point where $\theta = \dfrac{\pi}{4}$.

(iv) Find the co-ordinates of the point where $\theta = \dfrac{\pi}{4}$.

(v) Show the curve and the normal on a sketch.

SOLUTION

(i) $x = 4\cos\theta \quad \Rightarrow \quad \dfrac{dx}{d\theta} = -4\sin\theta$

$y = 3\sin\theta \quad \Rightarrow \quad \dfrac{dy}{d\theta} = 3\cos\theta$

$$\frac{dy}{dx} = \frac{\frac{dy}{d\theta}}{\frac{dx}{d\theta}} = \frac{3\cos\theta}{-4\sin\theta}$$

$$= -\frac{3\cos\theta}{4\sin\theta}$$

(ii) The tangent and normal are perpendicular, so the gradient of the normal is

$$-\dfrac{1}{\frac{dy}{dx}} \quad \text{which is} \quad +\dfrac{4\sin\theta}{3\cos\theta}.$$

($m_1 m_2 = -1$ for perpendicular lines.)

Using $y - y_1 = m(x - x_1)$ and taking the point (x_1, y_1) as $(4\cos\theta, 3\sin\theta)$, the equation of the normal at the point $(4\cos\theta, 3\sin\theta)$ is

$$y - 3\sin\theta = \dfrac{4\sin\theta}{3\cos\theta}(x - 4\cos\theta)$$

$$\Rightarrow \quad 3y\cos\theta - 9\sin\theta\cos\theta = 4x\sin\theta - 16\sin\theta\cos\theta$$

$$\Rightarrow \quad 4x\sin\theta - 3y\cos\theta - 7\sin\theta\cos\theta = 0$$

(iii) When $\theta = \dfrac{\pi}{4}$, $\cos\theta = \dfrac{1}{\sqrt{2}}$ and $\sin\theta = \dfrac{1}{\sqrt{2}}$, so the equation of the normal is

$$4x \times \dfrac{1}{\sqrt{2}} - 3y \times \dfrac{1}{\sqrt{2}} - 7 \times \dfrac{1}{\sqrt{2}} \times \dfrac{1}{\sqrt{2}} = 0$$

$$\Rightarrow \quad 4\sqrt{2}x - 3\sqrt{2}y - 7 = 0$$

$$\Rightarrow \quad 4x - 3y - 4.95 = 0 \quad \text{(to 2 decimal places)}$$

(iv) The co-ordinates of the point where $\theta = \dfrac{\pi}{4}$ are

$$\left(4\cos\dfrac{\pi}{4},\ 3\sin\dfrac{\pi}{4}\right) = \left(4 \times \dfrac{1}{\sqrt{2}},\ 3 \times \dfrac{1}{\sqrt{2}}\right)$$

$$\approx (2.83,\ 2.12)$$

(v)

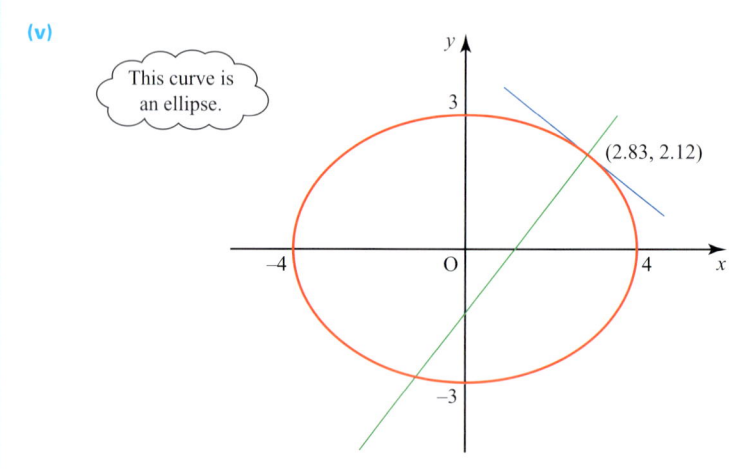

This curve is an ellipse.

Figure 4.14

Stationary points

When the equation of a curve is given parametrically, the easiest way to distinguish between stationary points is usually to consider the sign of $\frac{dy}{dx}$. If you use this method, you must be careful to ensure that you take points which are to the left and right of the stationary point, i.e. have x co-ordinates smaller and larger than those at the stationary point. These will not necessarily be points whose parameters are smaller and larger than those at the stationary point.

EXAMPLE 4.24 Find the stationary points of the curve with parametric equations $x = 2t + 1$, $y = 3t - t^3$, and distinguish between them.

SOLUTION

$x = 2t + 1 \implies \frac{dx}{dt} = 2$

$y = 3t - t^3 \implies \frac{dy}{dt} = 3 - 3t^2$

$\frac{dy}{dx} = \frac{\frac{dy}{dt}}{\frac{dx}{dt}} = \frac{3 - 3t^2}{2} = \frac{3(1 - t^2)}{2}$

Stationary points occur when $\frac{dy}{dx} = 0$:

$\implies t^2 = 1 \implies t = 1 \quad \text{or} \quad t = -1$

At $t = 1$: $\quad x = 3, y = 2$
At $t = 0.9$: $\quad x = 2.8$ (to the left); $\frac{dy}{dx} = 0.285$ (positive)
At $t = 1.1$: $\quad x = 3.2$ (to the right); $\frac{dy}{dx} = -0.315$ (negative)

There is a maximum at $(3, 2)$.

At $t = -1$: $\quad x = -1, y = -2$
At $t = -1.1$: $\quad x = -1.2$ (to the left); $\frac{dy}{dx} = -0.315$ (negative)
At $t = -0.9$: $\quad x = -0.8$ (to the right); $\frac{dy}{dx} = 0.285$ (positive)

There is a minimum at $(-1, -2)$.

An alternative method

Alternatively, to find $\frac{d^2y}{dx^2}$ when $\frac{dy}{dx}$ is expressed in terms of a parameter requires a further use of the chain rule:

$$\frac{d^2y}{dx^2} = \frac{d}{dx}\left(\frac{dy}{dx}\right) = \frac{d}{dt}\left(\frac{dy}{dx}\right) \times \frac{dt}{dx}.$$

EXERCISE 4E

1 For each of the following curves, find $\dfrac{dy}{dx}$ in terms of the parameter.

(i) $x = 3t^2$
$y = 2t^3$

(ii) $x = \theta - \cos\theta$
$y = \theta + \sin\theta$

(iii) $x = t + \dfrac{1}{t}$
$y = t - \dfrac{1}{t}$

(iv) $x = 3\cos\theta$
$y = 2\sin\theta$

(v) $x = (t+1)^2$
$y = (t-1)^2$

(vi) $x = \theta\sin\theta + \cos\theta$
$y = \theta\cos\theta - \sin\theta$

(vii) $x = e^{2t} + 1$
$y = e^t$

(viii) $x = \dfrac{t}{1+t}$
$y = \dfrac{t}{1-t}$

2 A curve has the parametric equations $x = \tan\theta$, $y = \tan 2\theta$. Find

(i) the value of $\dfrac{dy}{dx}$ when $\theta = \dfrac{\pi}{6}$

(ii) the equation of the tangent to the curve at the point where $\theta = \dfrac{\pi}{6}$

(iii) the equation of the normal to the curve at the point where $\theta = \dfrac{\pi}{6}$.

3 A curve has the parametric equations $x = t^2$, $y = 1 - \dfrac{1}{2t}$ for $t > 0$. Find

(i) the co-ordinates of the point P where the curve cuts the x axis
(ii) the gradient of the curve at this point
(iii) the equation of the tangent to the curve at P
(iv) the co-ordinates of the point where the tangent cuts the y axis.

4 A curve has parametric equations $x = at^2$, $y = 2at$, where a is constant. Find

(i) the equation of the tangent to the curve at the point with parameter t
(ii) the equation of the normal to the curve at the point with parameter t
(iii) the co-ordinates of the points where the normal cuts the x and y axes.

5 A curve has parametric equations $x = \cos\theta$, $y = \cos 2\theta$.

(i) Show that $\dfrac{dy}{dx} = 4\cos\theta$.

(ii) By writing $\dfrac{dy}{dx}$ in terms of x, show that $\dfrac{d^2y}{dx^2} - 4 = 0$.

6 The parametric equations of a curve are $x = at$, $y = \dfrac{b}{t}$, where a and b are constant. Find in terms of a, b and t

(i) $\dfrac{dy}{dx}$

(ii) the equation of the tangent to the curve at the general point $\left(at, \dfrac{b}{t}\right)$

(iii) the co-ordinates of the points X and Y where the tangent cuts the x and y axes.
(iv) Show that the area of triangle OXY is constant, where O is the origin.

7 The diagram shows a sketch of the curve given parametrically in terms of *t* by the equations $x = 4t$ and $y = 2t^2$ where *t* takes positive and negative values.

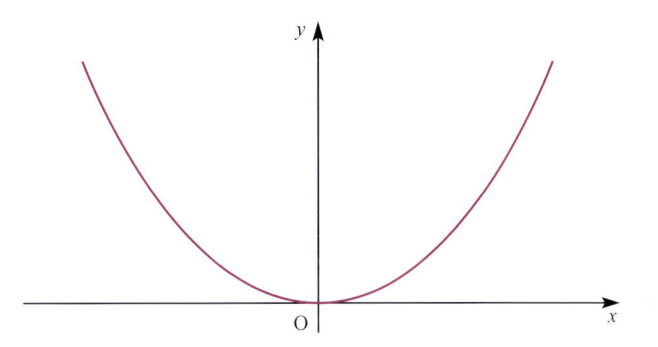

P is the point on the curve with parameter *t*.

(i) Show that the gradient at P is *t*.

(ii) Find and simplify the equation of the tangent at P.

The tangents at two points Q (with parameter t_1) and R (with parameter t_2) meet at S.

(iii) Find the co-ordinates of S.

(iv) In the case when $t_1 + t_2 = 2$ show that S lies on a straight line. Give the equation of the line.

[MEI, *adapted*]

8 The diagram shows a sketch of the curve given parametrically in terms of *t* by the equations $x = 1 - t^2$, $y = 2t + 1$.

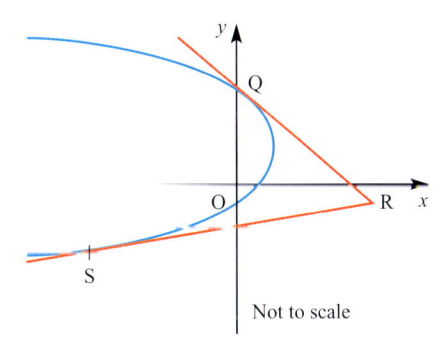

Not to scale

(i) Show that the point Q(0, 3) lies on the curve, stating the value of *t* corresponding to this point.

(ii) Show that, at the point with parameter *t*,
$$\frac{dy}{dx} = -\frac{1}{t}.$$

(iii) Find the equation of the tangent at Q.

(iv) Verify that the tangent at Q passes through the point R(4, −1).

(v) The other tangent from R to the curve touches the curve at the point S and has equation $3y - x + 7 = 0$. Find the co-ordinates of S.

[MEI]

9 The diagram shows a sketch of the curve with parametic equations $x = 1 - 2t$, $y = t^2$. The tangent and normal at P are also shown.

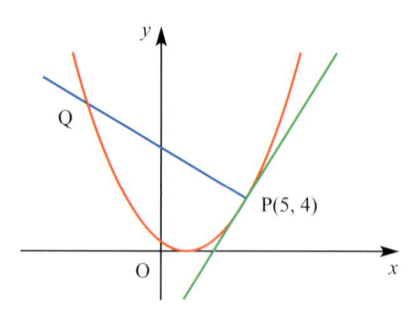

(i) Show that the point P(5, 4) lies on the curve by stating the value of t corresponding to this point.

(ii) Show that, at the point with parameter t, $\dfrac{dy}{dx} = -t$.

(iii) Find the equation of the tangent at P.

(iv) The normal at P cuts the curve again at Q. Find the co-ordinates of Q.

[MEI]

10 A particle P moves in a plane so that at time t its co-ordinates are given by $x = 4\cos t$, $y = 3\sin t$. Find

(i) $\dfrac{dy}{dx}$ in terms of t

(ii) the equation of the tangent to its path at time t

(iii) the values of t for which the particle is travelling parallel to the line $x + y = 0$.

11 (i) By differentiating $\dfrac{1}{\cos\theta}$, show that if $y = \sec\theta$ then $\dfrac{dy}{d\theta} = \sec\theta\tan\theta$.

(ii) The parametric equations of a curve are

$$x = 1 + \tan\theta, \quad y = \sec\theta,$$

for $-\tfrac{1}{2}\pi < \theta < \tfrac{1}{2}\pi$. Show that $\dfrac{dy}{dx} = \sin\theta$.

(iii) Find the co-ordinates of the point on the curve at which the gradient of the curve is $\tfrac{1}{2}$.

[Cambridge International AS & A Level Mathematics 9709, Paper 2 Q5 June 2005]

12 The parametric equations of a curve are

$$x = 3t + \ln(t - 1), \quad y = t^2 + 1, \quad \text{for } t > 1.$$

(i) Express $\dfrac{dy}{dx}$ in terms of t.

(ii) Find the co-ordinates of the only point on the curve at which the gradient of the curve is equal to 1.

[Cambridge International AS & A Level Mathematics 9709, Paper 2 Q3 June 2007]

13 The parametric equations of a curve are

$$x = 4\sin\theta, \quad y = 3 - 2\cos 2\theta,$$

where $-\tfrac{1}{2}\pi < \theta < \tfrac{1}{2}\pi$. Express $\dfrac{dy}{dx}$ in terms of θ, simplifying your answer as far as possible.

[Cambridge International AS & A Level Mathematics 9709, Paper 2 Q4 June 2009]

14 The parametric equations of a curve are

$$x = 1 - e^{-t}, \quad y = e^{t} + e^{-t}.$$

(i) Show that $\dfrac{dy}{dx} = e^{2t} - 1$.

(ii) Hence find the exact value of t at the point on the curve at which the gradient is 2.

[Cambridge International AS & A Level Mathematics 9709, Paper 22 Q4 November 2009]

15 The parametric equations of a curve are

$$x = 2\theta + \sin 2\theta, \quad y = 1 - \cos 2\theta.$$

Show that $\dfrac{dy}{dx} = \tan\theta$.

[Cambridge International AS & A Level Mathematics 9709, Paper 3 Q3 June 2006]

16 The parametric equations of a curve are

$$x = a\cos^3 t, \quad y = a\sin^3 t,$$

where a is a positive constant and $0 < t < \tfrac{1}{2}\pi$.

(i) Express $\dfrac{dy}{dx}$ in terms of t.

(ii) Show that the equation of the tangent to the curve at the point with parameter t is

$$x\sin t + y\cos t = a\sin t \cos t.$$

(iii) Hence show that, if this tangent meets the x axis at X and the y axis at Y, then the length of XY is always equal to a.

[Cambridge International AS & A Level Mathematics 9709, Paper 3 Q6 June 2009]

KEY POINTS

1. $y = kx^n \Rightarrow \dfrac{dy}{dx} = knx^{n-1}$ where k and n are real constants.

2. Chain rule: $\dfrac{dy}{dx} = \dfrac{dy}{du} \times \dfrac{du}{dx}$

3. Product rule (for $y = uv$): $\dfrac{dy}{dx} = v\dfrac{du}{dx} + u\dfrac{dv}{dx}$.

4. Quotient rule $\left(\text{for } y = \dfrac{u}{v}\right): \dfrac{dy}{dx} = \dfrac{v\dfrac{du}{dx} - u\dfrac{dv}{dx}}{v^2}$.

5. $\dfrac{dy}{dx} = \dfrac{1}{\dfrac{dx}{dy}}$.

6. $\dfrac{d}{dx}(\ln x) = \dfrac{1}{x}$

7. $\dfrac{d}{dx}(e^x) = e^x$

8. $\dfrac{d}{dx}(\sin kx) = k\cos kx$

   $\dfrac{d}{dx}(\cos kx) = -k\sin kx$

   $\dfrac{d}{dx}(\tan kx) = k\sec^2 kx$

9. An implicit function is one connecting x and y where y is not the subject. When you differentiate an implicit function:
   - differentiating y^2 with respect to x gives $2y\dfrac{dy}{dx}$
   - differentiating $4x^3y^2$ with respect to x gives $12x^2 \times y^2 + 4x^3 \times 2y\dfrac{dy}{dx}$.

   The derivative of any constant is 0.

10. In parametric equations the relationship between two variables is expressed by writing both of them in terms of a third variable or *parameter*.

11. To draw a graph from parametric equations, plot the points on the curve given by different values of the parameter.

12. $\dfrac{dy}{dx} = \dfrac{\dfrac{dy}{dt}}{\dfrac{dx}{dt}}$ provided that $\dfrac{dx}{dt} \neq 0$.

5 Integration

Every picture is worth a thousand words.

Traditional Chinese proverb

Integrals involving the exponential function

Since you know that

$$\frac{d}{dx}(e^{ax+b}) = ae^{ax+b},$$

you can see that

$$\int e^{ax+b}\,dx = \frac{1}{a}e^{ax+b} + c.$$

This increases the number of functions which you are able to integrate, as in the following example.

EXAMPLE 5.1

Find the following integrals.

(i) $\int e^{2x-3}\,dx$

(ii) $\int_1^5 6e^{3x}\,dx$

SOLUTION

(i) $\int e^{2x-3}\,dx = \tfrac{1}{2}e^{2x-3} + c$

(ii) $\int_1^5 6e^{3x}\,dx = \left[\frac{6e^{3x}}{3}\right]_1^5$

$= \left[2e^{3x}\right]_1^5$

$= 2(e^{15} - e^3)$

$= 6.54 \times 10^6$ (to 3 significant figures)

Integrals involving the natural logarithm function

You have already seen that

$$\int \frac{1}{x}\,dx = \ln x + c.$$

There are many other integrals that can be reduced to this form.

EXAMPLE 5.2 Evaluate $\int_2^5 \frac{1}{2x} dx$.

SOLUTION

$$\frac{1}{2}\int_2^5 \frac{1}{x} dx = \frac{1}{2}\Big[\ln x\Big]_2^5$$
$$= \frac{1}{2}(\ln 5 - \ln 2)$$
$$= 0.458 \quad \text{(to 3 significant figures)}$$

In this example the $\frac{1}{2}$ was taken outside the integral, allowing the standard result for $\frac{1}{x}$ to be used.

Since
$$y = \ln(ax+b) \Rightarrow \frac{dy}{dx} = \frac{a}{ax+b}$$

So
$$\int \frac{a}{ax+b} dx = \ln(ax+b) + c$$

c means 'an arbitrary constant' and so does not necessarily have the same value from one equation to another.

and
$$\int \frac{1}{ax+b} dx = \frac{1}{a}\ln(ax+b) + c$$

EXAMPLE 5.3 Find $\int_0^2 \frac{1}{5x+3} dx$.

SOLUTION

$$\int_0^2 \frac{1}{5x+3} dx = \Big[\frac{1}{5}\ln(5x+3)\Big]_0^2$$
$$= \frac{1}{5}\ln 13 - \frac{1}{5}\ln 3$$
$$= 0.293 \quad \text{(to 3 significant figures)}$$

Extending the domain for logarithmic integrals

The use of $\int \frac{1}{x} dx = \ln x + c$ has so far been restricted to cases where $x > 0$, since logarithms are undefined for negative numbers.

Look, however, at the area between $-b$ and $-a$ on the left-hand branch of the curve $y = \frac{1}{x}$ in figure 5.1. You can see that it is a real area, and that it must be possible to evaluate it.

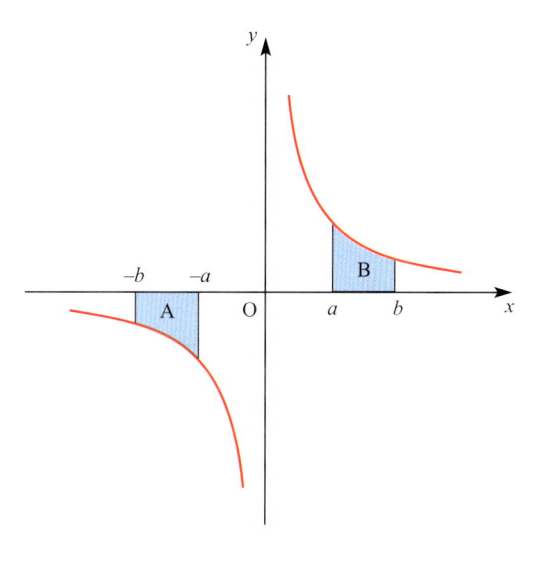

Figure 5.1

ACTIVITY 5.1

1. What can you say about the areas of the two shaded regions?
2. Try to prove your answer to part **1** before reading on.

Proof

Let $A = \int_{-b}^{-a} \frac{1}{x} \, dx$.

Now write the integral in terms of a new variable, u, where $u = -x$.
This gives new limits: $x = -b \Rightarrow u = b$
$\qquad\qquad\qquad\qquad\quad x = -a \Rightarrow u = a$.

$\frac{du}{dx} = -1 \Rightarrow dx = -du$.

So the integral becomes

$$A = \int_{b}^{a} \frac{1}{-u} (-du)$$

$$= \int_{b}^{a} \frac{1}{u} \, du$$

$$= [\ln a - \ln b]$$

$$= -[\ln b - \ln a] = -\text{area B}$$

So the area has the same size as that obtained if no notice is taken of the fact that the limits a and b have minus signs. However it has the opposite sign, as you would expect because the area is below the axis.

Consequently the restriction that $x > 0$ may be dropped, and the integral is written

$$\int \frac{1}{x} \, dx = \ln |x| + c.$$

Similarly, $\int \frac{f'(x)}{f(x)} \, dx = \ln |f(x)| + c.$

EXAMPLE 5.4 Find the value of $\int_5^7 \frac{1}{4-x} dx$.

SOLUTION

To make the top line into the differential of the bottom line, you write the integral in one of two ways.

$$-\int_5^7 \frac{-1}{4-x} dx = -[\ln|4-x|]_5^7 \qquad -\int_5^7 \frac{1}{x-4} dx = -[\ln|x-4|]_5^7$$
$$= -[(\ln|-3|) - (\ln|-1|)] \qquad\qquad = -[\ln 3 - \ln 1]$$
$$= -[\ln 3 - \ln 1] \qquad\qquad\qquad = -1.10 \text{ (to 3 s.f.)}$$
$$= -1.10 \text{ (to 3 s.f.)}$$

⚠ Since the curve $y = \frac{1}{x}$ is not defined at the discontinuity at $x = 0$ (see figure 5.2), it is not possible to integrate across this point.

Consequently in the integral $\int_p^q \frac{1}{x} dx$ both the limits p and q must have the same sign, either + or −. The integral is invalid otherwise.

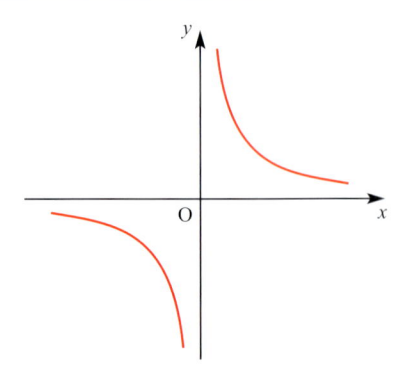

Figure 5.2

ⓟ The equation of a curve is $y = \frac{p_1(x)}{p_2(x)}$ where $p_1(x)$ and $p_2(x)$ are polynominals.

How can you tell from the equation whether the curve has a discontinuity?

How can you prove $y = x^2 - 2x + 3$ has no discontinuities?

EXERCISE 5A

1 Find the following indefinite integrals.

(i) $\int \frac{3}{x} dx$ (ii) $\int \frac{1}{4x} dx$ (iii) $\int \frac{1}{x-5} dx$ (iv) $\int \frac{1}{2x-9} dx$

2 Find the following indefinite integrals.

(i) $\int e^{3x} dx$ (ii) $\int e^{-4x} dx$ (iii) $\int e^{-\frac{x}{3}} dx$

(iv) $\int \frac{10}{e^{5x}} dx$ (v) $\int \frac{e^{3x}+4}{e^{2x}} dx$

3 Find the following definite integrals.
  Where appropriate give your answers to 3 significant figures.

  (i) $\int_0^4 4e^{2x}\,dx$

  (ii) $\int_1^3 \dfrac{4}{2x+1}\,dx$

  (iii) $\int_{-1}^1 (e^x + e^{-x})\,dx$

  (iv) $\int_{-2}^1 e^{3x-2}\,dx$

4 The graph of $y = x + \dfrac{4}{x}$ is shown below.

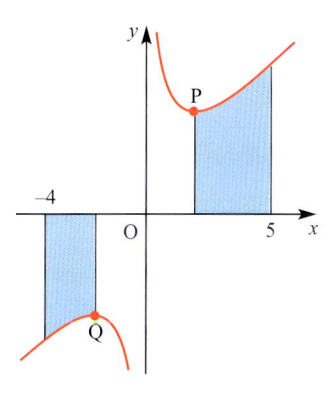

  (i) Find the co-ordinates of the minimum point, P, and the maximum point, Q.
  (ii) Find the area of each shaded region.

5 The diagram illustrates the graph of $y = e^x$. The point A has co-ordinates $(\ln 5, 0)$, B has co-ordinates $(\ln 5, 5)$ and C has co-ordinates $(0, 5)$.

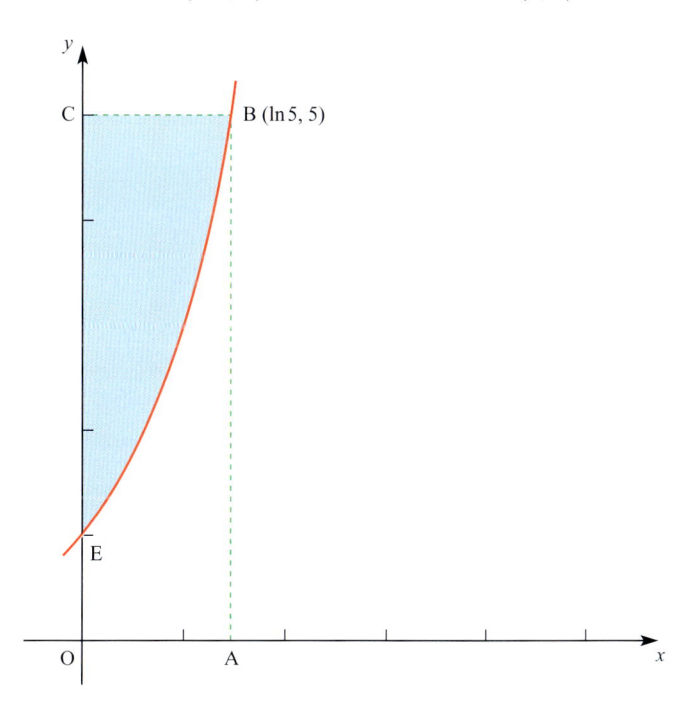

  (i) Find the area of the region OABE enclosed by the curve $y = e^x$, the x axis, the y axis and the line AB. Hence find the area of the shaded region EBC.

(ii) The graph of $y = e^x$ is transformed into the graph of $y = \ln x$. Describe this transformation geometrically.

(iii) Using your answers to parts **(i)** and **(ii)**, or otherwise, show that
$$\int_1^5 \ln x \, dx = 5\ln 5 - 4.$$

(iv) Deduce the values of

(a) $\int_1^5 \ln(x^3) \, dx$

(b) $\int_1^5 \ln(3x) \, dx.$

[MEI, *adapted*]

6 (i) Differentiate $\ln(2x+3)$.

(ii) Hence, or otherwise, show that
$$\int_{-1}^{3} \frac{1}{2x+3} dx = \ln 3.$$

(iii) Find the quotient and remainder when $4x^2 + 8x$ is divided by $2x+3$.

(iv) Hence show that
$$\int_{-1}^{3} \frac{4x^2 + 8x}{2x+3} dx = 12 - 3\ln 3.$$

[Cambridge International AS & A Level Mathematics 9709, Paper 2 Q7 June 2006]

7 A curve is such that $\dfrac{dy}{dx} = e^{2x} - 2e^{-x}$. The point $(0, 1)$ lies on the curve.

(i) Find the equation of the curve.

(ii) The curve has one stationary point. Find the x co-ordinate of this point and determine whether it is a maximum or a minimum point.

[Cambridge International AS & A Level Mathematics 9709, Paper 2 Q6 November 2005]

8 (i) Find the equation of the tangent to the curve $y = \ln(3x-2)$ at the point where $x = 1$.

(ii) (a) Find the value of the constant A such that
$$\frac{6x}{3x-2} \equiv 2 + \frac{A}{3x-2}.$$

(b) Hence show that $\int_2^6 \dfrac{6x}{3x-2} dx = 8 + \dfrac{8}{3}\ln 2.$

[Cambridge International AS & A Level Mathematics 9709, Paper 2 Q8 June 2009]

9 Find the exact value of the constant k for which $\int_1^k \dfrac{1}{2x-1} dx = 1.$

[Cambridge International AS & A Level Mathematics 9709, Paper 3 Q1 November 2007]

INVESTIGATIONS

e A series for e^x

The exponential function can be written as the infinite series

$$e^x = a_0 + a_1 x + a_2 x^2 + a_3 x^3 + a_4 x^4 + \ldots \quad \text{(for } x \in \mathbb{R}\text{)}$$

where $a_0, a_1, a_2, \ldots$ are numbers.

You can find the value of a_0 by substituting the value zero for x.

Since $e^0 = 1$, it follows that $1 = a_0 + 0 + 0 + 0 + \ldots$, and so $a_0 = 1$.

You can now write: $e^x = 1 + a_1 x + a_2 x^2 + a_3 x^3 + a_4 x^4 + \ldots$.

Now differentiate both sides: $e^x = a_1 + 2a_2 x + 3a_3 x^2 + 4a_4 x^3 + \ldots$,

and substitute $x = 0$ again: $1 = a_1 + 0 + 0 + 0 + \ldots$, and so $a_1 = 1$ also.

Now differentiate a second time, and again substitute $x = 0$. This time you find a_2. Continue this procedure until you can see the pattern in the values of $a_0, a_1, a_2, a_3, \ldots$.

When you have the series for e^x, substitute $x = 1$. The left-hand side is e^1 or e, and so by adding the terms on the right-hand side you obtain the value of e. You will find that the terms become small quite quickly, so you will not need to use very many to obtain the value of e correct to several decimal places.

If you are also studying statistics you will meet this series expansion of e^x in connection with the Poisson distribution.

e Compound interest

You win $100 000 in a prize draw and are offered two investment options.
A You are paid 100% interest at the end of 10 years, or
B You are paid 10% compound interest year by year for 10 years.
Under which scheme are you better off?

Clearly in scheme **A**, the ratio $R = \dfrac{\text{final money}}{\text{original money}}$ is $\dfrac{\$200\,000}{\$100\,000} = 2$.

What is the value of the ratio R in scheme **B**?

Suppose that you asked for the interest to be paid in 20 half-yearly instalments of 5% each (scheme **C**). What would be the value of R in this case?

Continue this process, investigating what happens to the ratio R when the interest is paid at increasingly frequent intervals.

Is there a limit to R as the time interval between interest payments tends to zero?

Integrals involving trigonometrical functions

Since $\dfrac{d}{dx}(\sin(ax+b)) = a\cos(ax+b)$

$\left(\int a\cos(ax+b)\,dx = \sin(ax+b) + c\right)$

it follows that $\displaystyle\int \cos(ax+b)\,dx = \dfrac{1}{a}\sin(ax+b) + c$

Similarly, since $\dfrac{d}{dx}(\cos(ax+b)) = -a\sin(ax+b)$

$\left(\int -a\sin(ax+b)\,dx = \cos(ax+b) + c\right)$

it also follows that $\displaystyle\int \sin(ax+b)\,dx = -\dfrac{1}{a}\cos(ax+b) + c$

Also $\dfrac{d}{dx}(\tan(ax+b)) = a\sec^2(ax+b)$

and so $\displaystyle\int \sec^2(ax+b)\,dx = \dfrac{1}{a}\tan(ax+b) + c$

EXAMPLE 5.5

Find

(i) $\displaystyle\int \sec^2 x\,dx$ (ii) $\displaystyle\int \sin 2x\,dx$ (iii) $\displaystyle\int \cos(3x-\pi)\,dx$.

SOLUTION

(i) $\displaystyle\int \sec^2 x\,dx = \tan x + c$

(ii) $\displaystyle\int \sin 2x\,dx = -\tfrac{1}{2}\cos 2x + c$

(iii) $\displaystyle\int \cos(3x-\pi)\,dx = \tfrac{1}{3}\sin(3x-\pi) + c$

EXAMPLE 5.6

Find the exact value of $\displaystyle\int_0^{\frac{\pi}{3}} (\sin 2x - \cos 4x)\,dx$.

SOLUTION

$\displaystyle\int_0^{\frac{\pi}{3}} (\sin 2x - \cos 4x)\,dx = \left[-\tfrac{1}{2}\cos 2x - \tfrac{1}{4}\sin 4x\right]_0^{\frac{\pi}{3}}$

$= \left[-\tfrac{1}{2}\cos\tfrac{2\pi}{3} - \tfrac{1}{4}\sin\tfrac{4\pi}{3}\right] - \left[-\tfrac{1}{2}\cos 0 - \tfrac{1}{4}\sin 0\right]$

$= \left[-\tfrac{1}{2}\times\left(-\tfrac{1}{2}\right) - \tfrac{1}{4}\times\left(-\tfrac{\sqrt{3}}{2}\right)\right] - \left[-\tfrac{1}{2}\times 1\right]$

$= \tfrac{1}{4} + \tfrac{\sqrt{3}}{8} + \tfrac{1}{2}$

$= \tfrac{\sqrt{3}}{8} + \tfrac{3}{4}$

$= \dfrac{6+\sqrt{3}}{8}$

Using trigonometrical identities in integration

Sometimes, when it is not immediately obvious how to integrate a function involving trigonometrical functions, it may help to rewrite the function using one of the trigonometrical identities.

EXAMPLE 5.7 Find $\int \sin^2 x \, dx$.

SOLUTION

Use the identity

$$\cos 2x = 1 - 2\sin^2 x.$$

(Remember that this is just one of the three expressions for $\cos 2x$.)

This identity may be rewritten as

$$\sin^2 x = \tfrac{1}{2}(1 - \cos 2x).$$

By putting $\sin^2 x$ in this form, you will be able to perform the integration.

$$\int \sin^2 x \, dx = \tfrac{1}{2} \int (1 - \cos 2x) \, dx$$
$$= \tfrac{1}{2}\left(x - \tfrac{1}{2}\sin 2x\right) + c$$
$$= \tfrac{1}{2}x - \tfrac{1}{4}\sin 2x + c$$

You can integrate $\cos^2 x$ in the same way, by using $\cos^2 x = \tfrac{1}{2}(\cos 2x + 1)$. Other even powers of $\sin x$ or $\cos x$ can also be integrated in a similar way, but you have to use the identity twice or more.

EXAMPLE 5.8 Find $\int \cos^4 x \, dx$.

SOLUTION

First express $\cos^4 x$ as $(\cos^2 x)^2$:

$$\cos^4 x = \left[\tfrac{1}{2}(\cos 2x + 1)\right]^2$$
$$= \tfrac{1}{4}(\cos^2 2x + 2\cos 2x + 1)$$

Next, apply the same identity to $\cos^2 2x$:

$$\cos^2 2x = \tfrac{1}{2}(\cos 4x + 1)$$

Hence $\cos^4 x = \tfrac{1}{4}\left(\tfrac{1}{2}\cos 4x + \tfrac{1}{2} + 2\cos 2x + 1\right)$
$$= \tfrac{1}{4}\left(\tfrac{1}{2}\cos 4x + 2\cos 2x + \tfrac{3}{2}\right)$$
$$= \tfrac{1}{8}\cos 4x + \tfrac{1}{2}\cos 2x + \tfrac{3}{8}$$

This can now be integrated.

$$\int \cos^4 x \, dx = \int \left(\tfrac{1}{8}\cos 4x + \tfrac{1}{2}\cos 2x + \tfrac{3}{8}\right) dx$$
$$= \tfrac{1}{32}\sin 4x + \tfrac{1}{4}\sin 2x + \tfrac{3}{8}x + c$$

EXERCISE 5B

1 Integrate the following with respect to x.

(i) $\sin x - 2\cos x$ (ii) $3\cos x + 2\sin x$ (iii) $5\sin x + 4\cos x$

(iv) $4\sec^2 x$ (v) $\sin(2x+1)$ (vi) $\cos(5x - \pi)$

(vii) $6\sec^2 2x$ (viii) $3\sec^2 3x - \sin 2x$ (ix) $4\sec^2 x - \cos 2x$

2 Find the exact value of the following.

(i) $\int_0^{\frac{\pi}{3}} \sin x \, dx$ (ii) $\int_0^{\frac{\pi}{4}} \sec^2 x \, dx$

(iii) $\int_{\frac{\pi}{6}}^{\frac{\pi}{3}} \cos x \, dx$ (iv) $\int_0^{\frac{2\pi}{3}} \sin 2x \, dx$

(v) $\int_0^{\frac{5\pi}{6}} \cos 3x \, dx$ (vi) $\int_{\frac{\pi}{8}}^{\frac{\pi}{6}} \sec^2 2x \, dx$

(vii) $\int_0^{\pi} \cos\left(2x + \frac{\pi}{2}\right) dx$ (viii) $\int_0^{\frac{\pi}{4}} (\sec^2 x + \cos 4x) \, dx$

(ix) $\int_0^{\frac{\pi}{6}} (\cos x + \sin 2x) \, dx$

3 (i) Show that $\sin x \cos x = \tfrac{1}{2}\sin 2x$.

   (ii) Hence find the exact value of $\int_0^{\frac{\pi}{3}} \sin x \cos x \, dx$.

4 Use a suitable trigonometric identity to help you find these.

(i) (a) $\int \cos^2 x \, dx$ (b) $\int_0^{\frac{\pi}{2}} \cos^2 x \, dx$

(ii) (a) $\int \sin^2 x \, dx$ (b) $\int_0^{\frac{\pi}{3}} \sin^2 x \, dx$

5 (i) By expanding $\sin(2x + x)$ and using double-angle formulae, show that

$$\sin 3x = 3\sin x - 4\sin^3 x.$$

   (ii) Hence show that

$$\int_0^{\frac{1}{3}\pi} \sin^3 x \, dx = \frac{5}{24}.$$

[Cambridge International AS & A Level Mathematics 9709, Paper 2 Q7 June 2005]

6 The diagram shows the part of the curve $y = \sin^2 x$ for $0 \leq x \leq \pi$.

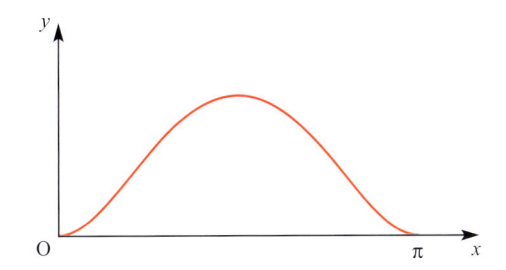

(i) Show that $\dfrac{dy}{dx} = \sin 2x$.

(ii) Hence find the x co-ordinates of the points on the curve at which the gradient of the curve is 0.5.

(iii) By expressing $\sin^2 x$ in terms of $\cos 2x$, find the area of the region bounded by the curve and the x axis between 0 and π.

[Cambridge International AS & A Level Mathematics 9709, Paper 2 Q7 November 2005]

7 (i) Express $\cos^2 x$ in terms of $\cos 2x$.

(ii) Hence show that

$$\int_0^{\frac{1}{3}\pi} \cos^2 x \, dx = \tfrac{1}{6}\pi + \tfrac{1}{8}\sqrt{3}.$$

(iii) By using an appropriate trigonometrical identity, deduce the exact value of

$$\int_0^{\frac{1}{3}\pi} \sin^2 x \, dx.$$

[Cambridge International AS & A Level Mathematics 9709, Paper 2 Q6 June 2007]

8 (i) Prove the identity

$$(\cos x + 3\sin x)^2 \equiv 5 - 4\cos 2x + 3\sin 2x.$$

(ii) Using the identity, or otherwise, find the exact value of

$$\int_0^{\frac{1}{4}\pi} (\cos x + 3\sin x)^2 \, dx.$$

[Cambridge International AS & A Level Mathematics 9709, Paper 2 Q7 November 2007]

9 (i) Show that $\int_0^{\frac{1}{4}\pi} \cos 2x \, dx = \tfrac{1}{2}$.

(ii) By using an appropriate trigonometrical identity, find the exact value of

$$\int_{\frac{1}{6}\pi}^{\frac{1}{3}\pi} 3\tan^2 x \, dx.$$

[Cambridge International AS & A Level Mathematics 9709, Paper 22 Q4 June 2010]

Numerical integration

There are times when you need to find the area under a graph but cannot do this by the integration methods you have met so far.

- The function may be one that cannot be integrated algebraically. (There are many such functions.)
- The function may be one that can be integrated algebraically but which requires a technique with which you are unfamiliar.
- It may be that you do not know the function in algebraic form, but just have a set of points (perhaps derived from an experiment).

In these circumstances you can always find an approximate answer using a numerical method, but you must:

(i) have a clear picture in your mind of the graph of the function, and how your method estimates the area beneath it

(ii) understand that a numerical answer without any estimate of its accuracy, or error bounds, is valueless.

The trapezium rule

In this chapter just one numerical method of integration is introduced, namely the *trapezium rule*. As an illustration of the rule, it is used to find the area under the curve $y = \sqrt{5x - x^2}$ for values of x between 0 and 4.

It is in fact possible to integrate this function algebraically, but not using the techniques that you have met so far.

> *Note*
>
> You should not use a numerical method when an algebraic (sometimes called analytic) technique is available to you. Numerical methods should be used only when other methods fail.

Figure 5.3 shows the area approximated by two trapezia of equal width.

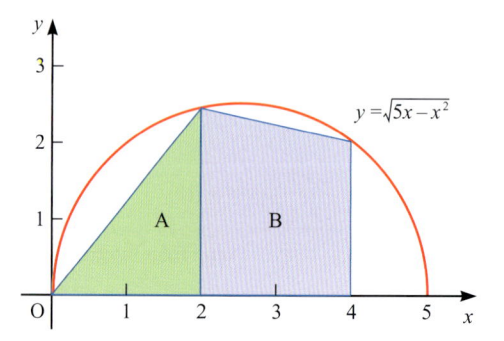

Figure 5.3

Remember the formula for the area of a trapezium, Area = $\frac{1}{2}h(a+b)$, where a and b are the lengths of the parallel sides and h the distance between them.

In the cases of the trapezia A and B, the parallel sides are vertical. The left-hand side of trapezium A has zero height, and so the trapezium is also a triangle.

When $x = 0$ $\Rightarrow$ $y = \sqrt{0} = 0$
When $x = 2$ $\Rightarrow$ $y = \sqrt{6} = 2.4495$ (to 4 d.p.)
When $x = 4,$ $\Rightarrow$ $y = \sqrt{4} = 2$

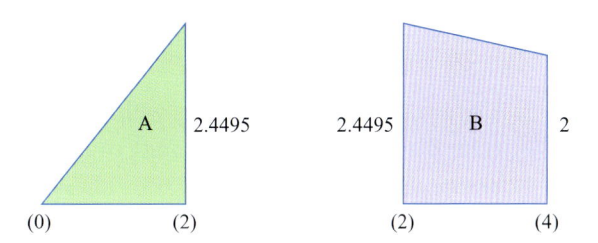

Figure 5.4

The area of trapezium A = $\frac{1}{2} \times 2 \times (0 + 2.4495) = 2.4495$

The area of trapezium B = $\frac{1}{2} \times 2 \times (2.4495 + 2) = \underline{4.4495}$

Total 6.8990

For greater accuracy you can use four trapezia, P, Q, R and S, each of width 1 unit as in figure 5.5. The area is estimated in just the same way.

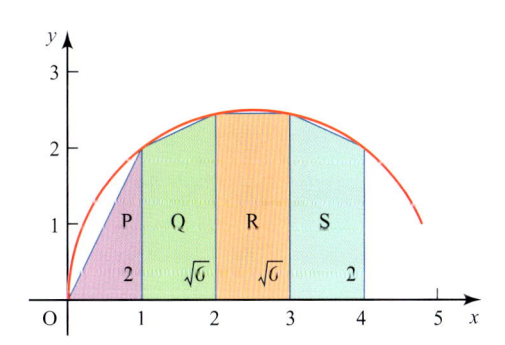

Figure 5.5

Trapezium P: $\frac{1}{2} \times 1 \times (0 + 2)$ = 1.0000
Trapezium Q: $\frac{1}{2} \times 1 \times (2 + 2.4495)$ = 2.2247
Trapezium R: $\frac{1}{2} \times 1 \times (2.4495 + 2.4495)$ = 2.4495
Trapezium S: $\frac{1}{2} \times 1 \times (2.4495 + 2)$ = $\underline{2.2247}$

Total 7.8990

These figures are given to 4 decimal places but the calculation has been done to more places on a calculator.

Accuracy

In this example, the first two estimates are 6.8989… and 7.8989… . You can see from figure 5.5 that the trapezia all lie underneath the curve, and so in this case the trapezium rule estimate of 7.8989… must be too small. You cannot, however, say by how much. To find that out you will need to take progressively more strips to find the value to which the estimate converges. Using 8 strips gives an estimate of 8.2407…, and 16 strips gives 8.3578… . The first figure, 8, looks reasonably certain but it is still not clear whether the second is 3, 4 or even 5. You need to take even more strips to be able to decide. In this example, the convergence is unusually slow because of the high curvature of the curve.

ACTIVITY 5.2 Use a graph-drawing program with the capability to calculate areas using trapezia. Calculate the area using progressively more strips and observe the convergence.

❓ It is possible to find this area without using calculus at all.

How can this be done? How close is the 16-strip estimate?

The procedure

In the previous example, the answer of 7.8990 from four strips came from adding the areas of the four trapezia P, Q, R and S:

$$\tfrac{1}{2} \times 1 \times (0 + 2) + \tfrac{1}{2} \times 1 \times (2 + 2.4495) + \tfrac{1}{2} \times 1 \times (2.4495 + 2.4495) + \tfrac{1}{2} \times 1 \times (2.4495 + 2)$$

and this can be written as

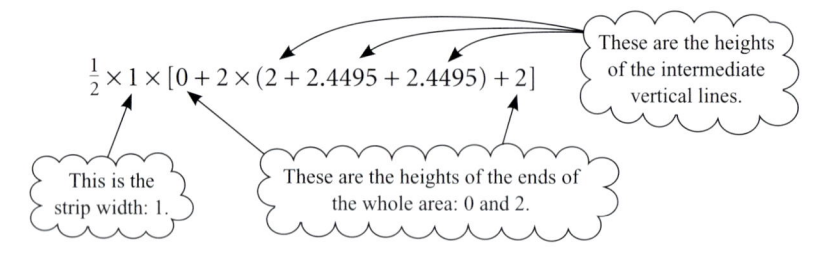

$$\tfrac{1}{2} \times 1 \times [0 + 2 \times (2 + 2.4495 + 2.4495) + 2]$$

- This is the strip width: 1.
- These are the heights of the ends of the whole area: 0 and 2.
- These are the heights of the intermediate vertical lines.

This is often stated in words as

$$\text{Area} \approx \tfrac{1}{2} \times \text{strip width} \times [\text{ends} + \text{twice middles}]$$

or in symbols, for n strips of width h

$$A \approx \tfrac{1}{2} \times h \times [y_0 + y_n + 2(y_1 + y_2 + \ldots + y_{n-1})].$$

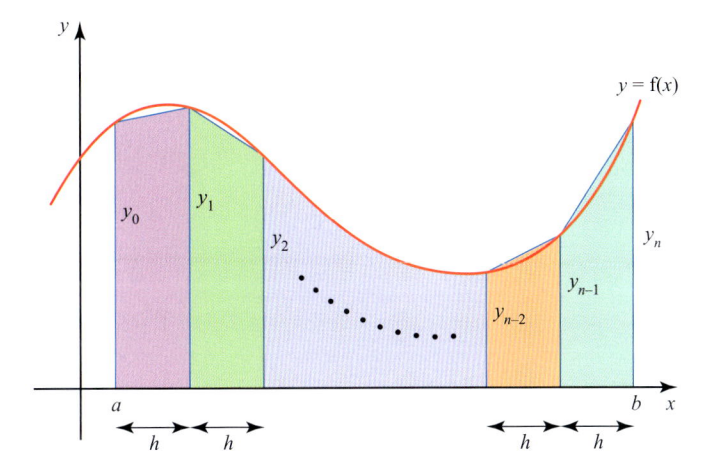

This is called the *trapezium rule* for width h (see figure 5.6).

Figure 5.6

? Look at the three graphs in figure 5.7, and in each case state whether the trapezium rule would underestimate or overestimate the area, or whether you cannot tell.

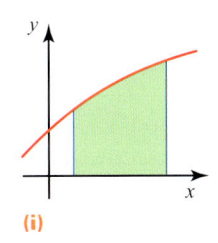

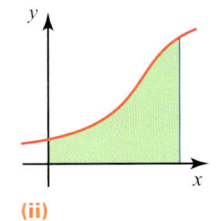

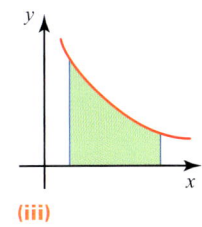

(i)   (ii)   (iii)

Figure 5.7

EXERCISE 5C

1 The speed v in ms^{-1} of a train is given at time t seconds in the following table.

t	0	10	20	30	40	50	60
v	0	5.0	6.7	8.2	9.5	10.6	11.6

The distance that the train has travelled is given by the area under the graph of the speed (vertical axis) against time (horizontal axis).

(i) Estimate the distance the train travels in this 1-minute period.
(ii) Give two reasons why your method cannot give a very accurate answer.

2 The definite integral $\int_0^1 \frac{1}{1+x^2}\,dx$ is known to equal $\frac{\pi}{4}$.

   (i) Using the trapezium rule for four strips, find an approximation for π.
   (ii) Repeat your calculation with 10 and 20 strips to obtain closer estimates.
   (iii) If you did not know the value of π, what value would you give it with confidence on the basis of your estimates in parts (i) and (ii)?

3 The table below gives the values of a function f(x) for different values of x.

x	0	0.5	1.0	1.5	2.0	2.5	3.0
f(x)	1.000	1.225	1.732	2.345	3.000	3.674	4.359

   (i) Apply the trapezium rule to the values in this table to obtain an approximation for $\int_0^3 f(x)\,dx$.
   (ii) By considering the shape of the curve $y = f(x)$, explain whether the approximation calculated in part (i) is likely to be an overestimate or an underestimate of the true area under the curve $y = f(x)$ between $x = 0$ and $x = 3$.

[MEI]

4 The graph of the function $y = \sqrt{2+x}$ (for $x \geqslant -2$) is given in the diagram.
The area of the shaded region ABCD is to be found.

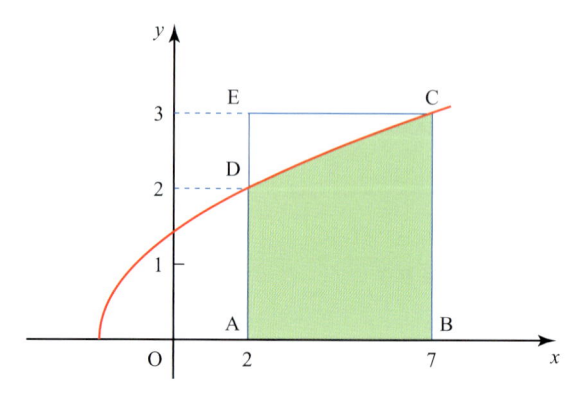

   (i) Make a table of values for y, for integer values of x from $x = 2$ to $x = 7$, giving each value of y correct to 4 decimal places.
   (ii) Use the trapezium rule with five strips, each 1 unit wide, to calculate an estimate for the area ABCD.
   State, giving a reason, whether your estimate is too large or too small.

Another method is to consider the area ABCD as the area of the rectangle ABCE minus the area of the region CDE.

   (iii) Show that the area CDE is given by $\int_2^3 (y^2 - 4)\,dy$.
   Calculate the exact value of this integral.
   (iv) Find the exact value of the area ABCD.
   Hence find the percentage error in using the trapezium rule.

[MEI, adapted]

5 The trapezium rule is used to estimate the value of $I = \int_0^{1.6} \sqrt{1+x^2}\, dx$.

  (i)   Draw the graph of $y = \sqrt{1+x^2}$ for $0 \leqslant x \leqslant 1.6$.
  (ii) Use strip widths of 0.8, 0.4, 0.2 and 0.1 to find approximations to the value of the integral.
  (iii) State the value of the integral to as many decimal places as you can justify.

6 The trapezium rule is used to estimate the value of $\int_0^1 \sqrt{\sin x}\, dx$.

  (i)   Draw the graph of $y = \sqrt{\sin x}$ for $0 \leqslant x \leqslant 1$.
  (ii) Use 1, 2, 4, 8 and 16 strips to find approximations to the value of the integral.
  (iii) State the value of the integral to as many decimal places as you can justify.

7 The trapezium rule is used to estimate the value of $\int_0^1 \dfrac{4}{1+x^2}\, dx$.

  (i)   Draw the graph of $y = \dfrac{4}{1+x^2}$ for $0 \leqslant x \leqslant 1$.
  (ii) Use strip widths of 1, 0.5, 0.25 and 0.125 to find approximations to the value of the integral.
  (iii) State the value of the integral to as many decimal places as you can justify.

8 A student uses the trapezium rule to estimate the value of $\int_0^2 (2 - \cos 2\pi x)\, dx$.

  (i)   Find approximations to the value of the integral by applying the trapezium rule using strip widths of, 2, 1, 0.5 and 0.25.
  (ii) Sketch the graph of $y = 2 - \cos 2\pi x$ for $0 \leqslant x \leqslant 2$.
         On copies of your graph shade the areas you have found in parts **(i)(a)** to **(d)**.
  (iii) Use integration to find the exact value of this integral.

9 The diagram shows the part of the curve $y = \dfrac{\ln x}{x}$ for $0 < x \leqslant 4$. The curve cuts the x-axis at A and its maximum point is M.

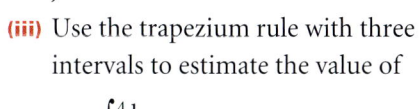

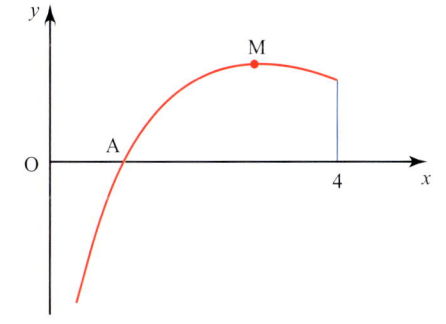

  (i)   Write down the co-ordinates of A.
  (ii) Show that the x co-ordinate of M is e, and write down the y co-ordinate of M in terms of e.
  (iii) Use the trapezium rule with three intervals to estimate the value of
  $$\int_1^4 \dfrac{\ln x}{x}\, dx,$$
  correct to 2 decimal places.
  (iv) State, with a reason, whether the trapezium rule gives an underestimate or an overestimate of the true value of the integral in part **(iii)**.

  [Cambridge International AS & A Level Mathematics 9709, Paper 2 Q6 June 2005]

10 The diagram shows the part of the curve $y = e^x \cos x$ for $0 \leq x \leq \frac{1}{2}\pi$. The curve meets the y axis at the point A. The point M is a maximum point.

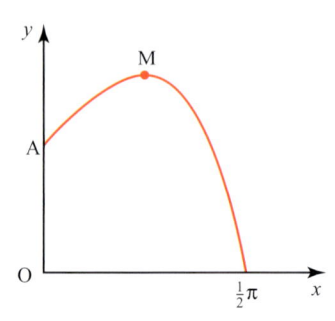

(i) Write down the co-ordinates of A.

(ii) Find the x co-ordinate of M.

(iii) Use the trapezium rule with three intervals to estimate the value of

$$\int_0^{\frac{1}{2}\pi} e^x \cos x \, dx,$$

giving your answer correct to 2 decimal places.

(iv) State, with a reason, whether the trapezium rule gives an underestimate or an overestimate of the true value of the integral in part **(iii)**.

[Cambridge International AS & A Level Mathematics 9709, Paper 2 Q7 June 2007]

11 The diagram shows the curve $y = x^2 e^{-x}$ and its maximum point M.

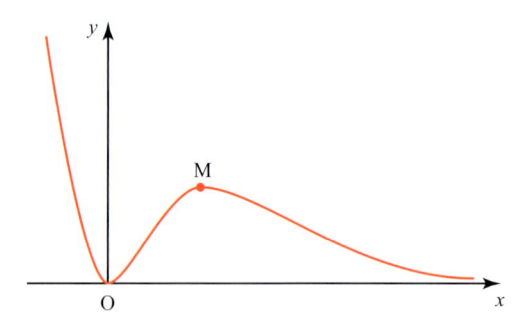

(i) Find the x co-ordinate of M.

(ii) Show that the tangent to the curve at the point where $x = 1$ passes through the origin.

(iii) Use the trapezium rule, with two intervals, to estimate the value of

$$\int_1^3 x^2 e^{-x} \, dx,$$

giving your answer correct to 2 decimal places.

[Cambridge International AS & A Level Mathematics 9709, Paper 2 Q8 November 2007]

12 The diagram shows a sketch of the curve $y = \dfrac{1}{1+x^3}$ for values of x from -0.6 to 0.6.

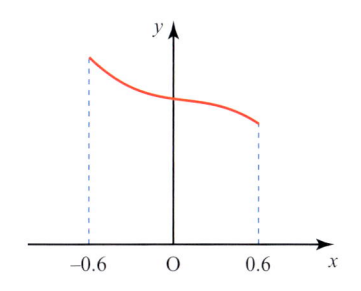

(i) Use the trapezium rule, with two intervals, to estimate the value of

$$\int_{-0.6}^{0.6} \dfrac{1}{1+x^3}\,dx,$$

giving your answer correct to 2 decimal places.

(ii) Explain, with reference to the diagram, why the trapezium rule may be expected to give a good approximation to the true value of the integral in this case.

[Cambridge International AS & A Level Mathematics 9709, Paper 3 Q2 June 2005]

KEY POINTS

1. $\displaystyle\int kx^n\,dx = \dfrac{kx^{n+1}}{n+1} + c$

2. $\displaystyle\int e^x\,dx = e^x + c$

   $\displaystyle\int e^{ax+b}\,dx = \dfrac{1}{a}e^{ax+b} + c$

3. $\displaystyle\int \dfrac{1}{x}\,dx = \ln|x| + c$

   $\displaystyle\int \dfrac{1}{ax+b}\,dx = \dfrac{1}{a}\ln|ax+b| + c$

4. $\displaystyle\int \cos(ax+b)\,dx = \dfrac{1}{a}\sin(ax+b) + c$

   $\displaystyle\int \sin(ax+b)\,dx = -\dfrac{1}{a}\cos(ax+b) + c$

   $\displaystyle\int \sec^2(ax+b)\,dx = \dfrac{1}{a}\tan(ax+b) + c$

5. You can use the trapezium rule, with n strips of width h, to find an approximate value for a definite integral as

$$A \approx \dfrac{h}{2}\big[y_0 + 2(y_1 + y_2 + \ldots + y_{n-1}) + y_n\big]$$

In words this is

Area $\approx \dfrac{1}{2} \times$ strip width $\times$ [ends + twice middles]

Numerical solution of equations

It is the true nature of mankind to learn from his mistakes.

Fred Hoyle

? Which of the following equations can be solved algebraically, and which cannot? For each equation find a solution, accurate or approximate.

(i) $x^2 - 4x + 3 = 0$ (ii) $x^2 + 10x + 8 = 0$ (iii) $x^5 - 5x + 3 = 0$
(iv) $x^3 - x = 0$ (v) $e^x = 4x$

You probably realised that the equations $x^5 - 5x + 3 = 0$ and $e^x = 4x$ cannot be solved algebraically. You may have decided to draw their graphs, either manually or using a graphic calculator or computer package, as in figure 6.1.

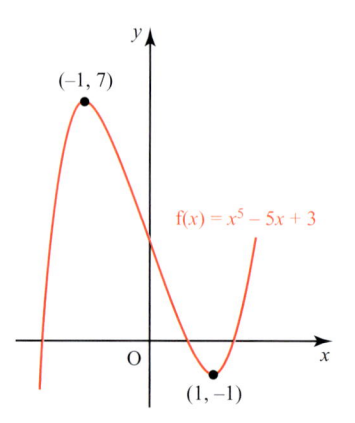

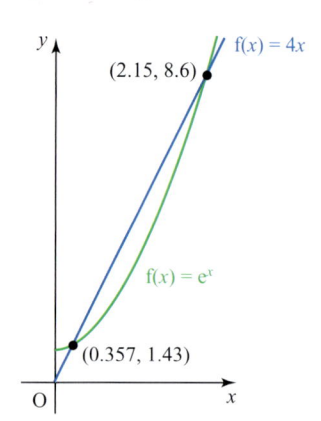

Figure 6.1

The graphs show you that

- $x^5 - 5x + 3 = 0$ has three roots, lying in the intervals $[-2, -1]$, $[0, 1]$ and $[1, 2]$.

- $e^x = 4x$ has two roots, lying in the intervals $[0, 1]$ and $[2, 3]$.

Note

An interval written as [*a*, *b*] means the interval between *a* and *b*, including *a* and *b*. This notation is used in this chapter. If *a* and b are not included, the interval is written (*a*, *b*). You may also elsewhere meet the notation]*a*, *b*[, indicating that *a* and *b* are not included.

The problem now is how to find the roots to any required degree of accuracy, and as efficiently as possible.

In many real problems, equations are obtained for which solutions using algebraic or analytic methods are not possible, but for which you nonetheless want to know the answers. In this chapter you will be introduced to numerical methods for solving such equations. In applying these methods, keep the following points in mind.

- Only use numerical methods when algebraic ones are not available. If you can solve an equation algebraically (e.g. a quadratic equation), that is the right method to use.

- Before starting to use a calculator or computer program, always start by drawing a sketch graph of the function whose equation you are trying to solve. This will show you how many roots the equation has and their approximate positions. It will also warn you of possible difficulties with particular methods. When using a graphic calculator or computer package ensure that the range of values of x is sufficiently large to, hopefully, find all the roots.

- Always give a statement about the accuracy of an answer (e.g. to 5 decimal places, or ± 0.000005). An answer obtained by a numerical method is worthless without this; the fact that at some point your calculator display reads, say, 1.6764705882 does not mean that all these figures are valid.

- Your statement about the accuracy must be obtained from within the numerical method itself. Usually you find a sequence of estimates of ever-increasing accuracy.

- Remember that the most suitable method for one equation may not be that for another.

Interval estimation – change-of-sign methods

Assume that you are looking for the roots of the equation $f(x) = 0$. This means that you want the values of x for which the graph of $y = f(x)$ crosses the x axis. As the curve crosses the x axis, $f(x)$ changes sign, so provided that $f(x)$ is a continuous function (its graph has no asymptotes or other breaks in it), once you have located an interval in which $f(x)$ changes sign, you know that that interval must contain a root. In both of the graphs in figure 6.2 (overleaf), there is a root lying between a and b.

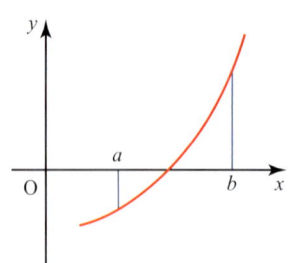

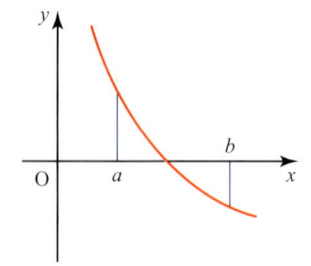

Figure 6.2

You have seen that $x^5 - 5x + 3 = 0$ has roots in the intervals $[-2, -1]$, $[0, 1]$ and $[1, 2]$. There are several ways of homing in on such roots systematically. Two of these are now described, using the search for the root in the interval $[0, 1]$ as an example.

Decimal search

In this method you first take increments in x of size 0.1 within the interval $[0, 1]$, working out the value of $f(x) = x^5 - 5x + 3$ for each one. You do this until you find a change of sign.

x	0.0	0.1	0.2	0.3	0.4	0.5	0.6	0.7
$f(x)$	3.00	2.50	2.00	1.50	1.01	0.53	0.08	−0.33

There is a sign change, and therefore a root, in the interval $[0.6, 0.7]$ since the function is continuous. Having narrowed down the interval, you can now continue with increments of 0.01 within the interval $[0.6, 0.7]$.

x	0.60	0.61	0.62
$f(x)$	0.08	0.03	−0.01

This shows that the root lies in the interval $[0.61, 0.62]$.

Alternative ways of expressing this information are that the root can be taken as 0.615 with a maximum error of ± 0.005, or the root is 0.6 (to 1 decimal place).

This process can be continued by considering $x = 0.611$, $x = 0.612$, … to obtain the root to any required number of decimal places.

? How many steps of decimal search would be necessary to find each of the values 0.012, 0.385 and 0.989, using $x = 0$ as a starting point?

When you use this procedure on a computer or calculator you should be aware that the machine is working in base 2, and that the conversion of many simple numbers from base 10 to base 2 introduces small rounding errors. This can lead to simple roots such as 2.7 being missed and only being found as 2.699 999.

e) Interval bisection

This method is similar to the decimal search, but instead of dividing each interval into ten parts and looking for a sign change, in this case the interval is divided into two parts – it is bisected.

Looking as before for the root in the interval [0, 1], you start by taking the mid-point of the interval, 0.5.

f(0.5) = 0.53, so f(0.5) > 0. Since f(1) < 0, the root is in [0.5, 1].

Now take the mid-point of this second interval, 0.75.

f(0.75) = −0.51, so f(0.75) < 0. Since f(0.5) > 0, the root is in [0.5, 0.75].

The mid-point of this further reduced interval is 0.625.

f(0.625) = −0.03, so the root is in the interval [0.5, 0.625].

The method continues in this manner until any required degree of accuracy is obtained. However, the interval bisection method is quite slow to converge to the root, and is cumbersome when performed manually.

ACTIVITY 6.1 Investigate how many steps of this method you need to achieve an accuracy of 1, 2, 3 and n decimal places, having started with an interval of length 1.

Error (or solution) bounds

Change-of-sign methods have the great advantage that they automatically provide bounds (the two ends of the interval) within which a root lies, so the maximum possible error in a result is known. Knowing that a root lies in the interval [0.61, 0.62] means that you can take the root as 0.615 with a maximum error of ± 0.005.

Problems with change-of-sign methods

There are a number of situations which can cause problems for change-of-sign methods if they are applied blindly, for example by entering the equation into a computer program without prior thought. In all cases you can avoid problems by first drawing a sketch graph, provided that you know what dangers to look out for.

The curve touches the x axis

In this case there is no change of sign, so change-of-sign methods are doomed to failure (see figure 6.3).

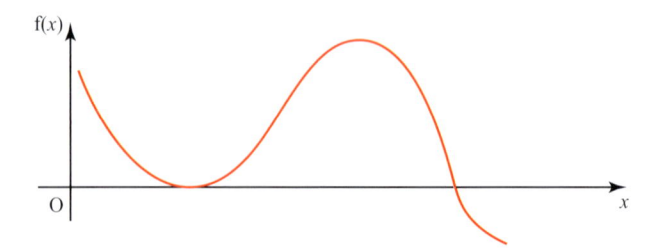

Figure 6.3

There are several roots close together

Where there are several roots close together, it is easy to miss a pair of them. The equation

$$f(x) = x^3 - 1.9x^2 + 1.11x - 0.189 = 0$$

has roots at 0.3, 0.7 and 0.9. A sketch of the curve of f(x) is shown in figure 6.4.

In this case f(0) < 0 and f(1) > 0, so you know there is a root between 0 and 1.

A decimal search would show that f(0.3) = 0, so that 0.3 is a root. You would be unlikely to search further in this interval.

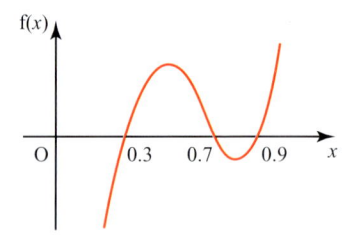

Figure 6.4

Interval bisection gives f(0.5) > 0, so you would search the interval [0, 0.5] and eventually arrive at the root 0.3, unaware of the existence of those at 0.7 and 0.9.

There is a discontinuity in f(x)

The curve $y = \dfrac{1}{x - 2.7}$ has a discontinuity at $x = 2.7$, as shown by the asymptote in figure 6.5.

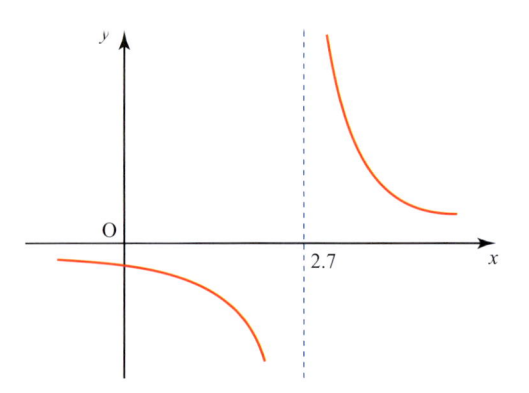

Figure 6.5

The equation $\dfrac{1}{x - 2.7} = 0$ has no root, but all change-of-sign methods will converge on a false root at $x = 2.7$.

None of these problems will arise if you start by drawing a sketch graph.

Note: Use of technology

It is important that you understand how each method works and are able, if necessary, to perform the calculations using only a scientific calculator. However, these repeated operations lend themselves to the use of a spreadsheet or a programmable calculator. Many packages, such as Autograph, will both perform the methods and illustrate them graphically.

EXERCISE 6A

1. (i) Show that the equation $x^3 + 3x - 5 = 0$ has no turning (stationary) points.
   (ii) Show with the aid of a sketch that the equation can have only one root, and that this root must be positive.
   (iii) Find the root, correct to 3 decimal places.

2. (i) How many roots has the equation $e^x - 3x = 0$?
   (ii) Find an interval of unit length containing each of the roots.
   (iii) Find each root correct to 2 decimal places.

3. (i) Sketch $y = 2^x$ and $y = x + 2$ on the same axes.
   (ii) Use your sketch to deduce the number of roots of the equation $2^x = x + 2$.
   (iii) Find each root, correct to 3 decimal places if appropriate.

4. Find all the roots of $x^3 - 3x + 1 = 0$, giving your answers correct to 2 decimal places.

5. Find the roots of $x^5 - 5x + 3 = 0$ in the intervals $[-2, -1]$ and $[1, 2]$, correct to 2 decimal places, using

   (i) decimal search
   (ii) interval bisection.

   Comment on the ease and efficiency with which the roots are approached by each method.

6. (i) Use a systematic search for a change of sign, starting with $x = -2$, to locate intervals of unit length containing each of the three roots of
   $$x^3 - 4x^2 - 3x + 8 = 0.$$
   (ii) Sketch the graph of $f(x) = x^3 - 4x^2 - 3x + 8$.
   (iii) Use the method of interval bisection to obtain each of the roots correct to 2 decimal places.
   (iv) Use your last intervals in part (iii) to give each of the roots in the form $a \pm (0.5)^n$ where a and n are to be determined.

7 The diagram shows a sketch of the graph of $f(x) = e^x - x^3$ without scales.

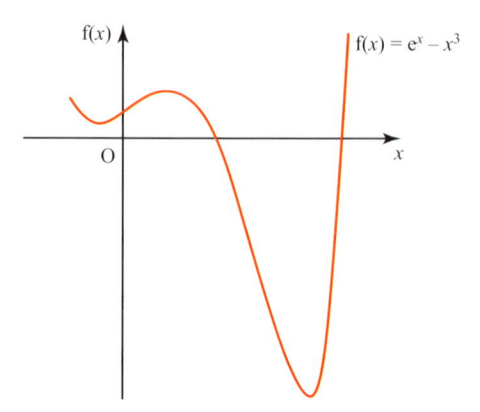

(i) Use a systematic search for a change of sign to locate intervals of unit length containing each of the roots.

(ii) Use a change-of-sign method to find each of the roots correct to 3 decimal places.

8 For each of the equations below

(a) sketch the curve

(b) write down any roots

(c) investigate what happens when you use a change-of-sign method with a starting interval of $[-0.3, 0.7]$.

(i) $y = \dfrac{1}{x}$ **(ii)** $y = \dfrac{x}{x^2 + 1}$ **(iii)** $y = \dfrac{x^2}{x^2 + 1}$

Fixed-point iteration

In fixed-point iteration you find a single value or point as your estimate for the value of x, rather than establishing an interval within which it must lie. This involves an *iterative process*, a method of generating a sequence of numbers by continued repetition of the same procedure. If the numbers obtained in this manner approach some limiting value, then they are said to *converge* to this value.

INVESTIGATION

Notice what happens in each of the following cases, and try to find some explanation for it.

(i) Set your calculator to the radian mode, enter zero if not automatically displayed and press the cosine key repeatedly.

(ii) Enter any positive number into your calculator and press the square root key repeatedly. Try this for both large and small numbers.

(iii) Enter any positive number into your calculator and press the sequence [+] [1] [=] [√] [=] repeatedly. Write down the number which appears each time you press [=]. The sequence generated appears to converge. You may recognise the number to which it appears to converge: it is called the Golden Ratio.

Rearranging the equation f(x) = 0 into the form x = F(x)

The first step, with an equation $f(x) = 0$, is to rearrange it into the form $x = F(x)$. Any value of x for which $x = F(x)$ is a root of the original equation, as shown in figure 6.6.

When $f(x) = x^2 - x - 2$, $f(x) = 0$ is the same as $x = x^2 - 2$.

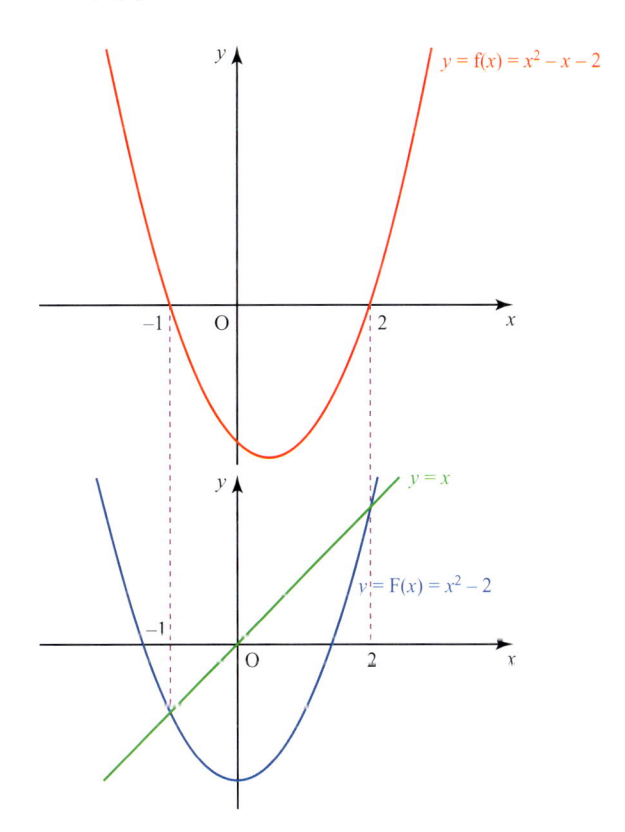

Figure 6.6

The equation $x^5 - 5x + 3 = 0$ which you met earlier can be rewritten in a number of ways. One of these is $5x = x^5 + 3$, giving

$$x = F(x) = \frac{x^5 + 3}{5}.$$

Figure 6.7 shows the graphs of $y = x$ and $y = F(x)$ in this case.

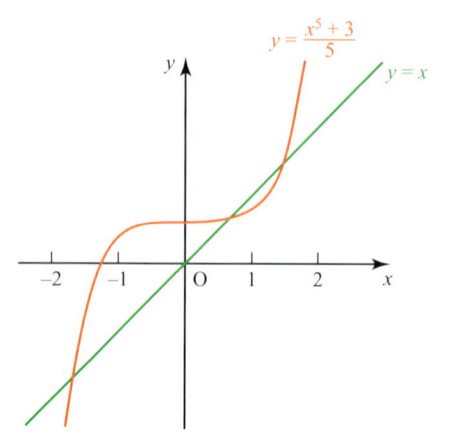

Figure 6.7

This provides the basis for the iterative formula

$$x_{n+1} = \frac{x_n^5 + 3}{5}.$$

Taking $x = 1$ as a starting point to find the root in the interval $[0, 1]$, successive approximations are:

$x_1 = 1$, $x_2 = 0.8$, $x_3 = 0.6655$, $x_4 = 0.6261$, $x_5 = 0.6192$,
$x_6 = 0.6182$, $x_7 = 0.6181$, $x_8 = 0.6180$, $x_9 = 0.6180$.

In this case the iteration has converged quite rapidly to the root for which you were looking.

? Another way of arranging $x^5 - 5x + 3 = 0$ is $x = \sqrt[5]{5x - 3}$. What other possible rearrangements can you find? How many are there altogether?

---

e The iteration process is easiest to understand if you consider the graph. Rewriting the equation $f(x) = 0$ in the form $x = F(x)$ means that instead of looking for points where the graph of $y = f(x)$ crosses the x axis, you are now finding the points of intersection of the curve $y = F(x)$ and the line $y = x$.

What you do	What it looks like on the graph
• Choose a value, x_1, of x	Take a starting point on the x axis
• Find the corresponding value of $F(x_1)$	Move vertically to the curve $y = F(x)$
• Take this value $F(x_1)$ as the new value of x, i.e. $x_2 = F(x_1)$	Move horizontally to the line $y = x$
• Find the value of $F(x_2)$ and so on	Move vertically to the curve

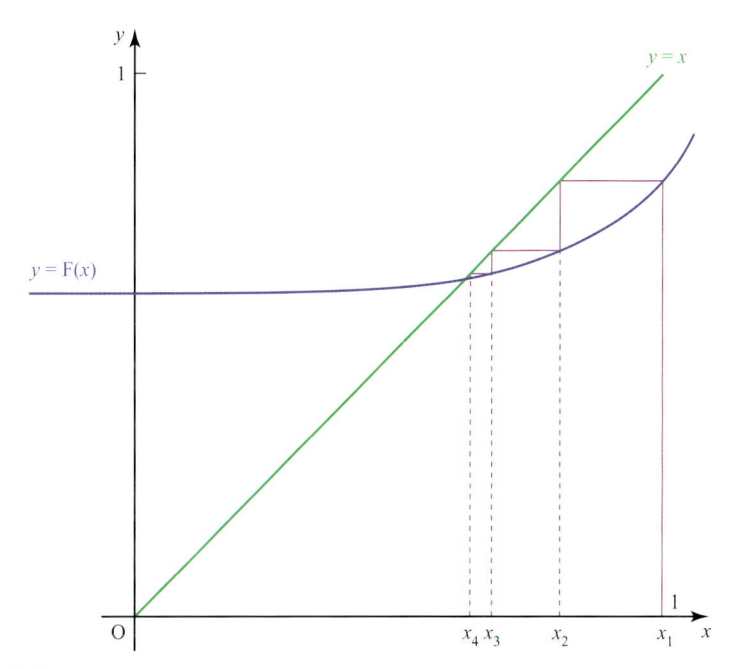

Figure 6.8

The effect of several repeats of this procedure is shown in figure 6.8. The successive steps look like a staircase approaching the root: this type of diagram is called a *staircase diagram*. In other examples, a *cobweb diagram* may be produced, as shown in figure 6.9.

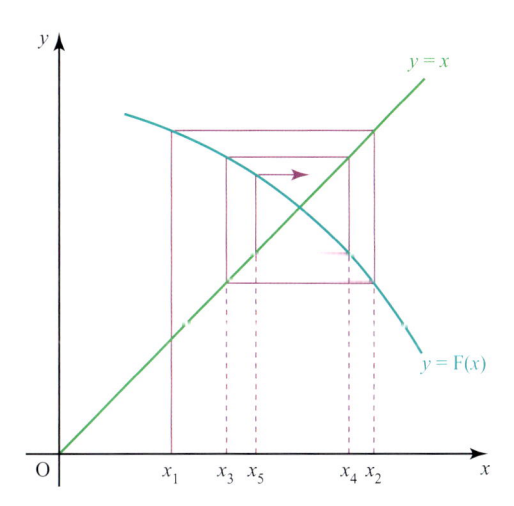

Figure 6.9

Successive approximations to the root are found by using the formula

$$x_{n+1} = F(x_n).$$

This is an example of an *iterative formula*. If the resulting values of x_n approach some limit, a, then $a = F(a)$, and so a is a *fixed point* of the iteration. It is also a root of the original equation $f(x) = 0$.

Note

In the staircase diagram, the values of x_n approach the root from one side, but in a cobweb diagram they oscillate about the root. From figures 6.8 and 6.9 it is clear that the error (the difference between a and x_n) is decreasing in both diagrams.

Accuracy of the method of rearranging the equation

Iterative procedures give you a sequence of point estimates. A staircase diagram, for example, might give the following.

$$1, 0.8, 0.6655, 0.6261, 0.6192$$

What can you say at this stage?

Looking at the pattern of convergence it seems as though the root lies between 0.61 and 0.62, but you cannot be absolutely certain from the available evidence. To be certain you must look for a change of sign.

$$f(0.61) = +0.034\ldots \quad f(0.62) = -0.0083\ldots$$

Ⓟ Explain why you can now be quite certain that your judgement is correct.

Note

Estimates from a cobweb diagram oscillate above and below the root and so naturally provide you with bounds.

Using different arrangements of the equation

So far only one possible arrangement of the equation $x^5 - 5x + 3 = 0$ has been used. What happens when you use a different arrangement, for example $x = \sqrt[5]{5x - 3}$, which leads to the iterative formula

$$x_{n+1} = \sqrt[5]{5x_n - 3}?$$

The resulting sequence of approximations is:

$x_1 = 1,$ $\quad x_2 = 1.1486\ldots,$ $\quad x_3 = 1.2236\ldots,$ $\quad x_4 = 1.2554\ldots,$
$x_5 = 1.2679\ldots,$ $\quad x_6 = 1.2727\ldots,$ $\quad x_7 = 1.2745\ldots,$ $\quad x_8 = 1.2752\ldots,$
$x_9 = 1.2755\ldots,$ $\quad x_{10} = 1.2756\ldots,$ $\quad x_{11} = 1.2756\ldots,$ $\quad x_{12} = 1.2756\ldots.$

⚠ In the calculations the full calculator values of x_n were used, but only the first 4 decimal places have been written down.

The process has clearly converged, but in this case not to the root for which you were looking: you have identified the root in the interval [1, 2]. If instead you had taken $x_1 = 0$ as your starting point and applied the second formula, you would have obtained a sequence converging to the value -1.6180, the root in the interval $[-2, -1]$.

e The choice of F(x)

A particular rearrangement of the equation $f(x) = 0$ into the form $x = F(x)$ will allow convergence to a root a of the equation, provided that $-1 < F'(a) < 1$ for values of x close to the root.

Look again at the two rearrangements of $x^5 - 5x + 3 = 0$ which were suggested. When you look at the graph of $y = F(x) = \sqrt[5]{5x - 3}$, as shown in figure 6.10, you can see that its gradient near A, the root you were seeking, is greater than 1. This makes $x_{n+1} = \sqrt[5]{5x_n - 3}$ an unsuitable iterative formula for finding the root in the interval [0, 1], as you saw earlier.

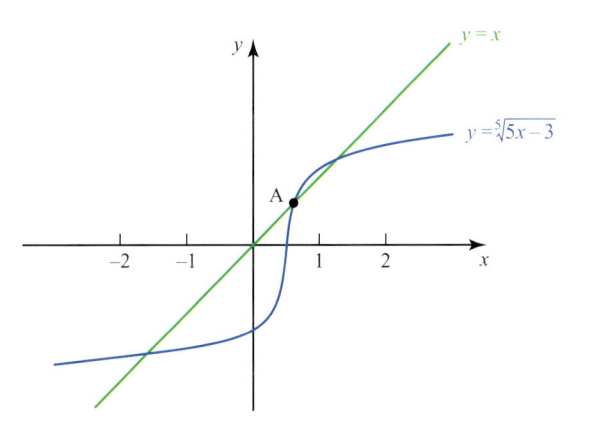

Figure 6.10

When an equation has two or more roots, a single rearrangement will not usually find all of them. This is demonstrated in figure 6.11.

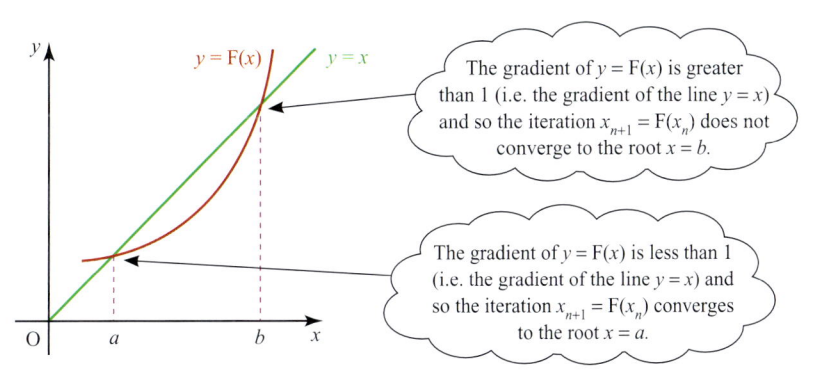

The gradient of $y = F(x)$ is greater than 1 (i.e. the gradient of the line $y = x$) and so the iteration $x_{n+1} = F(x_n)$ does not converge to the root $x = b$.

The gradient of $y = F(x)$ is less than 1 (i.e. the gradient of the line $y = x$) and so the iteration $x_{n+1} = F(x_n)$ converges to the root $x = a$.

Figure 6.11

ACTIVITY 6.2 Try using the iterative formula $x_{n+1} = \dfrac{x_n^5 + 3}{5}$ to find the roots in the intervals $[-2, -1]$ and $[1, 2]$. In both cases use each end point of the interval as a starting point. What happens?

Explain what you find by referring to a sketch of the curve $y = \dfrac{x^5 + 3}{5}$.

EXERCISE 6B

1. (i) Show that the equation $x^3 - x - 2 = 0$ has a root between 1 and 2.
   (ii) The equation is rearranged into the form $x = F(x)$, where
   $$F(x) = \sqrt[3]{x + 2}.$$
   Use the iterative formula suggested by this rearrangement to find the value of the root to 3 decimal places.

2. (i) Show that the equation $e^{-x} - x + 2 = 0$ has a root in the interval $[2, 3]$.
   (ii) The equation is rearranged into the form $= e^{-x} + 2$.
   Use the iterative formula suggested by this rearrangement to find the value of the root to 3 decimal places.

3. (i) Show that the equation $e^x + x - 6 = 0$ has a root in the interval $[1, 2]$.
   (ii) Show that this equation may be written in the form $x = \ln(6 - x)$.
   (iii) Use an iterative formula based on the equation $x = \ln(6 - x)$ to calculate the root correct to 3 decimal places.

4. (i) Sketch the curves $y = e^x$ and $y = x^2 + 2$ on the same graph.
   (ii) Use your sketch to explain why the equation $e^x - x^2 - 2 = 0$ has only one root.
   (iii) Rearrange this equation in the form $x = F(x)$.
   (iv) Use an iterative formula based on the equation found in part (iii) to calculate the root correct to 3 decimal places

5. (i) Show that $x^2 = \ln(x + 1)$ for $x = 0$ and for one other value of x.
   (ii) Use the method of fixed point iteration to find the second value to 3 decimal places.

6. (i) Sketch the graphs of $y = x$ and $y = \cos x$ on the same axes, for $0 \leq x \leq \dfrac{\pi}{2}$.
   (ii) Find the solution of the equation $x = \cos x$ to 5 decimal places.

7. The sequence of values given by the iterative formula
   $$x_{n+1} = \dfrac{3x_n}{4} + \dfrac{2}{x_n^3},$$
   with initial value $x_1 = 2$, converges to α.

   (i) Use this iteration to calculate α correct to 2 decimal places, showing the result of each iteration to 4 decimal places.
   (ii) State an equation which is satisfied by α and hence find the exact value of α.

   [Cambridge International AS & A Level Mathematics 9709, Paper 2 Q3 June 2005]

8 The sequence of values given by the iterative formula

$$x_{n+1} = \frac{2x_n}{3} + \frac{4}{x_n^2},$$

with initial value $x_1 = 2$, converges to α.

(i) Use this iterative formula to determine α correct to 2 decimal places, giving the result of each iteration to 4 decimal places.

(ii) State an equation that is satisfied by α and hence find the exact value of α.

[Cambridge International AS & A Level Mathematics 9709, Paper 2 Q2 November 2007]

9 (i) By sketching a suitable pair of graphs, show that the equation

$$\cos x = 2 - 2x,$$

where x is in radians, has only one root for $0 \leq x \leq \frac{1}{2}\pi$.

(ii) Verify by calculation that this root lies between 0.5 and 1.

(iii) Show that, if a sequence of values given by the iterative formula

$$x_{n+1} = 1 - \tfrac{1}{2}\cos x_n$$

converges, then it converges to the root of the equation in part (i).

(iv) Use this iterative formula, with initial value $x_1 = 0.6$, to determine this root correct to 2 decimal places. Give the result of each iteration to 4 decimal places.

[Cambridge International AS & A Level Mathematics 9709, Paper 2 Q7 November 2008]

10 The diagram shows the curve $y = x^2 \cos x$, for $0 \leq x \leq \frac{1}{2}\pi$, and its maximum point M.

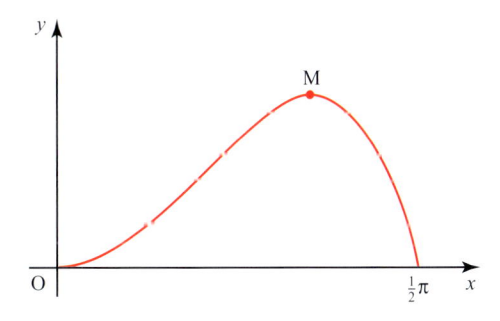

(i) Show by differentiation that the x co-ordinate of M satisfies the equation

$$\tan x = \frac{2}{x}.$$

(ii) Verify by calculation that this equation has a root (in radians) between 1 and 1.2.

(iii) Use the iterative formula $x_{n+1} = \tan^{-1}\left(\dfrac{2}{x_n}\right)$ to determine this root correct to 2 decimal places. Give the result of each iteration to 4 decimal places.

[Cambridge International AS & A Level Mathematics 9709, Paper 22 Q7 November 2009]

11 The diagram shows the curve $y = xe^{2x}$ and its minimum point M.

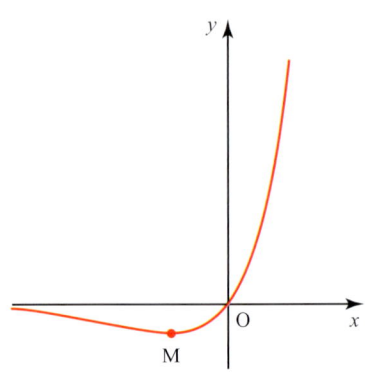

(i) Find the exact co-ordinates of M.

(ii) Show that the curve intersects the line $y = 20$ at the point whose x co-ordinate is the root of the equation

$$x = \tfrac{1}{2}\ln\left(\frac{20}{x}\right).$$

(iii) Use the iterative formula

$$x_{n+1} = \tfrac{1}{2}\ln\left(\frac{20}{x_n}\right),$$

with initial value $x_1 = 1.3$, to calculate the root correct to 2 decimal places, giving the result of each iteration to 4 decimal places.

[Cambridge International AS & A Level Mathematics 9709, Paper 2 Q7 June 2009]

12 (i) By sketching a suitable pair of graphs, show that the equation

$$\ln x = 2 - x^2$$

has only one root.

(ii) Verify by calculation that this root lies between $x = 1.3$ and $x = 1.4$.

(iii) Show that, if a sequence of values given by the iterative formula

$$x_{n+1} = \sqrt{(2 - \ln x_n)}$$

converges, then it converges to the root of the equation in part (i).

(iv) Use the iterative formula $x_{n+1} = \sqrt{(2 - \ln x_n)}$ to determine the root correct to 2 decimal places. Give the result of each iteration to 4 decimal places.

[Cambridge International AS & A Level Mathematics 9709, Paper 22 Q6 June 2010]

13 The equation $x^3 - 8x - 13 = 0$ has one real root.

(i) Find the two consecutive integers between which this root lies.

(ii) Use the iterative formula
$$x_{n+1} = (8x_n + 13)^{\frac{1}{3}}$$
to determine this root correct to 2 decimal places. Give the result of each iteration to 4 decimal places.

[Cambridge International AS & A Level Mathematics 9709, Paper 32 Q2 November 2009]

14 The equation $x^3 - 2x - 2 = 0$ has one real root.

(i) Show by calculation that this root lies between $x = 1$ and $x = 2$.

(ii) Prove that, if a sequence of values given by the iterative formula
$$x_{n+1} = \frac{2x_n^3 + 2}{3x_n^2 - 2}$$
converges, then it converges to this root.

(iii) Use this iterative formula to calculate the root correct to 2 decimal places. Give the result of each iteration to 4 decimal places.

[Cambridge International AS & A Level Mathematics 9709, Paper 3 Q4 June 2009]

KEY POINTS

1 When f(x) is a continuous function, if f(a) and f(b) have opposite signs, there will be at least one root of f(x) = 0 in the interval [a, b].

2 When an interval [a, b] containing a root has been found, this interval may be reduced systematically by decimal search or interval bisection.

3 Fixed-point iteration may be used to solve an equation f(x) = 0. You can sometimes find a root by rearranging the equation f(x) = 0 into the form x = F(x) and using the iteration $x_{n+1} = F(x_n)$.

e 4 Successive iterations will converge to the root a provided that $-1 < F'(a) < 1$ for values of x close to the root.

Pure Mathematics 3

7 Further algebra

At the age of twenty-one he wrote a treatise upon the Binomial Theorem. ... On the strength of it, he won the Mathematical Chair at one of our smaller Universities.

Sherlock Holmes on Professor Moriarty in 'The Final Problem' by Sir Arthur Conan Doyle

How would you find $\sqrt{101}$ correct to 3 decimal places, without using a calculator?

Many people are able to develop a very high degree of skill in mental arithmetic, particularly those whose work calls for quick reckoning. There are also those who have quite exceptional innate skills. Shakuntala Devi, pictured right, is famous for her mathematical speed. On one occasion she found the 23rd root of a 201-digit number in her head, beating a computer by 12 seconds. On another occasion she multiplied 7 686 369 774 870 by 2 465 099 745 779 in just 28 seconds.

While most mathematicians do not have Shakuntala Devi's high level of talent with numbers, they do acquire a sense of when something looks right or wrong. This often involves finding approximate values of numbers, such as $\sqrt{101}$, using methods that are based on series expansions, and these are the subject of the first part of this chapter.

INVESTIGATION

Using your calculator, write down the values of $\sqrt{1.02}$, $\sqrt{1.04}$, $\sqrt{1.06}$, ..., giving your answers correct to 2 decimal places. What do you notice?

Use your results to complete the following, giving the value of the constant k.

$$\sqrt{1.02} = (1 + 0.02)^{\frac{1}{2}} \approx 1 + 0.02k$$
$$\sqrt{1.04} = (1 + 0.04)^{\frac{1}{2}} \approx 1 + 0.04k$$

What is the largest value of x such that $\sqrt{1+x} \approx 1 + kx$ is true for the same value of k?

The general binomial expansion

In *Pure Mathematics 1* Chapter 3 you met the binomial expansion in the form

$$(1+x)^n = 1 + \binom{n}{1}x + \binom{n}{2}x^2 + \binom{n}{3}x^3 + \ldots + \binom{n}{r}x^r + \ldots$$

which holds when n is any positive integer (or zero), that is $n \in \mathbb{N}$.

> This is a short way of writing 'n is a *natural number*'. A natural is any positive integer or zero.

This may also be written as

$$(1+x)^n = 1 + nx + \frac{n(n-1)}{2!}x^2 + \frac{n(n-1)(n-2)}{3!}x^3 + \ldots$$
$$+ \frac{n(n-1)(n-2)\ldots(n-r+1)}{r!}x^r + \ldots$$

which, being the same expansion as above, also holds when $n \in \mathbb{N}$.

The general binomial theorem states that this second form, that is

$$(1+x)^n = 1 + nx + \frac{n(n-1)}{2!}x^2 + \frac{n(n-1)(n-2)}{3!}x^3 + \ldots$$
$$+ \frac{n(n-1(n-2)\ldots(n-r+1)}{r!}x^r + \ldots$$

is true when **n is any real number**, but there are two important differences to note when $n \notin \mathbb{N}$.

> This is a short way of writing 'n is not a natural number'.

- The series is infinite (or non-terminating).

- The expansion of $(1+x)^n$ is valid only if $|x| < 1$.

Proving this result is beyond the scope of an A-level course but you can assume that it is true.

Consider now the coefficients in the binomial expansion:

$$1, \quad n, \quad \frac{n(n-1)}{2!}, \quad \frac{n(n-1(n-2)}{3!}, \quad \frac{n(n-1)(n-2)(n-3)}{4!}, \quad \ldots$$

When $n = 0$, we get 1 0 0 0 0 .. (infinitely many zeros)
$n = 1$ 1 1 0 0 0 ... ditto
$n = 2$ 1 2 1 0 0 ... ditto
$n = 3$ 1 3 3 1 0 ... ditto
$n = 4$ 1 4 6 4 1 ... ditto

so that, for example

$$(1+x)^2 = 1 + 2x + x^2 + 0x^3 + 0x^4 + 0x^5 + \ldots$$
$$(1+x)^3 = 1 + 3x + 3x^2 + x^3 + 0x^4 + 0x^5 + \ldots$$
$$(1+x)^4 = 1 + 4x + 6x^2 + 4x^3 + x^4 + 0x^5 + \ldots$$

Of course, it is usual to discard all the zeros and write these binomial coefficients in the familiar form of Pascal's triangle:

$$
\begin{array}{ccccccccc}
& & & & 1 & & & & \\
& & & 1 & & 1 & & & \\
& & 1 & & 2 & & 1 & & \\
& 1 & & 3 & & 3 & & 1 & \\
1 & & 4 & & 6 & & 4 & & 1
\end{array}
$$

and the expansions as

$$(1 + x)^2 = 1 + 2x + x^2$$
$$(1 + x)^3 = 1 + 3x + 3x^2 + x^3$$
$$(1 + x)^4 = 1 + 4x + 6x^2 + 4x^3 + x^4$$

However, for other values of n (where $n \notin \mathbb{N}$) there are no zeros in the row of binomial coefficients and so we obtain an infinite sequence of non-zero terms. For example:

$n = -3$ gives $1 \quad -3 \quad \dfrac{(-3)(-4)}{2!} \quad \dfrac{(-3)(-4)(-5)}{3!} \quad \dfrac{(-3)(-4)(-5)(-6)}{4!} \quad \ldots$

that is $1 \quad -3 \quad 6 \quad -10 \quad 15 \quad \ldots$

$n = \tfrac{1}{2}$ gives $1 \quad \tfrac{1}{2} \quad \dfrac{\left(\tfrac{1}{2}\right)\left(-\tfrac{1}{2}\right)}{2!} \quad \dfrac{\left(\tfrac{1}{2}\right)\left(-\tfrac{1}{2}\right)\left(-\tfrac{3}{2}\right)}{3!} \quad \dfrac{\left(\tfrac{1}{2}\right)\left(-\tfrac{1}{2}\right)\left(-\tfrac{3}{2}\right)\left(-\tfrac{5}{2}\right)}{4!} \quad \ldots$

that is $1 \quad \tfrac{1}{2} \quad -\tfrac{1}{8} \quad \tfrac{1}{16} \quad -\tfrac{5}{128} \quad \ldots$

so that $(1 + x)^{-3} = 1 - 3x + 6x^2 - 10x^3 + 15x^4 + \ldots$

and $(1 + x)^{\tfrac{1}{2}} = 1 + \tfrac{1}{2}x - \tfrac{1}{8}x^2 + \tfrac{1}{16}x^3 - \tfrac{5}{128}x^4 + \ldots$

But remember: these two expansions are valid only if $|x| < 1$.

Show that the expansion of $(1 + x)^{\tfrac{1}{2}}$ is not valid when $x = 8$.

These examples confirm that there will be an infinite sequence of non-zero coefficients when $n \notin \mathbb{N}$.

In the investigation at the beginning of this chapter you showed that

$$\sqrt{1 + x} \approx 1 + \tfrac{1}{2}x$$

is a good approximation for small values of x. Notice that these are the first two terms of the binomial expansion for $n = \tfrac{1}{2}$. If you include the third term, the approximation is

$$\sqrt{1 + x} \approx 1 + \tfrac{1}{2}x - \tfrac{1}{8}x^2.$$

Take $y = 1 + \frac{1}{2}x$, $y = 1 + \frac{1}{2}x - \frac{1}{8}x^2$ and $y = \sqrt{1+x}$.

They are shown in the graph in figure 7.1 for values of x between -1 and 1.

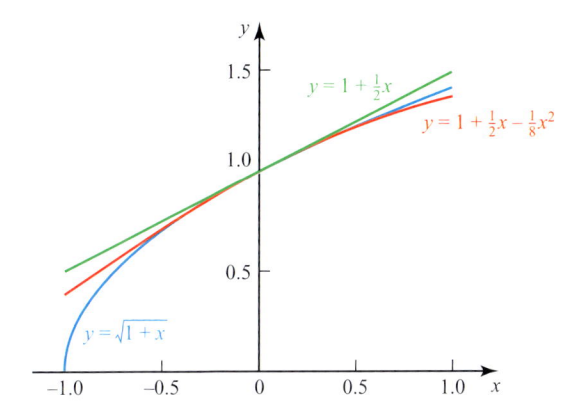

Figure 7.1

INVESTIGATION

For $n = \frac{1}{2}$ the first three terms of the binomial expansion are $1 + \frac{1}{2}x - \frac{1}{8}x^2$. Use your calculator to verify the approximate result

$$\sqrt{1+x} \approx 1 + \frac{1}{2}x - \frac{1}{8}x^2$$

for 'small' values of x.

What values of x can be considered as 'small' if you want the result to be correct to 2 decimal places?

Now take $n = -3$. Using the coefficients found earlier suggests the approximate result

$$(1+x)^{-3} \approx 1 - 3x + 6x^2.$$

Comment on values of x for which this approximation is correct to 2 decimal places.

When $|x| < 1$, the magnitudes of x^2, x^3, x^4, x^5, ... form a decreasing geometric sequence. In this case, the binomial expansion converges (just as a geometric progression converges for $-1 < r < 1$, where r is the common ratio) and has a sum to infinity.

ACTIVITY 7.1 Compare the geometric progression $1 - x + x^2 - x^3 + ...$ with the series obtained by putting $n = -1$ in the binomial expansion. What do you notice?

To summarise: when n is not a positive integer or zero, the binomial expansion of $(1+x)^n$ becomes an infinite series, and is only valid when some restriction is placed on the values of x.

The binomial theorem states that for any value of n:

$$(1+x)^n = 1 + nx + \frac{n(n-1)}{2!}x^2 + \frac{n(n-1)(n-2)}{3!}x^3 + \ldots$$

where

- if $n \in \mathbb{N}$, x may take any value;
- if $n \notin \mathbb{N}$, $|x| < 1$.

Note

The full statement is the binomial *theorem*, and the right-hand side is referred to as the binomial *expansion*.

EXAMPLE 7.1 Expand $(1-x)^{-2}$ as a series of ascending powers of x up to and including the term in x^3, stating the set of values of x for which the expansion is valid.

SOLUTION

$$(1+x)^n = 1 + nx + \frac{n(n-1)}{2!}x^2 + \frac{n(n-1)(n-2)}{3!}x^3 + \ldots$$

Replacing n by -2, and x by $(-x)$ gives

$$(1+(-x))^{-2} = 1 + (-2)(-x) + \frac{(-2)(-3)}{2!}(-x)^2 + \frac{(-2)(-3)(-4)}{3!}(-x)^3 + \ldots$$

when $|-x| < 1$

> It is important to put brackets round the term $-x$, since, for example, $(-x)^2$ is not the same as $-x^2$.

which leads to

$$(1-x)^{-2} \approx 1 + 2x + 3x^2 + 4x^3 \quad \text{when } |x| < 1.$$

Note

In this example the coefficients of the powers of x form a recognisable sequence, and it would be possible to write down a general term in the expansion. The coefficient is always one more than the power, so the rth term would be rx^{r-1}. Using sigma notation, the infinite series could be written as

$$\sum_{r=1}^{\infty} rx^{r-1}$$

EXAMPLE 7.2 Find a quadratic approximation for $\dfrac{1}{\sqrt{1+2t}}$ and state for which values of t the expansion is valid.

SOLUTION

$$\frac{1}{\sqrt{1+2t}} = \frac{1}{(1+2t)^{\frac{1}{2}}} = (1+2t)^{-\frac{1}{2}}$$

The binomial theorem states that

$$(a+x)^n = 1 + nx + \frac{n(n-1)}{2!}x^2 + \frac{n(n-1)(n-2)}{3!}x^3 + \ldots$$

Replacing n by $-\tfrac{1}{2}$ and x by $2t$ gives

$$(1+2t)^{-\frac{1}{2}} = 1 + \left(-\tfrac{1}{2}\right)(2t) + \frac{\left(-\tfrac{1}{2}\right)\left(-\tfrac{3}{2}\right)}{2!}(2t)^2 + \ldots \quad \text{when } |2t| < 1$$

$$\Rightarrow \quad (1+2t)^{-\frac{1}{2}} \approx 1 - t + \tfrac{3}{2}t^2 \quad \text{when } |t| < \tfrac{1}{2}$$

> Remember to put brackets round the term $2t$, since $(2t)^2$ is not the same as $2t^2$.

INVESTIGATION

Example 7.1 showed how using the binomial expansion for $(1-x)^{-2}$ gave a sequence of coefficients of powers of x which was easily recognisable, so that the particular binomial expansion could be written using sigma notation.

Investigate whether a recognisable pattern is formed by the coefficients in the expansions of $(1-x)^n$ for any other negative integers n.

The equivalent binomial expansion of $(a+x)^n$ when n is not a positive integer is rather unwieldy. It is easier to start by taking a outside the brackets:

$$(a+x)^n = a^n\left(1 + \frac{x}{a}\right)^n$$

The first entry inside the bracket is now 1 and so the first few terms of the expansion are

$$(a+x)^n = a^n\left[1 + n\left(\frac{x}{a}\right) + \frac{n(n-1)}{2!}\left(\frac{x}{a}\right)^2 + \frac{n(n-1)(n-2)}{3!}\left(\frac{x}{a}\right)^3 + \ldots\right]$$

for $\left|\dfrac{x}{a}\right| < 1$.

Note

Since the bracket is raised to the power n, any quantity you take out must be raised to the power n too, as in the following example.

EXAMPLE 7.3 Expand $(2 + x)^{-3}$ as a series of ascending powers of x up to and including the term in x^2, stating the values of x for which the expansion is valid.

SOLUTION

$$(2 + x)^{-3} = \frac{1}{(2 + x)^3}$$

$$= \frac{1}{2^3\left(1 + \frac{x}{2}\right)^3}$$

$$= \frac{1}{8}\left(1 + \frac{x}{2}\right)^{-3}$$

> Notice that this is the same as $2^{-3}\left(1 + \frac{x}{2}\right)^{-3}$.

Take the binomial expansion

$$(1 + x)^n = 1 + nx + \frac{n(n-1)}{2!}x^2 + \frac{n(n-1)(n-2)}{3!}x^3 + \ldots$$

and replace n by -3 and x by $\frac{x}{2}$ to give

$$\frac{1}{8}\left(1 + \frac{x}{2}\right)^{-3} = \frac{1}{8}\left[1 + (-3)\left(\frac{x}{2}\right) + \frac{(-3)(-4)}{2!}\left(\frac{x}{2}\right)^2 + \ldots\right] \quad \text{when } \left|\frac{x}{2}\right| < 1$$

$$\approx \frac{1}{8} - \frac{3x}{16} + \frac{3x^2}{16} \quad \text{when } |x| < 2$$

? The chapter began by asking how you would find $\sqrt{101}$ to 3 decimal places without using a calculator. How would you find it?

EXAMPLE 7.4 Find a quadratic approximation for $\frac{(2 + x)}{(1 - x^2)}$, stating the values of x for which the expansion is valid.

SOLUTION

$$\frac{(2 + x)}{(1 - x^2)} = (2 + x)(1 - x^2)^{-1}$$

Take the binomial expansion

$$(1 + x)^n = 1 + nx + \frac{n(n-1)}{2!}x^2 + \frac{n(n-1)(n-2)}{3!}x^3 + \ldots$$

and replace n by -1 and x by $(-x^2)$ to give

$$\left(1 + (-x^2)\right)^{-1} = 1 + (-1)(-x^2) + \frac{(-1)(-2)(-x^2)^2}{2!} + \ldots \quad \text{when } |-x^2| < 1$$

$$(1 - x^2)^{-1} = 1 + x^2 + \ldots \quad \text{when } |x^2| < 1, \text{ i.e. when } |x| < 1.$$

Multiply both sides by $(2 + x)$ to obtain $(2 + x)(1 - x^2)^{-1}$:

$$(2 + x)(1 - x^2)^{-1} \approx (2 + x)(1 + x^2)$$
$$\approx 2 + x + 2x^2 \quad \text{when } |x| < 1.$$

The term in x^3 has been omitted because the question asked for a quadratic approximation.

Sometimes two or more binomial expansions may be used together. If these impose different restrictions on the values of x, you need to decide which is the strictest.

EXAMPLE 7.5

Find a and b such that

$$\frac{1}{(1 - 2x)(1 + 3x)} \approx a + bx$$

and state the values of x for which the expansions you use are valid.

SOLUTION

$$\frac{1}{(1 - 2x)(1 + 3x)} = (1 - 2x)^{-1}(1 + 3x)^{-1}$$

Using the binomial expansion:

$$(1 - 2x)^{-1} \approx 1 + (-1)(-2x) \quad \text{for } |-2x| < 1$$
$$\text{and} \quad (1 + 3x)^{-1} \approx 1 + (-1)(3x) \quad \text{for } |3x| < 1$$
$$\Rightarrow \quad (1 - 2x)^{-1}(1 + 3x)^{-1} \approx (1 + 2x)(1 - 3x)$$
$$\approx 1 - x \quad \text{(ignoring higher powers of } x\text{)}$$

giving $a = 1$ and $b = -1$.

For the result to be valid, both $|2x| < 1$ and $|3x| < 1$ need to be satisfied.

$$|2x| < 1 \quad \Rightarrow \quad -\tfrac{1}{2} < x < \tfrac{1}{2}$$
$$\text{and} \quad |3x| < 1 \quad \Rightarrow \quad -\tfrac{1}{3} < x < \tfrac{1}{3}$$

Both of these restrictions are satisfied if $-\tfrac{1}{3} < x < \tfrac{1}{3}$. This is the stricter restriction.

Note

The binomial expansion may also be used when the first term is the variable. For example:

$$(x + 2)^{-1} \text{ may be written as } (2 + x)^{-1} = 2^{-1}\left(1 + \frac{x}{2}\right)^{-1}$$
$$\text{and} \quad (2x - 1)^{-3} = [(-1)(1 - 2x)]^{-3}$$
$$= (-1)^{-3}(1 - 2x)^{-3}$$
$$= -(1 - 2x)^{-3}$$

? What happens when you try to rearrange $\sqrt{x-1}$ so that the binomial expansion can be used?

EXERCISE 7A

1 For each of the expressions below

(a) write down the first three non-zero terms in their expansions as a series of ascending powers of x
(b) state the values of x for which the expansion is valid
(c) substitute $x = 0.1$ in both the expression and its expansion and calculate the percentage error, where

$$\text{percentage error} = \frac{\text{absolute error} \times 100}{\text{true value}}\%$$

(i) $(1+x)^{-2}$ (ii) $\dfrac{1}{1+2x}$ (iii) $\sqrt{1-x^2}$

(iv) $\dfrac{1+2x}{1-2x}$ (v) $(3+x)^{-1}$ (vi) $(1-x)\sqrt{4+x}$

(vii) $\dfrac{x+2}{x-3}$ (viii) $\dfrac{1}{\sqrt{3x+4}}$ (ix) $\dfrac{1+2x}{(2x-1)^2}$

(x) $\dfrac{1+x^2}{1-x^2}$ (xi) $\sqrt[3]{1+2x^2}$ (xii) $\dfrac{1}{(1+2x)(1+x)}$

2 (i) Write down the expansion of $(1+x)^3$.
(ii) Find the first four terms in the expansion of $(1-x)^{-4}$ in ascending powers of x. For what values of x is this expansion valid?
(iii) When the expansion is valid

$$\frac{(1+x)^3}{(1-x)^4} = 1 + 7x + ax^2 + bx^3 + \ldots.$$

Find the values of a and b.

[MEI]

3 (i) Write down the expansion of $(2-x)^4$.
(ii) Find the first four terms in the expansion of $(1+2x)^{-3}$ in ascending powers of x. For what range of values of x is this expansion valid?
(iii) When the expansion is valid

$$\frac{(2-x)^4}{(1+2x)^3} = 16 + ax + bx^2 + \ldots.$$

Find the values of a and b.

[MEI]

4 Write down the expansions of the following expressions in ascending powers of x, as far as the term containing x^3. In each case state the values of x for which the expansion is valid.

(i) $(1-x)^{-1}$ (ii) $(1+2x)^{-2}$ (iii) $\dfrac{1}{(1-x)(1+2x)^2}$

[MEI]

5 (i) Show that $\dfrac{1}{\sqrt{4-x}} = \dfrac{1}{2}\left(1 - \dfrac{x}{4}\right)^{-\frac{1}{2}}$.

(ii) Write down the first three terms in the binomial expansion of $\left(1 - \dfrac{x}{4}\right)^{-\frac{1}{2}}$ in ascending powers of x, stating the range of values of x for which this expansion is valid.

(iii) Find the first three terms in the expansion of $\dfrac{2(1+x)}{\sqrt{4-x}}$ in ascending powers of x, for small values of x.

[MEI]

6 (i) Expand $(1 + y)^{-1}$, where $-1 < y < 1$, as a series in powers of y, giving the first four terms.

(ii) Hence find the first four terms of the expansion of $\left(1 + \dfrac{2}{x}\right)^{-1}$ where $-1 < \dfrac{2}{x} < 1$.

(iii) Show that $\left(1 + \dfrac{2}{x}\right)^{-1} = \dfrac{x}{x+2} = \dfrac{x}{2}\left(1 + \dfrac{x}{2}\right)^{-1}$.

(iv) Find the first four terms of the expansion of $\dfrac{x}{2}\left(1 + \dfrac{x}{2}\right)^{-1}$ where $-1 < \dfrac{x}{2} < 1$.

(v) State the conditions on x under which your expansions for $\left(1 + \dfrac{2}{x}\right)^{-1}$ and $\dfrac{x}{2}\left(1 + \dfrac{x}{2}\right)^{-1}$ are valid and explain briefly why your expansions are different.

[MEI]

7 Expand $(2 + 3x)^{-2}$ in ascending powers of x, up to and including the term in x^2, simplifying the coefficients.

[Cambridge International AS & A Level Mathematics 9709, Paper 3 Q1 June 2007]

8 Expand $(1 + x)\sqrt{(1 - 2x)}$ in ascending powers of x, up to and including the term in x^2, simplifying the coefficients.

[Cambridge International AS & A Level Mathematics 9709, Paper 3 Q2 November 2008]

9 When $(1 + 2x)(1 + ax)^{\frac{2}{3}}$, where a is a constant, is expanded in ascending powers of x, the coefficient of the term in x is zero.

(i) Find the value of a.

(ii) When a has this value, find the term in x^3 in the expansion of $(1 + 2x)(1 + ax)^{\frac{2}{3}}$, simplifying the coefficient.

[Cambridge International AS & A Level Mathematics 9709, Paper 3 Q5 June 2009]

Review of algebraic fractions

If f(x) and g(x) are polynomials, the expression $\frac{f(x)}{g(x)}$ is an *algebraic fraction* or *rational function*. It may also be called a *rational expression*. There are many occasions in mathematics when a problem reduces to the manipulation of algebraic fractions, and the rules for this are exactly the same as those for numerical fractions.

Simplifying fractions

To simplify a fraction, you look for a factor common to both the numerator (top line) and the denominator (bottom line) and cancel by it.

For example, in arithmetic

$$\frac{15}{20} = \frac{5 \times 3}{5 \times 4} = \frac{3}{4}$$

and in algebra

$$\frac{6a}{9a^2} = \frac{2 \times 3 \times a}{3 \times 3 \times a \times a} = \frac{2}{3a}$$

Notice how you must *factorise* both the numerator and denominator before cancelling, since it is only possible to cancel by a *common factor*. In some cases this involves putting brackets in.

$$\frac{2a+4}{a^2-4} = \frac{2(a+2)}{(a+2)(a-2)} = \frac{2}{(a-2)}$$

Multiplying and dividing fractions

Multiplying fractions involves cancelling any factors common to the numerator and denominator. For example:

$$\frac{10a}{3b^2} \times \frac{9ab}{25} = \frac{2 \times 5 \times a}{3 \times b \times b} \times \frac{3 \times 3 \times a \times b}{5 \times 5} = \frac{6a^2}{5b}$$

As with simplifying, it is often necessary to factorise any algebraic expressions first.

$$\frac{a^2+3a+2}{9} \times \frac{12}{a+1} = \frac{(a+1)(a+2)}{3 \times 3} \times \frac{3 \times 4}{(a+1)}$$

$$= \frac{(a+2)}{3} \times \frac{4}{1}$$

$$= \frac{4(a+2)}{3}$$

Remember that when one fraction is divided by another, you change ÷ to × and invert the fraction which follows the ÷ symbol. For example:

$$\frac{12}{x^2-1} \div \frac{4}{x+1} = \frac{12}{(x+1)(x-1)} \times \frac{(x+1)}{4}$$

$$= \frac{3}{(x-1)}$$

Addition and subtraction of fractions

To add or subtract two fractions they must be replaced by equivalent fractions, both of which have the same denominator.

For example:

$$\frac{2}{3} + \frac{1}{4} = \frac{8}{12} + \frac{3}{12} = \frac{11}{12}$$

Similarly, in algebra:

$$\frac{2x}{3} + \frac{x}{4} = \frac{8x}{12} + \frac{3x}{12} = \frac{11x}{12}$$

and $\quad \dfrac{2}{3x} + \dfrac{1}{4x} = \dfrac{8}{12x} + \dfrac{3}{12x} = \dfrac{11}{12x}$

Notice how you only need $12x$ here, not $12x^2$.

You must take particular care when the subtraction of fractions introduces a sign change. For example:

$$\frac{4x - 3}{6} - \frac{2x + 1}{4} = \frac{2(4x - 3) - 3(2x + 1)}{12}$$

$$= \frac{8x - 6 - 6x - 3}{12}$$

$$= \frac{2x - 9}{12}$$

Notice how in addition and subtraction, the new denominator is the *lowest common multiple* of the original denominators. When two denominators have no *common factor*, their product gives the new denominator. For example:

$$\frac{2}{y + 3} + \frac{3}{y - 2} = \frac{2(y - 2) + 3(y + 3)}{(y + 3)(y - 2)}$$

$$= \frac{2y - 4 + 3y + 9}{(y + 3)(y - 2)}$$

$$= \frac{5y + 5}{(y + 3)(y - 2)}$$

$$= \frac{5(y + 1)}{(y + 3)(y - 2)}$$

It may be necessary to factorise denominators in order to identify common factors, as shown here.

$$\frac{2b}{a^2 - b^2} - \frac{3}{a + b} = \frac{2b}{(a + b)(a - b)} - \frac{3}{(a + b)}$$

$$= \frac{2b - 3(a - b)}{(a + b)(a - b)}$$

$$= \frac{5b - 3a}{(a + b)(a - b)}$$

$(a + b)$ is a common factor.

EXERCISE 7B

Simplify the expressions in questions **1** to **10**.

1. $\dfrac{6a}{b} \times \dfrac{a}{9b^2}$

2. $\dfrac{5xy}{3} \div 15xy^2$

3. $\dfrac{x^2 - 9}{x^2 - 9x + 18}$

4. $\dfrac{5x - 1}{x + 3} \times \dfrac{x^2 + 6x + 9}{5x^2 + 4x - 1}$

5. $\dfrac{4x^2 - 25}{4x^2 + 20x + 25}$

6. $\dfrac{a^2 + a - 12}{5} \times \dfrac{3}{4a - 12}$

7. $\dfrac{4x^2 - 9}{x^2 + 2x + 1} \div \dfrac{2x - 3}{x^2 + x}$

8. $\dfrac{2p + 4}{5} \div (p^2 - 4)$

9. $\dfrac{a^2 - b^2}{2a^2 + ab - b^2}$

10. $\dfrac{x^2 + 8x + 16}{x^2 + 6x + 9} \times \dfrac{x^2 + 2x - 3}{x^2 + 4x}$

In questions **11** to **24** write each of the expressions as a single fraction in its simplest form.

11. $\dfrac{1}{4x} + \dfrac{1}{5x}$

12. $\dfrac{x}{3} - \dfrac{(x + 1)}{4}$

13. $\dfrac{a}{a + 1} + \dfrac{1}{a - 1}$

14. $\dfrac{2}{x - 3} + \dfrac{3}{x - 2}$

15. $\dfrac{x}{x^2 - 4} - \dfrac{1}{x + 2}$

16. $\dfrac{p^2}{p^2 - 1} - \dfrac{p^2}{p^2 + 1}$

17. $\dfrac{2}{a + 1} - \dfrac{a}{a^2 + 1}$

18. $\dfrac{2y}{(y + 2)^2} - \dfrac{4}{y + 4}$

19. $x + \dfrac{1}{x + 1}$

20. $\dfrac{2}{b^2 + 2b + 1} - \dfrac{3}{b + 1}$

21. $\dfrac{2}{3(x - 1)} + \dfrac{3}{2(x + 1)}$

22. $\dfrac{6}{5(x + 2)} + \dfrac{2x}{(x + 2)^2}$

23. $\dfrac{2}{a + 2} - \dfrac{a - 2}{2a^2 + a - 6}$

24. $\dfrac{1}{x - 2} + \dfrac{1}{x} + \dfrac{1}{x + 2}$

Partial fractions

Sometimes, it is easier to deal with two or three simple separate fractions than it is to handle one more complicated one.

For example:

$$\dfrac{1}{(1 + 2x)(1 + x)}$$

may be written as

$$\dfrac{2}{(1 + 2x)} - \dfrac{1}{(1 + x)}.$$

- When $\frac{1}{(1+2x)(1+x)}$ is written as $\frac{2}{(1+2x)} - \frac{1}{(1+x)}$ you can then do binomial expansions on the two fractions, and so find an expansion for the original fraction.

- When integrating, it is easier to work with a number of simple fractions than a combined one. For example, the only analytic method for integrating $\frac{1}{(1+2x)(1+x)}$ involves first writing it as $\frac{2}{(1+2x)} - \frac{1}{(1+x)}$. You will meet this application in Chapter 8.

This process of taking an expression such as $\frac{1}{(1+2x)(1+x)}$ and writing it in the form $\frac{2}{(1+2x)} - \frac{1}{(1+x)}$ is called expressing the algebraic fraction in *partial fractions*.

When finding partial fractions you must always assume the most general numerator possible, and the method for doing this is illustrated in the following examples.

Type 1: Denominators of the form (ax + b)(cx + d)(ex + f)

EXAMPLE 7.6 Express $\frac{4+x}{(1+x)(2-x)}$ as a sum of partial fractions.

SOLUTION

Assume

$$\frac{4+x}{(1+x)(2-x)} \equiv \frac{A}{1+x} + \frac{B}{2-x}$$

Remember: a linear denominator ⇒ a constant numerator if the fraction is to be a proper fraction.

Multiplying both sides by $(1+x)(2-x)$ gives

$$4 + x \equiv A(2-x) + B(1+x). \qquad ①$$

This is an identity; it is true for all values of x.

There are two possible ways in which you can find the constants A and B. You can either

- substitute *any two* values of x in ① (two values are needed to give two equations to solve for the two unknowns A and B); or

- equate the constant terms to give one equation (this is the same as putting $x = 0$) and the coefficients of x to give another.

Sometimes one method is easier than the other, and in practice you will often want to use a combination of the two.

Method 1: Substitution

Although you can substitute any two values of x, the easiest to use are $x = 2$ and $x = -1$, since each makes the value of one bracket zero in the identity.

$$4 + x \equiv A(2 - x) + B(1 + x)$$

$$x = 2 \quad \Rightarrow \quad 4 + 2 = A(2 - 2) + B(1 + 2)$$

$$6 = 3B \quad \Rightarrow \quad B = 2$$

$$x = -1 \quad \Rightarrow \quad 4 - 1 = A(2 + 1) + B(1 - 1)$$

$$3 = 3A \quad \Rightarrow \quad A = 1$$

Substituting these values for A and B gives

$$\frac{4 + x}{(1 + x)(2 - x)} \equiv \frac{1}{1 + x} + \frac{2}{2 - x}$$

Method 2: Equating coefficients

In this method, you write the right-hand side of

$$4 + x \equiv A(2 - x) + B(1 + x)$$

as a polynomial in x, and then compare the coefficients of the various terms.

$$4 + x \equiv 2A - Ax + B + Bx$$

$$4 + 1x \equiv (2A + B) + (-A + B)x$$

Equating the constant terms: $\quad 4 = 2A + B$

Equating the coefficients of x: $\quad 1 = -A + B$

> These are simultaneous equations in A and B.

Solving these simultaneous equations gives $A = 1$ and $B = 2$ as before.

❓ In each of these methods the identity ($\equiv$) was later replaced by equality ($=$). Why was this done?

In some cases it is necessary to factorise the denominator before finding the partial fractions.

EXAMPLE 7.7 Express $\dfrac{x(5x + 7)}{(2x + 1)(x^2 - 1)}$ as a sum of partial fractions.

SOLUTION

$$\frac{x(5x + 7)}{(2x + 1)(x^2 - 1)} = \frac{x(5x + 7)}{(2x + 1)(x + 1)(x - 1)}$$

> Start by factorising the denominator fully, replacing $(x^2 - 1)$ with $(x + 1)(x - 1)$.

There are three factors in the denominator, so write

$$\frac{x(5x+7)}{(2x+1)(x+1)(x-1)} \equiv \frac{A}{2x+1} + \frac{B}{x+1} + \frac{C}{x-1}$$

Multiplying both sides by $(2x+1)(x+1)(x-1)$ gives

$$x(5x+7) \equiv A(x+1)(x-1) + B(2x+1)(x-1) + C(2x+1)(x+1)$$

Substituting $x = 1$ gives: $12 = 6C$

$\Rightarrow \qquad\qquad\qquad C = 2$

Substituting $x = -1$ gives: $-2 = 2B$

$\Rightarrow \qquad\qquad\qquad B = -1$

Notice how a combination of the two methods is used.

Equating coefficients of x^2 gives: $5 = A + 2B + 2C$

As $B = -1$ and $C = 2$: $\qquad 5 = A - 2 + 4$

$\Rightarrow \qquad\qquad\qquad A = 3$

Hence $\dfrac{x(5x+7)}{(2x+1)(x+1)(x-1)} \equiv \dfrac{3}{2x+1} - \dfrac{1}{x+1} + \dfrac{2}{x-1}$

In the next example the orders of the numerator (top line) and the denominator (bottom line) are the same.

EXAMPLE 7.8

Express $\dfrac{6-x^2}{4-x^2}$ as a sum of partial fractions.

SOLUTION

Start by dividing the numerator by the denominator. In this case the quotient is 1 and the remainder is 2.

So $\qquad \dfrac{6-x^2}{4-x^2} = 1 + \dfrac{2}{4-x^2}$

You can also use this method when the order of the numerator is greater than that of the denominator.

Now find $\dfrac{2}{4-x^2}$.

$$\frac{2}{4-x^2} \equiv \frac{2}{(2+x)(2-x)} \equiv \frac{A}{2+x} + \frac{B}{2-x}$$

Multiplying both sides by $(2+x)(2-x)$ gives

$$2 \equiv A(2-x) + B(2+x)$$
$$2 \equiv (2A + 2B) + x(B - A)$$

Equating constant terms: $\qquad 2 = 2A + 2B$

so $\qquad\qquad\qquad A + B = 1 \qquad$ ①

Equating coefficients of x: $\qquad 0 = B - A$, so $B = A$

Substituting in ① gives $\qquad A = B = \dfrac{1}{2}$

Using these values

$$\frac{2}{(2+x)(2-x)} \equiv \frac{\frac{1}{2}}{2+x} + \frac{\frac{1}{2}}{2-x} \equiv \frac{1}{2(2+x)} + \frac{1}{2(2-x)}$$

So $\quad \dfrac{6-x^2}{4-x^2} \equiv 1 + \dfrac{1}{2(2+x)} + \dfrac{1}{2(2-x)}$

EXERCISE 7C

Write the expressions in questions **1** to **15** as a sum of partial fractions.

1. $\dfrac{5}{(x-2)(x+3)}$
2. $\dfrac{1}{x(x+1)}$
3. $\dfrac{6}{(x-1)(x-4)}$
4. $\dfrac{x+5}{(x-1)(x+2)}$
5. $\dfrac{3x}{(2x-1)(x+1)}$
6. $\dfrac{4}{x^2-2x}$
7. $\dfrac{2}{(x-1)(3x-1)}$
8. $\dfrac{x-1}{x^2-3x-4}$
9. $\dfrac{x+2}{2x^2-x}$
10. $\dfrac{7}{2x^2+x-6}$
11. $\dfrac{2x-1}{2x^2+3x-20}$
12. $\dfrac{2x+5}{18x^2-8}$
13. $\dfrac{6x^2+22x+18}{(x+1)(x+2)(x+3)}$
14. $\dfrac{4x^2-25x-3}{(2x+1)(x-1)(x-3)}$
15. $\dfrac{5x^2+13x+10}{(2x+3)(x^2-4)}$

Type 2: Denominators of the form $(ax+b)(cx^2+d)$

EXAMPLE 7.9

Express $\dfrac{2x+3}{(x-1)(x^2+4)}$ as a sum of partial fractions.

SOLUTION

You need to assume a numerator of order 1 for the partial fraction with a denominator of x^2+4, which is of order 2.

$$\frac{2x+3}{(x-1)(x^2+4)} \equiv \frac{A}{x-1} + \frac{Bx+C}{x^2+4}$$

$Bx+C$ is the most general numerator of order 1.

Multiplying both sides by $(x-1)(x^2+4)$ gives

$$2x+3 \equiv A(x^2+4) + (Bx+C)(x-1) \qquad ①$$

$x=1 \implies 5=5A \implies A=1$

The other two unknowns, B and C, are most easily found by equating coefficients. Identity ① may be rewritten as

$$2x+3 \equiv (A+B)x^2 + (-B+C)x + (4A-C)$$

Equating coefficients of x^2: $\quad 0 = A+B \implies B=-1$

Equating constant terms: $\quad 3 = 4A-C \implies C=1$

This gives

$$\frac{2x+3}{(x-1)(x^2+4)} \equiv \frac{1}{x-1} + \frac{1-x}{x^2+4}$$

Type 3: Denominators of the form $(ax + b)(cx + d)^2$

The factor $(cx + d)^2$ is of order 2, so it would have an order 1 numerator in the partial fractions. However, in the case of a repeated factor there is a simpler form.

Consider $$\frac{4x + 5}{(2x + 1)^2}$$

This can be written as $$\frac{2(2x + 1) + 3}{(2x + 1)^2}$$

$$\equiv \frac{2(2x + 1)}{(2x + 1)^2} + \frac{3}{(2x + 1)^2}$$

$$\equiv \frac{2}{(2x + 1)} + \frac{3}{(2x + 1)^2}$$

> **Note**
>
> In this form, both the numerators are constant.

In a similar way, any fraction of the form $\dfrac{px + q}{(cx + d)^2}$ can be written as

$$\frac{A}{(cx + d)} + \frac{B}{(cx + d)^2}$$

When expressing an algebraic fraction in partial fractions, you are aiming to find the simplest partial fractions possible, so you would want the form where the numerators are constant.

EXAMPLE 7.10 Express $\dfrac{x + 1}{(x - 1)(x - 2)^2}$ as a sum of partial fractions.

SOLUTION

Let $$\frac{x + 1}{(x - 1)(x - 2)^2} \equiv \frac{A}{(x - 1)} + \frac{B}{(x - 2)} + \frac{C}{(x - 2)^2}$$

Notice that you only need $(x - 2)^2$ here and not $(x - 2)^3$.

Multiplying both sides by $(x - 1)(x - 2)^2$ gives

$$x + 1 \equiv A(x - 2)^2 + B(x - 1)(x - 2) + C(x - 1)$$

$x = 1$ (so that $x - 1 = 0$) $\Rightarrow$ $2 = A(-1)^2$ $\Rightarrow$ $A = 2$
$x = 2$ (so that $x - 2 = 0$) $\Rightarrow$ $3 = C$

Equating coefficients of x^2: $\Rightarrow$ $0 = A + B$ $\Rightarrow$ $B = -2$

This gives

$$\frac{x + 1}{(x - 1)(x - 2)^2} \equiv \frac{2}{x - 1} - \frac{2}{x - 2} + \frac{3}{(x - 2)^2}$$

EXAMPLE 7.11 Express $\dfrac{5x^2 - 3}{x^2(x+1)}$ as a sum of partial fractions.

SOLUTION

Let $\dfrac{5x^2 - 3}{x^2(x+1)} \equiv \dfrac{A}{x} + \dfrac{B}{x^2} + \dfrac{C}{x+1}$

Multiplying both sides by $x^2(x+1)$ gives

$$5x^2 - 3 \equiv Ax(x+1) + B(x+1) + Cx^2$$

$x = 0 \quad \Rightarrow \quad -3 = B$
$x = -1 \quad \Rightarrow \quad +2 = C$

Equating coefficients of x^2: $\quad +5 = A + C \quad \Rightarrow \quad A = 3$

This gives

$$\dfrac{5x^2 - 3}{x^2(x+1)} \equiv \dfrac{3}{x} - \dfrac{3}{x^2} + \dfrac{2}{x+1}$$

EXERCISE 7D

1 Express each of the following fractions as a sum of partial fractions.

(i) $\dfrac{4}{(1-3x)(1-x)^2}$

(ii) $\dfrac{4 + 2x}{(2x-1)(x^2+1)}$

(iii) $\dfrac{5 - 2x}{(x-1)^2(x+2)}$

(iv) $\dfrac{2x+1}{(x-2)(x^2+4)}$

(v) $\dfrac{2x^2 + x + 4}{(2x^2 - 3)(x+2)}$

(vi) $\dfrac{x^2 - 1}{x^2(2x+1)}$

(vii) $\dfrac{x^2 + 3}{x(3x^2 - 1)}$

(viii) $\dfrac{2x^2 + x + 2}{(2x^2 + 1)(x+1)}$

(ix) $\dfrac{4x^2 - 3}{x(2x-1)^2}$

2 Given that

$$\dfrac{x^2 + 2x + 7}{(2x+3)(x^2+4)} \equiv \dfrac{A}{(2x+3)} + \dfrac{Bx + C}{(x^2+4)}$$

find the values of the constants A, B and C.

[MEI, *part*]

3 Calculate the values of the constants A, B and C for which

$$\dfrac{x^2 - 4x + 23}{(x-5)(x^2+3)} \equiv \dfrac{A}{(x-5)} + \dfrac{Bx + C}{(x^2+3)}$$

[MEI, *part*]

Using partial fractions with the binomial expansion

One of the most common reasons for writing an expression in partial fractions is to enable binomial expansions to be applied, as in the following example.

EXAMPLE 7.12 Express $\dfrac{2x+7}{(x-1)(x+2)}$ in partial fractions and hence find the first three terms of its binomial expansion, stating the values of x for which this is valid.

SOLUTION

$$\frac{2x+7}{(x-1)(x+2)} \equiv \frac{A}{(x-1)} + \frac{B}{(x+2)}$$

Multiplying both sides by $(x-1)(x+2)$ gives

$$2x + 7 \equiv A(x+2) + B(x-1)$$

$x = 1 \quad \Rightarrow \quad 9 = 3A \quad \Rightarrow \quad A = 3$
$x = -2 \quad \Rightarrow \quad 3 = -3B \quad \Rightarrow \quad B = -1$

This gives

$$\frac{2x+7}{(x-1)(x+2)} \equiv \frac{3}{(x-1)} - \frac{1}{(x+2)}$$

In order to obtain the binomial expansion, each bracket must be of the form $(1 \pm \ldots)$, giving

$$\frac{2x+7}{(x-1)(x+2)} \equiv \frac{-3}{(1-x)} - \frac{1}{2\left(1+\dfrac{x}{2}\right)}$$

$$\equiv -3(1-x)^{-1} - \frac{1}{2}\left(1+\frac{x}{2}\right)^{-1} \qquad \text{①}$$

The two binomial expansions are

$$(1-x)^{-1} = 1 + (-1)(-x) + \frac{(-1)(-2)}{2!}(-x)^2 + \ldots \qquad \text{for } |x| < 1$$

$$\approx 1 + x + x^2$$

and $\left(1+\dfrac{x}{2}\right)^{-1} = 1 + (-1)\left(\dfrac{x}{2}\right) + \dfrac{(-1)(-2)}{2!}\left(\dfrac{x}{2}\right)^2 + \ldots \qquad \text{for } \left|\dfrac{x}{2}\right| < 1$

$$\approx 1 - \frac{x}{2} + \frac{x^2}{4}$$

Substituting these in ① gives

$$\frac{2x+7}{(x-1)(x+2)} \approx -3(1 + x + x^2) - \frac{1}{2}\left(1 - \frac{x}{2} + \frac{x^2}{4}\right)$$

$$= -\frac{7}{2} - \frac{11x}{4} - \frac{25x^2}{8}$$

The expansion is valid when $|x| < 1$ and $\left|\dfrac{x}{2}\right| < 1$. The stricter of these is $|x| < 1$.

INVESTIGATION

Find a binomial expansion for the function

$$f(x) = \frac{1}{(1+2x)(1-x)}$$

and state the values of x for which it is valid

(i) by writing it as $(1+2x)^{-1}(1-x)^{-1}$
(ii) by writing it as $[1+(x-2x^2)]^{-1}$ and treating $(x-2x^2)$ as one term
(iii) by first expressing f(x) as a sum of partial fractions.

Decide which method you find simplest for the following cases.
(a) When a linear approximation for f(x) is required.
(b) When a quadratic approximation for f(x) is required.
(c) When the coefficient of x^n is required.

EXERCISE 7E

1 Find the first three terms in ascending powers of x in the binomial expansion of the following fractions.

(i) $\dfrac{4}{(1-3x)(1-x)^2}$

(ii) $\dfrac{4+2x}{(2x-1)(x^2+1)}$

(iii) $\dfrac{5-2x}{(x-1)^2(x+2)}$

(iv) $\dfrac{2x+1}{(x-2)(x^2+4)}$

2 (i) Express $\dfrac{7-4x}{(2x-1)(x+2)}$ in partial fractions as $\dfrac{A}{(2x-1)} + \dfrac{B}{(x+2)}$ where A and B are to be found.

(ii) Find the expansion of $\dfrac{1}{(1-2x)}$ in the form $a + bx + cx^2 + \ldots$ where a, b and c are to be found.
Give the range of values of x for which this expansion is valid.

(iii) Find the expansion of $\dfrac{1}{(2+x)}$ as far as the term containing x^2.
Give the range of values of x for which this expansion is valid.

(iv) Hence find a quadratic approximation for $\dfrac{7-4x}{(2x-1)(x+2)}$ when $|x|$ is small.
Find the percentage error in this approximation when $x = 0.1$.

[MEI]

3 (i) Expand $(2-x)(1+x)$.
Hence express $\dfrac{3x}{2+x-x^2}$ in partial fractions.

(ii) Use the binomial expansion of the partial fractions in part (i) to show that

$$\frac{3x}{2+x-x^2} = \frac{3}{2}x - \frac{3}{4}x^2 + \ldots$$

State the range of values of x for which this result is valid.

[MEI, part]

4 (i) Given that $f(x) = \dfrac{8x - 6}{(1 - x)(3 - x)}$, express $f(x)$ in partial fractions.

Hence show that
$$f'(x) = (1 - x)^{-2} - \left(1 - \dfrac{x}{3}\right)^{-2}.$$

(ii) Using the results in part **(i)**, or otherwise, find the x co-ordinates of the stationary points on the graph of $y = f(x)$.

(iii) Use the binomial expansion, together with the result in part **(i)**, to expand $f'(x)$ in powers of x up to and including the term in x^2.

(iv) Show that, when $f'(x)$ is expanded in powers of x, the coefficients of all the powers of x are positive.

[MEI]

5 (i) Express $\dfrac{10}{(2 - x)(1 + x^2)}$ in partial fractions.

(ii) Hence, given that $|x| < 1$, obtain the expansion of $\dfrac{10}{(2 - x)(1 + x^2)}$ in ascending powers of x, up to and including the term in x^3, simplifying the coefficients.

[Cambridge International AS & A Level Mathematics 9709, Paper 3 Q9 June 2006]

6 (i) Express $\dfrac{3x^2 + x}{(x + 2)(x^2 + 1)}$ in partial fractions.

(ii) Hence obtain the expansion of $\dfrac{3x^2 + x}{(x + 2)(x^2 + 1)}$ in ascending powers of x, up to and including the term in x^3.

[Cambridge International AS & A Level Mathematics 9709, Paper 3 Q9 November 2005]

7 (i) Express $\dfrac{2 - x + 8x^2}{(1 - x)(1 + 2x)(2 + x)}$ in partial fractions.

(ii) Hence obtain the expansion of $\dfrac{2 - x + 8x^2}{(1 - x)(1 + 2x)(2 + x)}$ in ascending powers of x, up to and including the term in x^2.

[Cambridge International AS & A Level Mathematics 9709, Paper 3 Q9 November 2007]

KEY POINTS

1. The general binomial expansion for $n \in \mathbb{R}$ is

$$(1 + x)^n = 1 + nx + \frac{n(n-1)}{2!}x^2 + \frac{n(n-1)(n-2)}{3!}x^3 + \ldots.$$

   In the special case when $n \in \mathbb{N}$, the series expansion is finite and valid for all x.

   When $n \notin \mathbb{N}$, the series expansion is non-terminating (infinite) and valid only if $|x| < 1$.

2. When $n \notin \mathbb{N}$, $(a + x)^n$ should be written as $a^n\left(1 + \frac{x}{a}\right)^n$ before obtaining the binomial expansion.

3. When multiplying algebraic fractions, you can only cancel when the same factor occurs in both the numerator and the denominator.

4. When adding or subtracting algebraic fractions, you first need to find a common denominator.

5. The easiest way to solve any equation involving fractions is usually to multiply both sides by a quantity which will eliminate the fractions.

6. A proper algebraic fraction with a denominator which factorises can be decomposed into a sum of proper partial fractions.

7. The following forms of partial fraction should be used.

$$\frac{px + q}{(ax + b)(cx + d)(ex + f)} \equiv \frac{A}{ax + b} + \frac{B}{cx + d} + \frac{C}{ex + f}$$

$$\frac{px^2 + qr + r}{(ax + b)(cx^2 + d)} \equiv \frac{A}{ax + b} + \frac{Bx + C}{cx^2 + d}$$

$$\frac{px^2 + qx + r}{(ax + b)(cx + d)^2} \equiv \frac{A}{ax + b} + \frac{B}{cx + d} + \frac{C}{(cx + d)^2}$$

8 Further integration

The mathematical process has a reality and virtue in itself, and once discovered it constitutes a new and independent factor.

Winston Churchill (1876–1965)

Figure 8.1 shows the graph of $y = \sqrt{x}$.

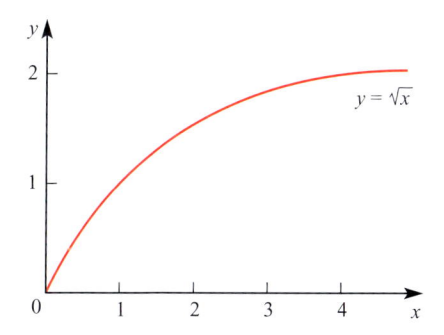

Figure 8.1

? How does it allow you to find the shaded area in the graph in figure 8.2?

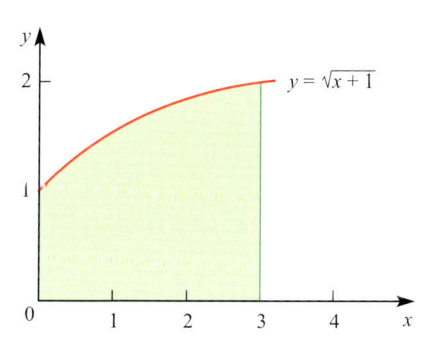

Figure 8.2

Integration by substitution

The graph of $y = \sqrt{x-1}$ is shown in figure 8.3.

The shaded area is given by

$$\int_1^5 \sqrt{x-1}\,dx = \int_1^5 (x-1)^{\frac{1}{2}}\,dx.$$

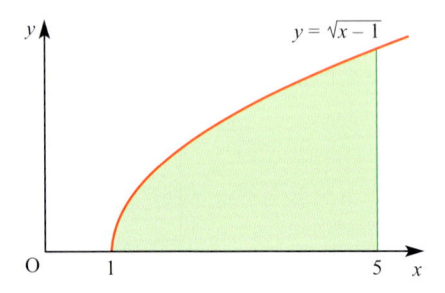

Figure 8.3

You may remember how to investigate this by inspection. However, you can also transform the integral into a simpler one by using the substitution $u = x - 1$ to get $\int_a^b u^{\frac{1}{2}}\,du$.

When you make this substitution it means that you are now integrating with respect to a new variable, namely u. The limits of the integral, and the 'dx', must be written in terms of u.

The new limits are given by $\quad x = 1 \quad \Rightarrow \quad u = 1 - 1 = 0$
and $\quad x = 5 \quad \Rightarrow \quad u = 5 - 1 = 4$.

Since $u = x - 1$, $\dfrac{du}{dx} = 1$.

Even though $\dfrac{du}{dx}$ is not a fraction, it is usual to treat it as one in this situation (see the warning below), and to write the next step as '$du = dx$'.

The integral now becomes:

$$\int_{u=0}^{u=4} u^{\frac{1}{2}}\,du = \left[\frac{u^{\frac{3}{2}}}{\frac{3}{2}}\right]_0^4$$

$$= \left[\frac{2u^{\frac{3}{2}}}{3}\right]_0^4$$

$$= 5\tfrac{1}{3}$$

178

This method of integration is known as *integration by substitution*. It is a very powerful method which allows you to integrate many more functions. Since you are changing the variable from x to u, the method is also referred to as *integration by change of variable*.

⚠ The last example included the statement '$du = dx$'. Some mathematicians are reluctant to write such statements on the grounds that du and dx may only be used in the form $\frac{du}{dx}$, i.e. as a gradient. This is not in fact true; there is a well-defined branch of mathematics which justifies such statements but it is well beyond the scope of this book. In the meantime it may help you to think of it as shorthand for 'in the limit as $\delta x \to 0$, $\frac{\delta u}{\delta x} \to 1$, and so $\delta u = \delta x$'.

EXAMPLE 8.1

Evaluate $\int_1^3 (x+1)^3 \, dx$ by making a suitable substitution.

SOLUTION

Let $u = x + 1$.

Converting the limits: $\quad x = 1 \implies u = 1 + 1 = 2$
$\quad\quad\quad\quad\quad\quad\quad\quad\;\; x = 3 \implies u = 3 + 1 = 4$

Converting dx to du:

$$\frac{du}{dx} = 1 \implies du = dx.$$

$$\int_1^3 (x+1)^3 \, dx = \int_2^4 u^3 \, du$$
$$= \left[\frac{u^4}{4}\right]_2^4$$
$$= \frac{4^4}{4} - \frac{2^4}{4}$$
$$= 60$$

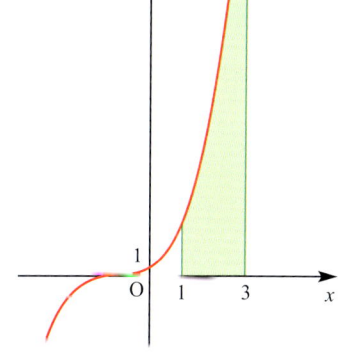

Figure 8.4

❓ Can integration by substitution be described as the reverse of the chain rule?

EXAMPLE 8.2 Evaluate $\int_3^4 2x(x^2-4)^{\frac{1}{2}}\,dx$ by making a suitable substitution.

SOLUTION

Notice that $2x$ is the derivative of the expression in the brackets, $x^2 - 4$, and so $u = x^2 - 4$ is a natural substitution to try.

This gives $\dfrac{du}{dx} = 2x \Rightarrow du = 2x\,dx$

Converting the limits: $x = 3 \Rightarrow u = 9 - 4 = 5$
$x = 4 \Rightarrow u = 16 - 4 = 12$

So the integral becomes:

$$\int_3^4 (x^2-4)^{\frac{1}{2}} 2x\,dx = \int_5^{12} u^{\frac{1}{2}}\,du$$

$$= \left[\frac{2u^{\frac{3}{2}}}{3}\right]_5^{12}$$

$$= 20.3 \text{ (to 3 significant figures)}$$

Note

In the last example there were two expressions multiplied together; the second expression is raised to a power. The two expressions are in this case related, since the first expression, $2x$, is the derivative of the expression in brackets, $x^2 - 4$. It was this relationship that made the integration possible.

EXAMPLE 8.3 Find $\int x(x^2+2)^3\,dx$ by making an appropriate substitution.

SOLUTION

Since this is an indefinite integral there are no limits to change, and the final answer will be a function of x.

Let $u = x^2 + 2$, then:

$\dfrac{du}{dx} = 2x \Rightarrow \tfrac{1}{2}du = x\,dx$ ◀ *You only have $x\,dx$ in the integral, not $2x\,dx$.*

So $\int x(x^2+2)^3\,dx = \int (x^2+2)^3 x\,dx$

$$= \int u^3 \times \tfrac{1}{2}\,du$$

$$= \frac{u^4}{8} + c$$

$$= \frac{(x^2+2)^4}{8} + c$$

> Always remember, when finding an indefinite integral by substitution, to substitute back at the end. The original integral was in terms of x, so your final answer must be too.

EXAMPLE 8.4 By making a suitable substitution, find $\int x\sqrt{x-2}\,dx$.

SOLUTION

This question is not of the same type as the previous ones since x is not the derivative of $(x-2)$. However, by making the substitution $u = x - 2$ you can still make the integral into one you can do.

Let $u = x - 2$, then:

$$\frac{du}{dx} = 1 \implies du = dx$$

There is also an x in the integral so you need to write down an expression for x in terms of u. Since $u = x - 2$ it follows that $x = u + 2$.

In the original integral you can now replace $\sqrt{x-2}$ by $u^{\frac{1}{2}}$, dx by du and x by $u+2$.

$$\int x\sqrt{x-2}\,dx = \int (u+2)u^{\frac{1}{2}}\,du$$
$$= \int \left(u^{\frac{3}{2}} + 2u^{\frac{1}{2}}\right)du$$
$$= \tfrac{2}{5}u^{\frac{5}{2}} + \tfrac{4}{3}u^{\frac{3}{2}} + c$$

Replacing u by $x - 2$ and tidying up gives $\tfrac{2}{15}(3x+4)(x-2)^{\frac{3}{2}} + c$.

ACTIVITY 8.1 Complete the algebraic steps involved in tidying up the answer above.

EXERCISE 8A

1 Find the following indefinite integrals by making the suggested substitution. Remember to give your final answer in terms of x.

(i) $\int 3x^2(x^3+1)^7\,dx$, $u = x^3 + 1$

(ii) $\int 2x(x^2+1)^5\,dx$, $u = x^2 + 1$

(iii) $\int 3x^2(x^3-2)^4\,dx$, $u = x^3 - 2$

(iv) $\int x\sqrt{2x^2-5}\,dx$, $u = 2x^2 - 5$

(v) $\int x\sqrt{2x+1}\,dx$, $u = 2x + 1$

(vi) $\int \dfrac{x}{\sqrt{x+9}}\,dx$, $u = x + 9$

2 Evaluate each of the following definite integrals by using a suitable substitution. Give your answer to 3 significant figures where appropriate.

(i) $\int_1^5 x^2(x^3+1)^2\,dx$

(ii) $\int_{-1}^2 2x(x-3)^5\,dx$

(iii) $\int_1^5 x\sqrt{x-1}\,dx$

3 Find the area of the shaded region for each of the following graphs.

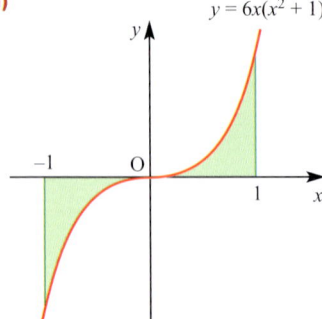

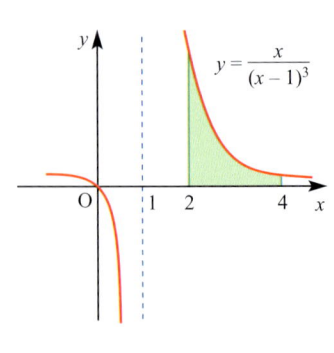

(i) $y = 6x(x^2 + 1)^3$

(ii) $y = \dfrac{x}{(x-1)^3}$

4 The sketch shows part of the graph of $y = x\sqrt{1+x}$.

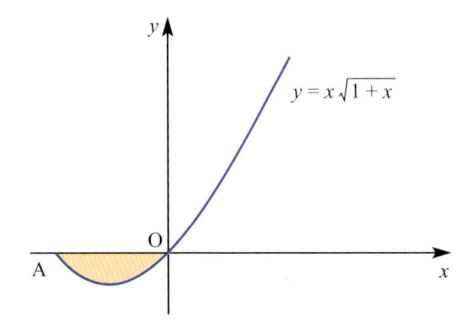

(i) Find the co-ordinates of point A and the range of values of x for which the function is defined.

(ii) Show that the area of the shaded region is $\dfrac{4}{15}$.

You may find the substitution $u = 1 + x$ useful.

[MEI]

5 (i) By substituting $u = 1 + x$ or otherwise, find

   (a) $\displaystyle\int (1+x)^3 \, dx$

   (b) $\displaystyle\int_{-1}^{1} x(1+x)^3 \, dx$.

(ii) By substituting $t = 1 + x^2$ or otherwise, evaluate $\displaystyle\int_0^1 x\sqrt{1+x^2} \, dx$.

[MEI]

6 (i) Integrate with respect to x.

   (a) $\dfrac{4}{\sqrt{x}} + \dfrac{3}{x^3}$

   (b) $6x(1+x^2)^{\frac{1}{2}}$

(ii) Show that the substitution $x = u^2$ transforms $\displaystyle\int_1^4 \dfrac{(1+\sqrt{x})^3}{\sqrt{x}} \, dx$ into an integral of the form $\displaystyle\int_a^b k(1+u)^3 \, du$.

State the values of k, a and b.

Evaluate this integral.

[MEI, adapted]

Integrals involving exponentials and natural logarithms

In Chapter 5 you met integrals involving logarithms and exponentials. That work is extended here using integration by substitution.

EXAMPLE 8.5 By making a suitable substitution, find $\int_0^4 2x e^{x^2} \, dx$.

SOLUTION

$$\int_0^4 2x e^{x^2} \, dx = \int_0^4 e^{x^2} 2x \, dx$$

Since $2x$ is the derivative of x^2, let $u = x^2$.

$$\frac{du}{dx} = 2x \implies du = 2x \, dx$$

The new limits are given by $\quad x = 0 \implies u = 0$
and $\quad x = 4 \implies u = 16$

The integral can now be written as

$$\int_0^{16} e^u \, du = \left[e^u \right]_0^{16}$$
$$= e^{16} - e^0$$
$$= 8.89 \times 10^6 \quad \text{(to 3 significant figures)}$$

EXAMPLE 8.6 Evaluate $\int_1^5 \frac{2x}{x^2 + 3} \, dx$.

SOLUTION

In this case, substitute $u = x^2 + 3$, so that

$$\frac{du}{dx} = 2x \implies du = 2x \, dx$$

The new limits are given by
$x = 1 \implies u = 4$
and $x = 5 \implies u = 28$

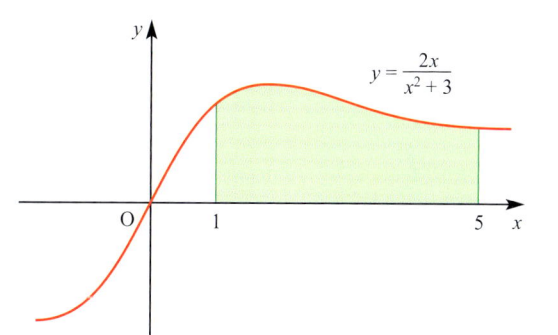

Figure 8.5

$$\int_1^5 \frac{2x}{x^2 + 3} \, dx = \int_4^{28} \frac{1}{u} \, du$$
$$= \left[\ln u \right]_4^{28}$$
$$= \ln 28 - \ln 4$$
$$= 1.95 \quad \text{(to 3 significant figures)}$$

The last example is of the form $\int \frac{f'(x)}{f(x)} dx$, where $f(x) = x^2 + 3$. In such cases the substitution $u = f(x)$ transforms the integral into $\int \frac{1}{u} du$. The answer is then $\ln u + c$ or $\ln(f(x)) + c$ (assuming that $u = f(x)$ is positive). This result may be stated as the working rule below.

If you obtain the top line when you differentiate the bottom line, the integral is the natural logarithm of the bottom line. So,

$$\int \frac{f'(x)}{f(x)} dx = \ln |f(x)| + c.$$

EXAMPLE 8.7

Evaluate $\int_1^2 \frac{5x^4 + 2x}{x^5 + x^2 + 4} dx$.

SOLUTION

You can work this out by substituting $u = x^5 + x^2 + 4$ but, since differentiating the bottom line gives the top line, you could apply the rule above and just write:

$$\int_1^2 \frac{5x^4 + 2x}{x^5 + x^2 + 4} dx = \left[\ln(x^5 + x^2 + 4) \right]_1^2$$

$$= \ln 40 - \ln 6$$

$$= 1.90 \quad \text{(to 2 significant figures)}$$

In the next example some adjustment is needed to get the top line into the required form.

EXAMPLE 8.8

Evaluate $\int_0^1 \frac{x^5}{x^6 + 7} dx$.

SOLUTION

The differential of $x^6 + 7$ is $6x^5$, so the integral is rewritten as $\frac{1}{6} \int_0^1 \frac{6x^5}{x^6 + 7} dx$.

Integrating this gives $\frac{1}{6} \left[\ln(x^6 + 7) \right]_0^1$ or 0.022 (to 2 significant figures).

EXERCISE 8B

1 Find the following indefinite integrals.

(i) $\int \frac{2x}{x^2 + 1} dx$

(ii) $\int \frac{2x + 3}{3x^2 + 9x - 1} dx$

(iii) $\int 12x^2 e^{x^3} dx$

2 Find the following definite integrals.
Where appropriate give your answers to 3 significant figures.

(i) $\int_2^3 2x e^{-x^2} dx$

(ii) $\int_2^4 \frac{x - 3}{x^2 - 6x + 9} dx$

3 The sketch shows the graph of $y = xe^{x^2}$.

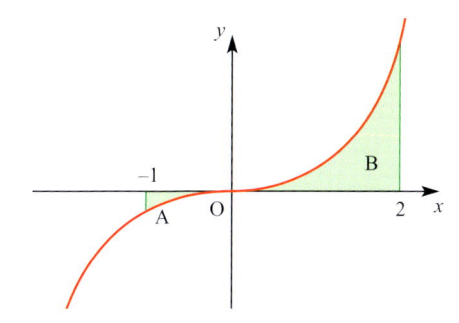

 (i) Find the area of region A.
 (ii) Find the area of region B.
 (iii) Hence write down the total area of the shaded region.

4 The graph of $y = \dfrac{x+2}{x^2 + 4x + 3}$ is shown below.

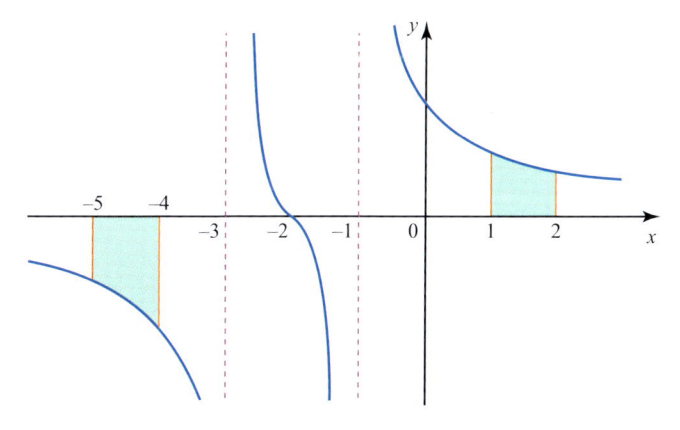

Find the area of each shaded region.

5 A curve has the equation $y = (x+3)e^{-x}$.

 (i) Find $\dfrac{dy}{dx}$.
 (ii) Hence find $\displaystyle\int \dfrac{x+2}{e^x}\,dx$.
 (iii) Find the x and y co-ordinates of the stationary point S on the curve.
 (iv) Calculate $\dfrac{d^2y}{dx^2}$ at the point S.
 What does its value indicate about the stationary point?
 (v) Show that the substitution $u = e^x$ converts $\displaystyle\int \dfrac{2 + \ln u}{u^2}\,du$ into $\displaystyle\int \dfrac{2+x}{e^x}\,dx$.
 (vi) Hence evaluate $\displaystyle\int_1^e \dfrac{2 + \ln u}{u^2}\,du$.

[MEI, adapted]

6 (i) Use a substitution, such as $u^2 = 2x - 3$, to find $\int 2x\sqrt{2x-3}\,dx$.

(ii) Differentiate $x^{\frac{1}{2}}\ln x$ with respect to x. Hence find $\int \dfrac{2 + \ln x}{\sqrt{x}}\,dx$.

(iii) The function f(x) has the property $f'(x) = e^{-x^2}$.
  - **(a)** Find $f''(x)$.
  - **(b)** Differentiate $f(x^3)$ with respect to x.

[MEI]

7 (i) Find the following integrals.

  (a) $\displaystyle\int_1^6 \dfrac{1}{2x+3}\,dx$

  (b) $\displaystyle\int \dfrac{x}{\sqrt{9+x^2}}\,dx$ (Use the substitution $v = \sqrt{9+x^2}$, or otherwise.)

(ii) (a) Show that $\dfrac{d}{dx}(e^{-x^2}) = -2xe^{-x^2}$.

The sketch below shows the curve with equation $y = xe^{-x^2}$.

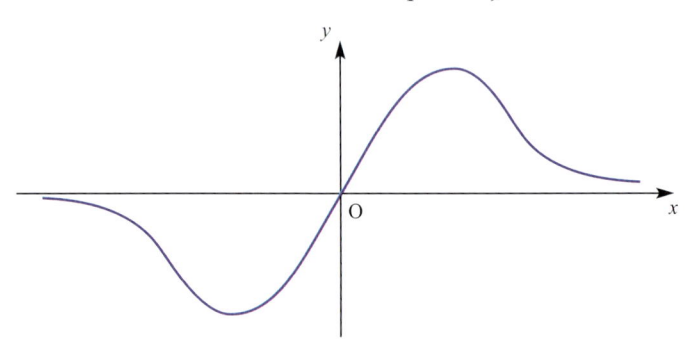

(b) Differentiate xe^{-x^2} and find the co-ordinates of the two stationary points on the curve.

(c) Find the area of the region between the curve and the x axis for $0 \leq x \leq 0.4$.

[MEI]

8 (i) Sketch the curve with equation $y = \dfrac{e^x}{e^x + 1}$ for values of x between 0 and 2.

(ii) Find the area of the region enclosed by this curve, the axes and the line $x = 2$.

(iii) Find the value of $\displaystyle\int_1^e \dfrac{2t}{t^2+1}\,dt$.

(iv) Compare your answers to parts **(ii)** and **(iii)**. Explain this result.

9 (i) Differentiate with respect to x
  - **(a)** e^{-2x^2}
  - **(b)** xe^{-2x^2}.

You are given that $f(x) = xe^{-2x^2}$.

(ii) Find $\displaystyle\int_0^k f(x)\,dx$ in terms of k.

(iii) Show that $f''(x) = 4xe^{-2x^2}(4x^2 - 3)$.

(iv) Show that there is just one stationary point on the curve $y = f(x)$ for positive x. State its co-ordinates and determine its nature.

[MEI]

10 The diagram shows part of the curve $y = \dfrac{x}{x^2+1}$ and its maximum point M. The shaded region R is bounded by the curve and by the lines $y = 0$ and $x = p$.

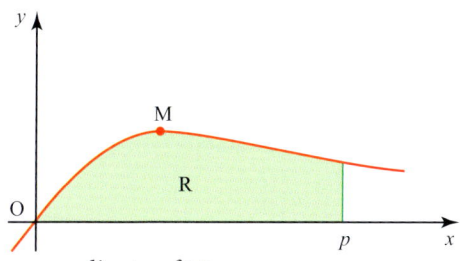

(i) Calculate the x co-ordinate of M.

(ii) Find the area of R in terms of p.

(iii) Hence calculate the value of p for which the area of R is 1, giving your answer correct to 3 significant figures.

[Cambridge International AS & A Level Mathematics 9709, Paper 3 Q9 June 2005]

11 Let $I = \displaystyle\int_1^4 \dfrac{1}{x(4-\sqrt{x})}\,dx$.

(i) Use the substitution $u = \sqrt{x}$ to show that $I = \displaystyle\int_1^2 \dfrac{2}{u(4-u)}\,du$.

(ii) Hence show that $I = \tfrac{1}{2}\ln 3$.

[Cambridge International AS & A Level Mathematics 9709, Paper 3 Q7 June 2007]

Integrals involving trigonometrical functions

In Chapter 5 you met integrals involving trigonometrical functions. That work is extended here using integration by substitution.

EXAMPLE 8.9 Find $\displaystyle\int 2x\cos(x^2+1)\,dx$.

SOLUTION

Make the substitution $u = x^2 + 1$. Then differentiate.

$$\dfrac{du}{dx} = 2x \quad \Rightarrow \quad 2x\,dx = du$$

$$\int 2x\cos(x^2+1)\,dx = \int \cos u\,du$$
$$= \sin u + c$$
$$= \sin(x^2+1) + c$$

Notice that the last example involves two expressions multiplied together, namely $2x$ and $\cos(x^2+1)$. These two expressions are related by the fact that $2x$ is the derivative of $x^2 + 1$. Because of this relationship, the substitution $u = x^2 + 1$ may be used to perform the integration. You can apply this method to other integrals involving trigonometrical functions, as in the next example.

EXAMPLE 8.10

Find $\int_0^{\frac{\pi}{2}} \cos x \sin^2 x \, dx$.

> Remember that $\sin^2 x$ means the same as $(\sin x)^2$.

SOLUTION

This integral is the product of two expressions, $\cos x$ and $(\sin x)^2$.

Now $(\sin x)^2$ is a function of $\sin x$, and $\cos x$ is the derivative of $\sin x$, so you should use the substitution $u = \sin x$.

Differentiating:

$$\frac{du}{dx} = \cos x \quad \Rightarrow \quad du = \cos x \, dx.$$

The limits of integration need to be changed as well:

$$x = 0 \quad \Rightarrow \quad u = 0$$
$$x = \frac{\pi}{2} \quad \Rightarrow \quad u = 1$$

Therefore
$$\int_0^{\frac{\pi}{2}} \cos x \sin^2 x \, dx = \int_0^1 u^2 \, du$$
$$= \left[\frac{u^3}{3}\right]_0^1$$
$$= \frac{1}{3}$$

EXAMPLE 8.11

Find $\int \cos^3 x \, dx$.

SOLUTION

First write $\cos^3 x = \cos x \cos^2 x$.

Now remember that

$$\cos^2 x + \sin^2 x = 1 \quad \Rightarrow \quad \cos^2 x = 1 - \sin^2 x.$$

This gives

$$\cos^3 x = \cos x (1 - \sin^2 x)$$
$$= \cos x - \cos x \sin^2 x$$

The first part of this expression, $\cos x$, is easily integrated to give $\sin x$.

The second part is more complicated, but you can see that it is of a type that you have met already, as it is a product of two expressions, one of which is a function of $\sin x$ and the other of which is the derivative of $\sin x$. This can be integrated either by making the substitution $u = \sin x$ or simply in your head (by inspection).

$$\int \cos^3 x \, dx = \int (\cos x - \cos x \sin^2 x) \, dx$$
$$= \sin x - \frac{1}{3} \sin^3 x + c$$

EXAMPLE 8.12

Find

(i) $\int \cot x \, dx$

(ii) $\int_{\frac{\pi}{6}}^{\frac{\pi}{3}} \tan x \, dx$

SOLUTION

(i) Rewrite $\cot x$ as $\dfrac{\cos x}{\sin x}$.

$$\int \cot x \, dx = \int \dfrac{\cos x}{\sin x} \, dx$$

Now you can use the substitution $u = \sin x$.

$$\dfrac{du}{dx} = \cos x \Rightarrow du = \cos x \, dx$$

$$\int \dfrac{\cos x}{\sin x} \, dx = \int \dfrac{1}{\sin x} \times \cos x \, dx = \int \dfrac{1}{u} \, du$$

$$= \ln|u| + c = \ln|\sin x| + c$$

> The 'top line' is the derivative of the 'bottom line'.

You may have noticed that the integral $\int \dfrac{\cos x}{\sin x} \, dx$ is in the form $\int \dfrac{f'(x)}{f(x)} \, dx = \ln|f(x)| + c$, and so you could have written the answer down directly.

(ii) $\displaystyle\int_{\pi/6}^{\pi/3} \tan x \, dx = \int_{\pi/6}^{\pi/3} \dfrac{\sin x}{\cos x} \, dx$

Adjusting the numerator to make it the derivative of the denominator gives:

$$\int_{\pi/6}^{\pi/3} \dfrac{\sin x}{\cos x} \, dx = -\int_{\pi/6}^{\pi/3} \dfrac{-\sin x}{\cos x} \, dx$$

$$= \left[-\ln|\cos x| \right]_{\pi/6}^{\pi/3}$$

$$= \left[-\ln \dfrac{1}{2} \right] - \left[-\ln \dfrac{\sqrt{3}}{2} \right]$$

$$= -\ln \dfrac{1}{2} + \ln \dfrac{\sqrt{3}}{2}$$

$$= \ln \sqrt{3}$$

$$= \tfrac{1}{2} \ln 3$$

> $\cos \dfrac{\pi}{3} = \dfrac{1}{2}$ and $\cos \dfrac{\pi}{6} = \dfrac{\sqrt{3}}{2}$

> Use the laws of logs: $\ln \dfrac{\sqrt{3}}{2} - \ln \dfrac{1}{2} = \ln \left(\dfrac{\sqrt{3}}{2} \div \dfrac{1}{2} \right)$

> Use the laws of logs: $\ln \sqrt{3} = \ln 3^{\frac{1}{2}} = \tfrac{1}{2} \ln 3$

Note

You may find that as you gain practice in this type of integration you become able to work out the integral without writing down the substitution. However, if you are unsure, it is best to write down the whole process.

EXERCISE 8C

1 Integrate the following by using the substitution given, or otherwise.

(i) $\cos 3x$ $u = 3x$

(ii) $\sin(1 - x)$ $u = 1 - x$

(iii) $\sin x \cos^3 x$ $u = \cos x$

(iv) $\dfrac{\sin x}{2 - \cos x}$ $u = 2 - \cos x$

(v) $\tan x$ $u = \cos x$ (write $\tan x$ as $\dfrac{\sin x}{\cos x}$)

(vi) $\sin 2x (1 + \cos 2x)^2$ $u = 1 + \cos 2x$

2 Use a suitable substitution to integrate the following.

(i) $2x \sin(x^2)$ (ii) $\cos x \, e^{\sin x}$

(iii) $\dfrac{\tan x}{\cos^2 x}$ (iv) $\dfrac{\cos x}{\sin^2 x}$

3 Evaluate the following definite integrals by using suitable substitutions.

(i) $\displaystyle\int_0^{\frac{\pi}{2}} \cos\left(2x - \dfrac{\pi}{2}\right) \, dx$ (ii) $\displaystyle\int_0^{\frac{\pi}{4}} \cos x \sin^3 x \, dx$

(iii) $\displaystyle\int_0^{\sqrt{\pi}} x \sin(x^2) \, dx$ (iv) $\displaystyle\int_0^{\frac{\pi}{4}} \dfrac{e^{\tan x}}{\cos^2 x} \, dx$

(v) $\displaystyle\int_0^{\frac{\pi}{4}} \dfrac{1}{\cos^2 x (1 + \tan x)} \, dx$

4 (i) Use the substitution $x = \tan\theta$ to show that
$$\int \dfrac{1 - x^2}{(1 + x^2)^2} \, dx = \int \cos 2\theta \, d\theta.$$

(ii) Hence find the value of
$$\int_0^1 \dfrac{1 - x^2}{(1 + x^2)^2} \, dx.$$

[Cambridge International AS & A Level Mathematics 9709, Paper 3 Q4 June 2005]

5 (i) Express $\cos\theta + (\sqrt{3})\sin\theta$ in the form $R\cos(\theta - \alpha)$, where $R > 0$ and $0 < \alpha < \tfrac{1}{2}\pi$, giving the exact values of R and α.

(ii) Hence show that $\displaystyle\int_0^{\frac{1}{2}\pi} \dfrac{1}{(\cos\theta + (\sqrt{3})\sin\theta)^2} \, d\theta = \dfrac{1}{\sqrt{3}}.$

[Cambridge International AS & A Level Mathematics 9709, Paper 3 Q5 June 2007]

6 (i) Use the substitution $x = \sin^2\theta$ to show that
$$\int \sqrt{\left(\dfrac{x}{1-x}\right)} \, dx = \int 2\sin^2\theta \, d\theta.$$

(ii) Hence find the exact value of
$$\int_0^{\frac{1}{4}} \sqrt{\left(\dfrac{x}{1-x}\right)} \, dx.$$

[Cambridge International AS & A Level Mathematics 9709, Paper 3 Q6 November 2005]

The use of partial fractions in integration

❓ Why is it not possible to use any of the integration techniques you have learnt so far to find $\displaystyle\int \dfrac{2}{x^2 - 1} \, dx$?

Partial fractions reminder

In Chapter 7 you met partial fractions. Here is a reminder of the work you did there.

Since $x^2 - 1$ can be factorised to give $(x+1)(x-1)$, you can write the expression to be integrated as partial fractions.

$$\frac{2}{x^2 - 1} = \frac{A}{x - 1} + \frac{B}{x + 1}$$

$$2 \equiv A(x + 1) + B(x - 1)$$

This is true for all values of x. It is an identity and to emphasise this point we use the identity symbol $\equiv$.

Let $x = 1$ $2 = 2A$ $\Rightarrow$ $A = 1$
Let $x = -1$ $2 = -2B$ $\Rightarrow$ $B = -1$

Substituting these values for A and B gives

$$\frac{2}{x^2 - 1} = \frac{1}{x - 1} - \frac{1}{x + 1}.$$

The integral then becomes

$$\int \frac{2}{x^2 - 1}\,dx = \int \frac{1}{x - 1}\,dx - \int \frac{1}{x + 1}\,dx.$$

Now the two integrals on the right can be recognised as logarithms.

$$\int \frac{2}{x^2 - 1}\,dx = \ln|x - 1| - \ln|x + 1| + c$$
$$= \ln\left|\frac{x - 1}{x + 1}\right| + c$$

Here you worked with the simplest type of partial fraction, in which there are two different linear factors in the denominator. This type will always result in two fractions both of which can be integrated to give logarithmic expressions.
Now look at the other types of partial fraction.

A repeated factor in the denominator

EXAMPLE 8.13

Find $\displaystyle\int \frac{x + 4}{(2x - 1)(x + 1)^2}\,dx.$

SOLUTION

First write the expression as partial fractions:

$$\frac{x + 4}{(2x - 1)(x + 1)^2} = \frac{A}{(2x - 1)} + \frac{B}{(x + 1)} + \frac{C}{(x + 1)^2}$$

where
$$x + 4 \equiv A(x + 1)^2 + B(2x - 1)(x + 1) + C(2x - 1).$$

Let $x = -1$ $3 = -3C$ $\Rightarrow$ $C = -1$
Let $x = \tfrac{1}{2}$ $\tfrac{9}{2} = A\left(\tfrac{3}{2}\right)^2$ $\Rightarrow$ $\tfrac{9}{2} = \tfrac{9}{4}A$ $\Rightarrow$ $A = 2$
Let $x = 0$ $4 = A - B - C$ $\Rightarrow$ $B = A - C - 4 = 2 + 1 - 4 = -1$

Substituting these values for A, B and C gives

$$\frac{x+4}{(2x-1)(x+1)^2} = \frac{2}{(2x-1)} - \frac{1}{(x+1)} - \frac{1}{(x+1)^2}$$

Now that the expression is in partial fractions, each part can be integrated separately.

$$\int \frac{x+4}{(2x-1)(x+1)^2} \, dx = \int \frac{2}{(2x-1)} \, dx - \int \frac{1}{(x+1)} \, dx - \int \frac{1}{(x+1)^2} \, dx$$

The first two integrals give logarithmic expressions as you saw above. The third, however, is of the form u^{-2} and therefore can be integrated by using the substitution $u = x + 1$, or by inspection (i.e. in your head).

$$\int \frac{x+4}{(2x-1)(x+1)^2} \, dx = \ln|2x-1| - \ln|x+1| + \frac{1}{x+1} + c$$

$$= \ln\left|\frac{2x-1}{x+1}\right| + \frac{1}{x+1} + c$$

A quadratic factor in the denominator

EXAMPLE 8.14 Find $\displaystyle\int \frac{x-2}{(x^2+2)(x+1)} \, dx$.

SOLUTION

First write the expression as partial fractions:

$$\frac{x-2}{(x^2+2)(x+1)} = \frac{Ax+B}{(x^2+2)} + \frac{C}{(x+1)}$$

where

$$x - 2 \equiv (Ax + B)(x + 1) + C(x^2 + 2)$$

Rearranging gives

$$x - 2 \equiv (A + C)x^2 + (A + B)x + (B + 2C)$$

Equating coefficients:

$$\begin{aligned} x^2 &\Rightarrow & A + C &= 0 \\ x &\Rightarrow & A + B &= 1 \\ \text{constant terms} &\Rightarrow & B + 2C &= -2 \end{aligned}$$

Solving these gives $A = 1$, $B = 0$, $C = -1$.

Hence

$$\frac{x-2}{(x^2+2)(x+1)} = \frac{x}{(x^2+2)} - \frac{1}{(x+1)}$$

$$\int \frac{x-2}{(x^2+2)(x+1)}\,dx = \int \frac{x}{(x^2+2)}\,dx - \int \frac{1}{(x+1)}\,dx$$

$$= \frac{1}{2}\int \frac{2x}{(x^2+2)}\,dx - \int \frac{1}{(x+1)}\,dx$$

$$= \frac{1}{2}\ln|x^2+2| - \ln|x+1| + c$$

$$= \ln\left|\frac{\sqrt{x^2+2}}{x+1}\right| + c$$

> $\frac{1}{2}\ln|x^2+2| = \ln\sqrt{x^2+2}$
> Notice that (x^2+2) is positive for all values of x.

(e) *Note*

If B had not been zero, you would have had an expression of the form $\frac{Ax+B}{x^2+2}$ to integrate. This can be split into $\frac{Ax}{x^2+2} + \frac{B}{x^2+2}$.

The first part of this can be integrated as in Example 8.13, but the second part cannot be integrated by any method you have met so far. If you come across a situation where you need to find such an integral, you may choose to use the standard result:

$$\int \frac{1}{(x^2+a^2)}\,dx = \frac{1}{a}\tan^{-1}\left(\frac{x}{a}\right) + c.$$

EXERCISE 8D

1 Express the fractions in each of the following integrals as partial fractions, and hence perform the integration.

(i) $\displaystyle\int \frac{1}{(1-x)(3x-2)}\,dx$

(ii) $\displaystyle\int \frac{7x-2}{(x-1)^2(2x+3)}\,dx$

(iii) $\displaystyle\int \frac{x+1}{(x^2+1)(x-1)}\,dx$

(iv) $\displaystyle\int \frac{3x+3}{(x-1)(2x+1)}\,dx$

(v) $\displaystyle\int \frac{1}{x^2(1-x)}\,dx$

(vi) $\displaystyle\int \frac{1}{(x+1)(x+3)}\,dx$

(vii) $\displaystyle\int \frac{2x-4}{(x^2+4)(x+2)}\,dx$

(viii) $\displaystyle\int \frac{5x+1}{(x+2)(2x+1)^2}\,dx$

2 Express in partial fractions

$$f(x) = \frac{3x+4}{(x^2+4)(x-3)}$$

and hence find $\displaystyle\int_0^2 f(x)\,dx$.

[MEI, adapted]

3 Express $\dfrac{1}{x^2(2x+1)}$ in partial fractions. Hence show that

$$\int_1^2 \frac{dx}{x^2(2x+1)} = \tfrac{1}{2} + 2\ln\tfrac{5}{6}.$$

[MEI]

4 (i) (a) Express $\dfrac{3}{(1+x)(1-2x)}$ in partial fractions.

(b) Hence find

$$\int_0^{0.1} \frac{3}{(1+x)(1-2x)}\,dx$$

giving your answer to 5 decimal places.

(ii) (a) Find the first three terms in the binomial expansion of
$$3(1+x)^{-1}(1-2x)^{-1}.$$

(b) Use the first three terms of this expansion to find an approximation for
$$\int_0^{0.1} \frac{3}{(1+x)(1-2x)} \, dx.$$

(c) What is the percentage error in your answer to part (b)?

5 (i) Given that
$$\frac{x^2 - x - 24}{(x+2)(x-4)} \equiv A + \frac{B}{(x+2)} + \frac{C}{(x-4)},$$
find the values of the constants A, B and C.

(ii) Find $\int_1^3 \frac{x^2 - x - 24}{(x+2)(x-4)} \, dx.$

[MEI]

6 (i) Given that $f(x) = \dfrac{16 + 2x + 15x^2}{(1+x^2)(2-x)} \equiv \dfrac{A + Bx}{1+x^2} + \dfrac{C}{2-x}$, find the values of B and C and show that $A = 0$.

(ii) Find $\int_0^1 f(x) \, dx$ in an exact form.

(iii) Express $f(x)$ as a sum of powers of x up to and including the term in x^4. Determine the range of values of x for which this expansion of $f(x)$ is valid.

[MEI]

7 (i) Find the values of the constants A, B, C and D such that
$$\frac{2x^3 - 1}{x^2(2x-1)} \equiv A + \frac{B}{x} + \frac{C}{x^2} + \frac{D}{2x-1}.$$

(ii) Hence show that
$$\int_1^2 \frac{2x^3 - 1}{x^2(2x-1)} \, dx = \frac{3}{2} + \frac{1}{2} \ln\left(\frac{16}{27}\right).$$

[Cambridge International AS & A Level Mathematics 9709, Paper 32 Q10 June 2010]

8 Let $f(x) \equiv \dfrac{x^2 + 3x + 3}{(x+1)(x+3)}.$

(i) Express $f(x)$ in partial fractions.

(ii) Hence show that $\int_0^3 f(x) \, dx = 3 - \frac{1}{2} \ln 2.$

[Cambridge International AS & A Level Mathematics 9709, Paper 3 Q7 June 2008]

Integration by parts

There are still many integrations which you cannot yet do. In fact, many functions cannot be integrated at all, although virtually all functions can be differentiated. However, some functions can be integrated by techniques which you have not yet met. Integration by parts is one of those techniques.

EXAMPLE 8.15 Find $\int x \cos x \, dx$.

SOLUTION

The expression to be integrated is clearly a product of two simpler expressions, x and $\cos x$, so your first thought may be to look for a substitution to enable you to perform the integration. However, there are some expressions which are products but which cannot be integrated by substitution. This is one of them. You need a new technique to integrate such expressions.

Take the expression $x \sin x$ and differentiate it, using the product rule.

$$\frac{d}{dx}(x \sin x) = x \cos x + \sin x$$

Now integrate both sides. This has the effect of 'undoing' the differentiation, so

$$x \sin x = \int x \cos x \, dx + \int \sin x \, dx$$

Rearranging this gives

$$\int x \cos x \, dx = x \sin x - \int \sin x \, dx$$
$$= x \sin x - (-\cos x) + c$$
$$= x \sin x + \cos x + c$$

This has enabled you to find the integral of $x \cos x$.

The work in this example can be generalised into the method of integration by parts. Before coming on to that, do the following activity.

ACTIVITY 8.2 For each of the following

(a) differentiate using the product rule
(b) rearrange your expression to find an expression for the given integral I
(c) use this expression to find the given integral.

(i) $y = x \cos x$ $\quad I = \int x \sin x \, dx$
(ii) $y = x e^{2x}$ $\quad I = \int 2x e^{2x} \, dx$

❓ The work in Activity 8.2 has enabled you to work out some integrals which you could not previously have done, but you needed to be given the expressions to be differentiated first. Effectively you were given the answers.

Look at the expressions you found in part **(b)** of Activity 8.2.
Can you see any way of working out these expressions without starting by differentiating a given product?

The general result for integration by parts

The method just investigated can be generalised.

Look back at Example 8.14. Use u to stand for the function x, and v to stand for the function $\sin x$.

Using the product rule to differentiate the function uv,

$$\frac{d}{dx}(uv) = v\frac{du}{dx} + u\frac{dv}{dx}.$$

Integrating gives

$$uv = \int v\frac{du}{dx}\,dx + \int u\frac{dv}{dx}\,dx.$$

Rearranging gives

$$\int u\frac{dv}{dx}\,dx = uv - \int v\frac{du}{dx}\,dx.$$

This is the formula you use when you need to integrate by parts.

In order to use it, you have to split the function you want to integrate into two simpler functions. In Example 8.15 you split $x \cos x$ into the two functions x and $\cos x$. One of these functions will be called u and the other $\frac{dv}{dx}$, to fit the left-hand side of the expression. You will need to decide which will be which. Two considerations will help you.

- As you want to use $\frac{du}{dx}$ on the right-hand side of the expression, u should be a function which becomes a simpler function after differentiation. So in this case, u will be the function x.

- As you need v to work out the right-hand side of the expression, it must be possible to integrate the function $\frac{dv}{dx}$ to obtain v. In this case, $\frac{dv}{dx}$ will be the function $\cos x$.

So now you can find $\int x \cos x\,dx$.

Put $\quad u = x \quad\Rightarrow\quad \frac{du}{dx} = 1$

and $\quad \frac{dv}{dx} = \cos x \quad\Rightarrow\quad v = \sin x$

Substituting in

$$\int u\frac{dv}{dx}\,dx = uv - \int v\frac{du}{dx}\,dx$$

gives

$$\int x \cos x\,dx = x \sin x - \int 1 \times \sin x\,dx$$
$$= x \sin x - (-\cos x) + c$$
$$= x \sin x + \cos x + c$$

EXAMPLE 8.16

Find $\int 2x\,e^x\,dx$.

SOLUTION

First split $2xe^x$ into the two simpler expressions, $2x$ and e^x. Both can be integrated easily but, as $2x$ becomes a simpler expression after differentiation and e^x does not, take u to be $2x$.

$$u = 2x \quad \Rightarrow \quad \frac{du}{dx} = 2$$

$$\frac{dv}{dx} = e^x \quad \Rightarrow \quad v = e^x$$

Substituting in

$$\int u\frac{dv}{dx}\,dx = uv - \int v\frac{du}{dx}\,dx$$

gives

$$\int 2x\,e^x\,dx = 2x\,e^x - \int 2e^x\,dx$$
$$= 2x\,e^x - 2e^x + c$$

In some cases, the choices of u and v may be less obvious.

EXAMPLE 8.17

Find $\int x \ln x\,dx$.

SOLUTION

It might seem at first that u should be taken as x, because it becomes a simpler expression after differentiation.

$$u = x \quad \Rightarrow \quad \frac{du}{dx} = 1$$

$$\frac{dv}{dx} = \ln x$$

Now you need to integrate $\ln x$ to obtain v. Although it is possible to integrate $\ln x$, it has to be done by parts, as you will see in the next example. The wrong choice has been made for u and v, resulting in a more complicated integral.

So instead, let $u = \ln x$.

$$u = \ln x \quad \Rightarrow \quad \frac{du}{dx} = \frac{1}{x}$$

$$\frac{dv}{dx} = x \quad \Rightarrow \quad v = \tfrac{1}{2}x^2$$

Substituting in

$$\int u\frac{dv}{dx}\,dx = uv - \int v\frac{du}{dx}\,dx$$

gives

$$\int x \ln x\,dx = \tfrac{1}{2}x^2 \ln x - \int \frac{\tfrac{1}{2}x^2}{x}\,dx$$
$$= \tfrac{1}{2}x^2 \ln x - \int \tfrac{1}{2}x\,dx$$
$$= \tfrac{1}{2}x^2 \ln x - \tfrac{1}{4}x^2 + c$$

EXAMPLE 8.18 Find $\int \ln x \, dx$.

SOLUTION

You need to start by writing $\ln x$ as $1 \ln x$ and then use integration by parts.

As in the last example, let $u = \ln x$.

$$u = \ln x \quad \Rightarrow \quad \frac{du}{dx} = \frac{1}{x}$$

$$\frac{dv}{dx} = 1 \quad \Rightarrow \quad v = x$$

Substituting in

$$\int u \frac{dv}{dx} \, dx = uv - \int v \frac{du}{dx} \, dx$$

gives

$$\int 1 \ln x \, dx = x \ln x - \int x \times \frac{1}{x} \, dx$$
$$= x \ln x - \int 1 \, dx$$
$$= x \ln x - x + c$$

Using integration by parts twice

Sometimes it is necessary to use integration by parts twice or more to complete the integration successfully.

EXAMPLE 8.19 Find $\int x^2 \sin x \, dx$.

SOLUTION

First split $x^2 \sin x$ into two: x^2 and $\sin x$. As x^2 becomes a simpler expression after differentiation, take u to be x^2.

$$u = x^2 \quad \Rightarrow \quad \frac{du}{dx} = 2x$$

$$\frac{dv}{dx} = \sin x \quad \Rightarrow \quad v = -\cos x$$

Substituting in

$$\int u \frac{dv}{dx} \, dx = uv - \int v \frac{du}{dx} \, dx$$

gives

$$\int x^2 \sin x \, dx = -x^2 \cos x - \int -2x \cos x \, dx$$
$$= -x^2 \cos x + \int 2x \cos x \, dx \qquad ①$$

Now the integral of $2x\cos x$ cannot be found without using integration by parts again. It has to be split into the expressions $2x$ and $\cos x$ and, as $2x$ becomes a simpler expression after differentiation, take u to be $2x$.

$$u = 2x \quad\Rightarrow\quad \frac{du}{dx} = 2$$

$$\frac{dv}{dx} = \cos x \quad\Rightarrow\quad v = \sin x$$

Substituting in

$$\int u \frac{dv}{dx} dx = uv - \int v \frac{du}{dx} dx$$

gives

$$\int 2x \cos x \, dx = 2x \sin x - \int 2 \sin x \, dx$$
$$= 2x \sin x - (-2 \cos x) + c$$
$$= 2x \sin x + 2 \cos x + c$$

So in ① $\int x^2 \sin x \, dx = -x^2 \cos x + 2x \sin x + 2 \cos x + c$.

The technique of integration by parts is usually used when the two functions are of different types: polynomials, trigonometrical functions, exponentials, logarithms. There are, however, some exceptions, as in questions 3 and 4 of Exercise 8E.

Integration by parts is a very important technique which is needed in many other branches of mathematics. For example, integrals of the form $\int x\, f(x)\, dx$ are used in statistics to find the mean of a probability density function, and in mechanics to find the centre of mass of a shape. Integrals of the form $\int x^2 f(x)\, dx$ are used in statistics to find variance and in mechanics to find moments of inertia.

EXERCISE 8E

1 For each of these integrals

(a) write down the expression to be taken as u and the expression to be taken as $\frac{dv}{dx}$

(b) use the formula for integration by parts to complete the integration.

(i) $\int x e^x \, dx$ (ii) $\int x \cos 3x \, dx$

(iii) $\int (2x+1) \cos x \, dx$ (iv) $\int x e^{-2x} \, dx$

(v) $\int x e^{-x} \, dx$ (vi) $\int x \sin 2x \, dx$

2 Use integraton by parts to integrate

(i) $x^3 \ln x$ (ii) $3x e^{3x}$

(iii) $2x \cos 2x$ (iv) $x^2 \ln 2x$

3 Find $\int x \sqrt{1+x} \, dx$

(i) by using integration by parts

(ii) by using the substitution $u = 1 + x$.

4 Find $\int 2x(x-2)^4 \, dx$

   (i) by using integration by parts

   (ii) by using the substitution $u = x - 2$.

5 (i) By writing $\ln x$ as the product of $\ln x$ and 1, use integration by parts to find $\int \ln x \, dx$.

   (ii) Use the same method to find $\int \ln 3x \, dx$.

   (iii) Write down $\int \ln px \, dx$ where $p > 0$.

6 Find $\int x^2 e^x \, dx$.

7 Find $\int (2-x)^2 \cos x \, dx$.

Definite integration by parts

When you use the method of integration by parts on a definite integral, it is important to remember that the term uv on the right-hand side of the expression has already been integrated and so should be written in square brackets with the limits indicated.

$$\int_a^b u \frac{dv}{dx} \, dx = \left[uv \right]_a^b - \int_a^b v \frac{du}{dx} \, dx$$

EXAMPLE 8.20 Evaluate $\int_0^2 x e^x \, dx$.

SOLUTION

Put $u = x$ $\Rightarrow$ $\dfrac{du}{dx} = 1$

and $\dfrac{dv}{dx} = e^x$ $\Rightarrow$ $v = e^x$

Substituting in

$$\int_a^b u \frac{dv}{dx} \, dx = \left[uv \right]_a^b - \int_a^b v \frac{du}{dx} \, dx$$

gives

$$\int_0^2 x e^x \, dx = \left[x e^x \right]_0^2 - \int_0^2 e^x \, dx$$

$$= \left[x e^x \right]_0^2 - \left[e^x \right]_0^2$$

$$= (2e^2 - 0) - (e^2 - e^0)$$

$$= 2e^2 - e^2 + 1$$

$$= e^2 + 1$$

EXAMPLE 8.21 Find the area of the region between the curve $y = x \cos x$ and the x axis, between $x = 0$ and $x = \frac{\pi}{2}$.

SOLUTION

Figure 8.6 shows the region whose area is to be found.

To find the required area, you need to integrate the function $x \cos x$ between the limits 0 and $\frac{\pi}{2}$. You therefore need to work out

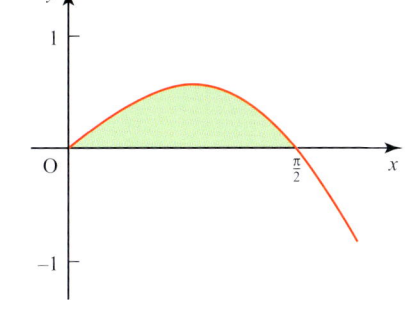

$$\int_0^{\frac{\pi}{2}} x \cos x \, dx.$$

Put $u = x$ $\Rightarrow$ $\frac{du}{dx} = 1$

and $\frac{dv}{dx} = \cos x$ $\Rightarrow$ $v = \sin x$

Figure 8.6

Substituting in

$$\int_a^b u \frac{dv}{dx} \, dx = \left[uv \right]_a^b - \int_a^b v \frac{du}{dx} \, dx.$$

gives

$$\int_0^{\frac{\pi}{2}} x \cos x \, dx = \left[x \sin x \right]_0^{\frac{\pi}{2}} - \int_0^{\frac{\pi}{2}} \sin x \, dx$$

$$= \left[x \sin x \right]_0^{\frac{\pi}{2}} - \left[-\cos x \right]_0^{\frac{\pi}{2}}$$

$$= \left[x \sin x + \cos x \right]_0^{\frac{\pi}{2}}$$

$$= \left(\frac{\pi}{2} + 0 \right) - \left(0 + 1 \right)$$

$$= \frac{\pi}{2} - 1$$

So the required area is $\left(\frac{\pi}{2} - 1 \right)$ square units.

EXERCISE 8F

1 Evaluate these definite integrals.

(i) $\int_0^1 x e^{3x} \, dx$

(ii) $\int_0^{\pi} (x - 1) \cos x \, dx$

(iii) $\int_0^2 (x + 1) e^x \, dx$

(iv) $\int_1^2 \ln 2x \, dx$

(v) $\int_0^{\frac{\pi}{2}} x \sin 2x \, dx$

(vi) $\int_1^4 x^2 \ln x \, dx$

2 (i) Find the co-ordinates of the points where the graph of $y = (2 - x)e^{-x}$ cuts the x and y axes.
  (ii) Hence sketch the graph of $y = (2 - x)e^{-x}$.
  (iii) Use integration by parts to find the area of the region between the x axis, the y axis and the graph $y = (2 - x)e^{-x}$.

3 (i) Sketch the graph of $y = x \sin x$ from $x = 0$ to $x = \pi$ and shade the region between the curve and the x axis.
  (ii) Find the area of this region using integration by parts.

4 Find the area of the region between the x axis, the line $x = 5$ and the graph $y = \ln x$.

5 Find the area of the region between the x axis and the graph $y = x \cos x$ from $x = 0$ to $x = \dfrac{\pi}{2}$.

6 Find the area of the region between the negative x axis and the graph $y = x\sqrt{x+1}$

  (i) using integration by parts
  (ii) using the substitution $u = x + 1$.

7 The sketch shows the curve with equation $y = x^2 \ln 2x$.

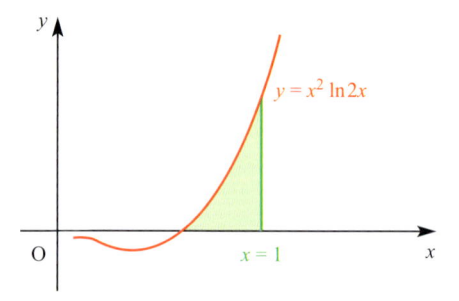

Find the x co-ordinate of the point where the curve cuts the x axis.
Hence calculate the area of the shaded region using the method of integration by parts applied to the product of $\ln 2x$ and x^2.
Give your answer correct to 3 decimal places.
[MEI]

8 Show that $\int_0^1 x^2 e^x \, dx = e - 2$.

Show that the use of the trapezium rule with five strips (six ordinates) gives an estimate that is about 3.8% too high.
Explain why approximate evaluation of this integral using the trapezium rule will always result in an overestimate, however many strips are used.
[MEI]

9 (i) Find $\int x \cos kx \, dx$, where k is a non-zero constant.

(ii) Show that
$$\cos(A - B) - \cos(A + B) = 2 \sin A \sin B.$$

Hence express $2 \sin 5x \sin 3x$ as the difference of two cosines.

(iii) Use the results in parts **(i)** and **(ii)** to show that
$$\int_0^{\frac{\pi}{4}} x \sin 5x \sin 3x \, dx = \frac{\pi - 2}{16}.$$
[MEI]

10 Use integration by parts to show that
$$\int_2^4 \ln x \, dx = 6 \ln 2 - 2.$$
[Cambridge International AS & A Level Mathematics 9709, Paper 3 Q3 November 2007]

11 The constant a is such that $\int_0^a x e^{\frac{1}{2}x} \, dx = 6$.

(i) Show that a satisfies the equation
$$x = 2 + e^{-\frac{1}{2}x}.$$

(ii) By sketching a suitable pair of graphs, show that this equation has only one root.

(iii) Verify by calculation that this root lies between 2 and 2.5.

(iv) Use an iterative formula based on the equation in part **(i)** to calculate the value of a correct to 2 decimal places. Give the result of each iteration to 4 decimal places.

[Cambridge International AS & A Level Mathematics 9709, Paper 3 Q9 November 2008]

12 The diagram shows the curve $y = e^{-\frac{1}{2}x} \sqrt{(1 + 2x)}$ and its maximum point M. The shaded region between the curve and the axes is denoted by R.

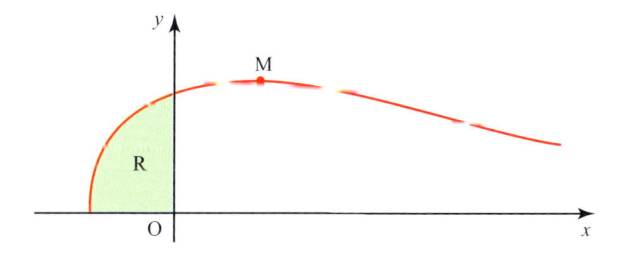

(i) Find the x co-ordinate of M.

(ii) Find by integration the volume of the solid obtained when R is rotated completely about the x axis. Give your answer in terms of π and e.

[Cambridge International AS & A Level Mathematics 9709, Paper 3 Q9 June 2008]

General integration

You now know several techniques for integration which can be used to integrate a wide variety of functions. One of the difficulties which you may now experience when faced with an integration is deciding which technique is appropriate! This section gives you some guidelines on this, as well as revising all the work on integration that you have done so far.

❓ Look at the integrals below and try to decide which technique you would use and, in the case of a substitution, which expression you would write as u. Do not attempt actually to carry out the integrations. Make a note of your decisions – you will return to these integrals later.

(i) $\displaystyle\int \frac{x-5}{x^2+2x-3}\,dx$
(ii) $\displaystyle\int \frac{x+1}{x^2+2x-3}\,dx$

(iii) $\displaystyle\int xe^x\,dx$
(iv) $\displaystyle\int xe^{x^2}\,dx$

(v) $\displaystyle\int \frac{2x+\cos x}{x^2+\sin x}\,dx$
(vi) $\displaystyle\int \cos x \sin^2 x\,dx$

Choosing an appropriate method of integration

You have now met the following standard integrals.

$f(x)$		$\int f(x)\,dx$
x^n	$(n \neq -1)$	$\dfrac{x^{n+1}}{n+1}$
$\dfrac{1}{x}$	$(x \neq 0)$	$\ln\lvert x \rvert$
e^x	$(x \in \mathbb{R})$	e^x
$\sin x$	$(x \in \mathbb{R})$	$-\cos x$
$\cos x$	$(x \in \mathbb{R})$	$\sin x$

If you are asked to integrate any of these standard functions, you may simply write down the answer.

For other integrations, the following table may help.

Type of expression to be integrated	Examples	Method of integration
Simple variations of any of the standard functions	$\cos(2x+1)$ e^{3x}	Substitution may be used, but it should be possible to do these by inspection.
Product of two expressions of the form $f'(x)g[f(x)]$ Note that $f'(x)$ means $\frac{d}{dx}[f(x)]$	$2xe^{x^2}$ $x^2(x^3+1)^6$	Substitution $u = f(x)$
Other products, particularly when one expression is a small positive integer power of x or a polynomial in x	xe^x $x^2 \sin x$	Integration by parts
Quotients of the form $\frac{f'(x)}{f(x)}$ or expressions which can easily be converted to this form	$\frac{x}{x^2+1}$ $\frac{\sin x}{\cos x}$	Substitution $u = f(x)$ or by inspection: $k \ln|f(x)| + c$, where k is known
Polynomial quotients which may be split into partial fractions	$\frac{x+1}{x(x-1)}$ $\frac{x-4}{x^2-x-2}$	Split into partial fractions and integrate term by term
Odd powers of $\sin x$ or $\cos x$	$\cos^3 x$	Use $\cos^2 x + \sin^2 x = 1$ and write in form $f'(x)g[f(x)]$
Even powers of $\sin x$ or $\cos x$	$\sin^2 x$ $\cos^4 x$	Use the double-angle formulae to transform the expression before integrating.

It is impossible to give an exhaustive list of possible types of integration, but the table above and that on the previous page cover the most common situations that you will meet.

ACTIVITY 8.3 Now look back at the integrals in the discussion point on the previous page and the decisions you made about which method of integration should be used for each one. Now find these integrals.

(i) $\int \dfrac{x-5}{x^2+2x-3}\, dx$

(ii) $\int \dfrac{x+1}{x^2+2x-3}\, dx$

(iii) $\int xe^x\, dx$

(iv) $\int xe^{x^2}\, dx$

(v) $\int \dfrac{2x+\cos x}{x^2+\sin x}\, dx$

(vi) $\int \cos x \sin^2 x\, dx$

EXERCISE 8G

1 Choose an appropriate method and integrate the following.
   You may find it helpful to discuss in class first which method to use.

   (i) $\int \cos(3x-1)\,dx$
   (ii) $\int \dfrac{2x+1}{(x^2+x-1)^2}\,dx$

   (iii) $\int e^{1-x}\,dx$
   (iv) $\int \cos 2x\,dx$

   (v) $\int \ln 2x\,dx$
   (vi) $\int \dfrac{x}{(x^2-1)^3}\,dx$

   (vii) $\int \sqrt{2x-3}\,dx$
   (viii) $\int \dfrac{4x-1}{(x-1)^2(x+2)}\,dx$

   (ix) $\int x^3 \ln x\,dx$
   (x) $\int \dfrac{5}{2x^2-7x+3}\,dx$

   (xi) $\int (x+1)e^{x^2+2x}\,dx$
   (xii) $\int \dfrac{\sin x - \cos x}{\sin x + \cos x}\,dx$

   (xiii) $\int x^2 \sin 2x\,dx$
   (xiv) $\int \sin^3 2x\,dx$

2 Evaluate the following definite integrals.

   (i) $\int_8^{24} \dfrac{dx}{\sqrt{3x-8}}$
   (ii) $\int_8^{24} \dfrac{dx}{3x-8}$
   (iii) $\int_8^{24} \dfrac{9x}{3x-8}\,dx$

   (iv) $\int_0^{\frac{\pi}{2}} \sin^3 x\,dx$
   (v) $\int_1^2 x^2 \ln x\,dx$

3 Evaluate $\int_0^2 \dfrac{x^2}{\sqrt{1+x^3}}\,dx$, using the substitution $u = 1 + x^3$, or otherwise.
   [MEI]

4 Find $\int_0^{\frac{\pi}{2}} \dfrac{\sin\theta}{\cos^4\theta}\,d\theta$ in terms of $\sqrt{2}$.
   [MEI]

5 Using the substitution $u = \ln x$, or otherwise, find $\int_1^2 \dfrac{\ln x}{x}\,dx$, giving your answer to 2 decimal places.
   [MEI, part]

6 Find $\int_0^{\frac{\pi}{2}} x\cos 2x\,dx$, expressing your answer in terms of π.
   [MEI]

7 (i) Find $\int xe^{-2x}\,dx$.
   (ii) Evaluate $\int_0^1 \dfrac{x}{(4+x^2)}\,dx$, giving your answer correct to 3 significant figures.
   [MEI]

8 (i) Find $\int \sin(2x-3)\,dx$.
   (ii) Use the method of integration by parts to evaluate $\int_0^2 xe^{2x}\,dx$.
   (iii) Using the substitution $t = x^2 - 9$, or otherwise, find $\int \dfrac{x}{x^2-9}\,dx$.
   [MEI]

9 Evaluate

(i) $\int_0^1 (2x^2 + 1)(2x^3 + 3x + 4)^{\frac{1}{2}} \, dx$

(ii) $\int_1^e \dfrac{\ln x}{x^3} \, dx.$

[MEI]

10 Find $\int_0^{\frac{\pi}{2}} \sin x \cos^3 x \, dx$ and $\int_0^1 t e^{-2t} \, dt.$

[MEI]

KEY POINTS

1 $\int kx^n \, dx = \dfrac{kx^{n+1}}{n+1} + c$ where k and n are constants but $n \neq -1$.

2 Substitution is often used to change a non-standard integral into a standard one.

3 $\int e^x \, dx = e^x + c$

$\int e^{ax+b} \, dx = \dfrac{1}{a} e^{ax+b} + c$

4 $\int \dfrac{1}{x} \, dx = \ln|x| + c$

$\int \dfrac{1}{ax+b} \, dx = \dfrac{1}{a} \ln|ax+b| + c$

5 $\int \dfrac{f'(x)}{f(x)} \, dx = \ln|f(x)| + c$

6 $\int \cos(ax+b) \, dx = \dfrac{1}{a} \sin(ax+b) + c$

$\int \sin(ax+b) \, dx = -\dfrac{1}{a} \cos(ax+b) + c$

$\int \sec^2(ax+b) \, dx = \dfrac{1}{a} \tan(ax+b) + c$

7 Using partial fractions often makes it possible to use logarithms to integrate the quotient of two polynomials.

8 Some products may be integrated by parts using the formulae

$\int u \dfrac{dv}{dx} \, dx = uv - \int v \dfrac{du}{dx} \, dx$

$\int_a^b u \dfrac{dv}{dx} \, dx = [uv]_a^b - \int_a^b v \dfrac{du}{dx} \, dx.$

9 Differential equations

The greater our knowledge increases, the more our ignorance unfolds.
John F. Kennedy

Suppose you are in a hurry to go out and want to drink a cup of hot tea before you go.

How long will you have to wait until it is cool enough to drink?

To solve this problem, you would need to know something about the rate at which liquids cool at different temperatures.

Figure 9.1 shows an example of the temperature of a liquid plotted against time.

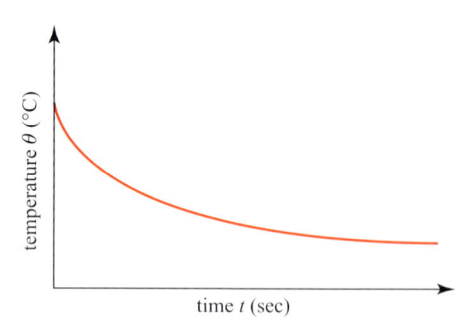

Figure 9.1

Notice that the graph is steepest at high temperatures and becomes less steep as the liquid cools. In other words, the rate of change of temperature is numerically greatest at high temperatures and gets numerically less as the temperature drops. The rate of change is always negative since the temperature is decreasing.

If you study physics, you may have come across Newton's law of cooling: The rate of cooling of a body is proportional to the difference in temperature of the body and that of the surrounding air.

The gradient of the temperature graph may be written as $\frac{d\theta}{dt}$, where θ is the temperature of the liquid and t is the time. The quantity $\frac{d\theta}{dt}$ tells us the rate at which the temperature of the liquid is increasing. As the liquid is cooling, $\frac{d\theta}{dt}$ will be negative, so the rate of cooling may be written as $-\frac{d\theta}{dt}$.

The difference in temperature of the liquid and that of the surrounding air may be written as $\theta - \theta_0$, where θ_0 is the temperature of the surrounding air. So Newton's law of cooling may be expressed mathematically as:

$$-\frac{d\theta}{dt} \propto (\theta - \theta_0)$$

or $\quad \dfrac{d\theta}{dt} = -k(\theta - \theta_0)$

where k is a positive constant.

Any equation, like this one, which involves a derivative, such as $\dfrac{d\theta}{dt}, \dfrac{dy}{dx}$ or $\dfrac{d^2y}{dx^2}$, is known as a *differential equation*. A differential equation which only involves a first derivative such as $\dfrac{dy}{dx}$ is called a *first-order differential equation*. One which involves a second derivative such as $\dfrac{d^2y}{dx^2}$ is called a *second-order differential equation*. A third-order differential equation involves a third derivative and so on.

In this chapter, you will be looking only at first-order differential equations such as the one above for Newton's law of cooling.

By the end of this chapter, you will be able to solve problems such as the tea cooling problem given at the beginning of this chapter, by using first-order differential equations.

Forming differential equations from rates of change

If you are given sufficient information about the rate of change of a quantity, such as temperature or velocity, you can work out a differential equation to model the situation, like the one above for Newton's law of cooling. It is important to look carefully at the wording of the problem which you are studying in order to write an equivalent mathematical statement. For example, if the altitude of an aircraft is being considered, the phrase 'the rate of change of height' might be used. This actually means 'the rate of change of height *with respect to time*' and could be written as $\dfrac{dh}{dt}$. However, you might be more interested in how the height of the aircraft changes according to the horizontal distance it has travelled. In this case, you would talk about 'the rate of change of height *with respect to horizontal distance*' and could write this as $\dfrac{dh}{dx}$, where x is the horizontal distance travelled.

Some of the situations you meet in this chapter involve motion along a straight line, and so you will need to know the meanings of the associated terms.

The position of an object (+5 in figure 9.2, overleaf) is its distance from the origin O in the direction you have chosen to define as being positive.

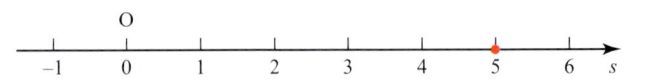

Figure 9.2

The rate of change of position of the object with respect to time is its velocity, and this can take positive or negative values according to whether the object is moving away from the origin or towards it.

$$v = \frac{ds}{dt}$$

The rate of change of an object's velocity with respect to time is called its acceleration, a.

$$a = \frac{dv}{dt}$$

Velocity and acceleration are vector quantities but in one-dimensional motion there is no choice in direction, only in sense (i.e. whether positive or negative). Consequently, as you may already have noticed, the conventional bold type for vectors is not used in this chapter.

EXAMPLE 9.1

An object is moving through a liquid so that the rate at which its velocity decreases is proportional to its velocity at any given instant. When it enters the liquid, it has a velocity of $5\,\text{m s}^{-1}$ and the velocity is decreasing at a rate of $1\,\text{m s}^{-2}$. Find the differential equation to model this situation.

SOLUTION

The rate of change of velocity means the rate of change of velocity with respect to time and so can be written as $\frac{dv}{dt}$. As it is decreasing, the rate of change must be negative, so

$$-\frac{dv}{dt} \propto v$$

or

$$\frac{dv}{dt} = -kv$$

where k is a positive constant.

When the object enters the liquid its velocity is $5\,\text{m s}^{-1}$, so $v = 5$, and the velocity is decreasing at the rate of $1\,\text{m s}^{-2}$, so

$$\frac{dv}{dt} = -1$$

Putting this information into the equation $\frac{dv}{dt} = -kv$ gives

$$-1 = -k \times 5 \quad \Rightarrow \quad k = \tfrac{1}{5}.$$

So the situation is modelled by the differential equation

$$\frac{dv}{dt} = -\frac{v}{5}$$

EXAMPLE 9.2 A model is proposed for the temperature gradient within a star, in which the temperature decreases with respect to the distance from the centre of the star at a rate which is inversely proportional to the square of the distance from the centre. Express this model as a differential equation.

SOLUTION

In this example the rate of change of temperature is not with respect to time but with respect to distance. If θ represents the temperature at a point in the star and r the distance from the centre of the star, the rate of change of temperature with respect to distance may be written as $-\dfrac{d\theta}{dr}$, so

$$-\frac{d\theta}{dr} \propto \frac{1}{r^2} \text{ or } \frac{d\theta}{dr} = -\frac{k}{r^2}$$

where k is a positive constant.

> **Note**
>
> This model must break down near the centre of the star, otherwise it would be infinitely hot there.

EXAMPLE 9.3 The area A of a square is increasing at a rate proportional to the length of its side s. The constant of proportionality is k. Find an expression for $\dfrac{ds}{dt}$.

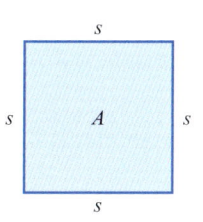

Figure 9.3

SOLUTION

The rate of increase of A with respect to time may be written as $\dfrac{dA}{dt}$.

As this is proportional to s, it may be written as

$$\frac{dA}{dt} = ks$$

where k is a positive constant.

You can use the chain rule to write down an expression for $\dfrac{ds}{dt}$ in terms of $\dfrac{dA}{dt}$.

$$\frac{ds}{dt} = \frac{ds}{dA} \times \frac{dA}{dt}$$

You now need an expression for $\frac{ds}{dA}$. Because A is a square

$$A = s^2$$
$$\Rightarrow \quad \frac{dA}{ds} = 2s$$
$$\Rightarrow \quad \frac{ds}{dA} = \frac{1}{2s}$$

Substituting the expressions for $\frac{ds}{dA}$ and $\frac{dA}{dt}$ into the expression for $\frac{ds}{dt}$

$$\Rightarrow \quad \frac{ds}{dt} = \frac{1}{2s} \times ks$$
$$\Rightarrow \quad \frac{ds}{dt} = \tfrac{1}{2}k$$

EXERCISE 9A

1 The differential equation
$$\frac{dv}{dt} = 5v^2$$
models the motion of a particle, where v is the velocity of the particle in m s^{-1} and t is the time in seconds. Explain the meaning of $\frac{dv}{dt}$ and what the differential equation tells you about the motion of the particle.

2 A spark from a firework is moving in a straight line at a speed which is inversely proportional to the square of the distance which the spark has travelled from the firework. Find an expression for the speed (i.e. the rate of change of distance travelled) of the spark.

3 The rate at which a sunflower increases in height is proportional to the natural logarithm of the difference between its final height H and its height h at a particular time. Find a differential equation to model this situation.

4 In a chemical reaction in which substance A is converted into substance B, the rate of increase of the mass of substance B is inversely proportional to the mass of substance B present. Find a differential equation to model this situation.

5 After a major advertising campaign, an engineering company finds that its profits are increasing at a rate proportional to the square root of the profits at any given time. Find an expression to model this situation.

6 The coefficient of restitution e of a squash ball increases with respect to the ball's temperature θ at a rate proportional to the temperature, for typical playing temperatures. (The coefficient of restitution is a measure of how elastic, or bouncy, the ball is. Its value lies between zero and one, zero meaning that the ball is not at all elastic and one meaning that it is perfectly elastic.) Find a differential equation to model this situation.

7 A cup of tea cools at a rate proportional to the temperature of the tea above that of the surrounding air. Initially, the tea is at a temperature of 95°C and is cooling at a rate of $0.5°C\,s^{-1}$. The surrounding air is at 15°C.
Find a differential equation to model this situation.

8 The rate of increase of bacteria is modelled as being proportional to the number of bacteria at any time during their initial growth phase.

When the bacteria number 2×10^6 they are increasing at a rate of 10^5 per day. Find a differential equation to model this situation.

9 The acceleration (i.e. the rate of change of velocity) of a moving object under a particular force is inversely proportional to the square root of its velocity. When the speed is $4\,m\,s^{-1}$ the acceleration is $2\,m\,s^{-2}$. Find a differential equation to model this situation.

10 The radius of a circular patch of oil is increasing at a rate inversely proportional to its area A. Find an expression for $\dfrac{dA}{dt}$.

11 A poker, 80 cm long, has one end in a fire. The temperature of the poker decreases with respect to the distance from that end at a rate proportional to that distance. Halfway along the poker, the temperature is decreasing at a rate of $10°C\,cm^{-1}$. Find a differential equation to model this situation.

12 A spherical balloon is allowed to deflate. The rate at which air is leaving the balloon is proportional to the volume V of air left in the balloon. When the radius of the balloon is 15 cm, air is leaving at a rate of $8\,cm^3\,s^{-1}$.
Find an expression for $\dfrac{dV}{dt}$.

13 A tank is shaped as a cuboid with a square base of side 10 cm. Water runs out through a hole in the base at a rate proportional to the square root of the height, h cm, of water in the tank. At the same time, water is pumped into the tank at a constant rate of $2\,cm^3\,s^{-1}$. Find an expression for $\dfrac{dh}{dt}$.

INVESTIGATION

Figure 9.4 shows the isobars (lines of equal pressure) on a weather map featuring a storm. The wind direction is almost parallel to the isobars and its speed is proportional to the pressure gradient.

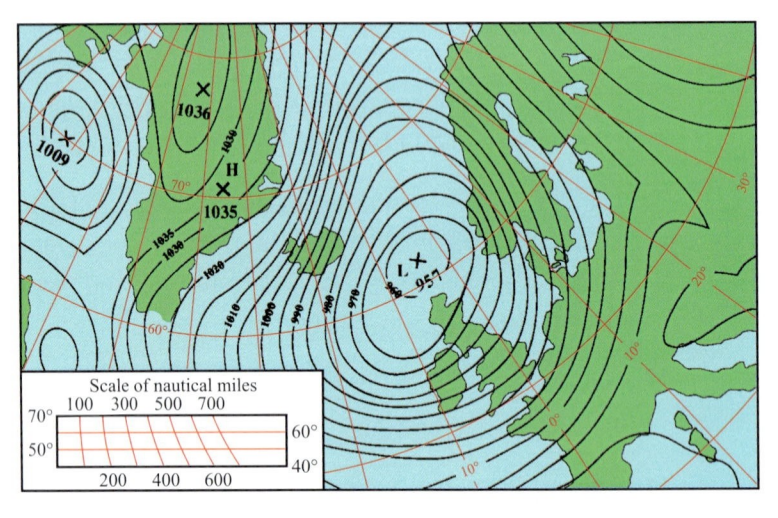

Figure 9.4

Draw a line from the point H to the point L. This runs approximately perpendicular to the isobars. It is suggested that along this line the pressure gradient (and so the wind speed) may be modelled by the differential equation

$$\frac{dp}{dx} = -a \sin bx$$

Suggest values for a and b, and comment on the suitability of this model.

Solving differential equations

The general solution of a differential equation

Finding an expression for f(x) from a differential equation involving derivatives of f(x) is called solving the equation.

Some differential equations may be solved simply by integration.

EXAMPLE 9.4 Solve the differential equation $\frac{dy}{dx} = 3x^2 - 2$.

SOLUTION

Integrating gives

$$y = \int (3x^2 - 2) \, dx$$
$$y = x^3 - 2x + c$$

Notice that when you solve a differential equation, you get not just one solution, but a whole family of solutions, as c can take any value. This is called the *general solution* of the differential equation. The family of solutions for the differential equation in the example above would be translations in the y direction of the curve $y = x^3 - 2x$. Graphs of members of the family of curves can be found in figure 9.5 on page 217.

The method of separation of variables

It is not difficult to solve a differential equation like the one in Example 9.4, because the right-hand side is a function of x only. So long as the function can be integrated, the equation can be solved.

Now look at the differential equation $\dfrac{dy}{dx} = xy$.

This cannot be solved directly by integration, because the right-hand side is a function of both x and y. However, as you will see in the next example, you can solve this and similar differential equations where the right-hand side consists of a function of x and a function of y multiplied together.

EXAMPLE 9.5 Find, for $y > 0$, the general solution of the differential equation $\dfrac{dy}{dx} = xy$.

SOLUTION

The equation may be rewritten as

$$\frac{1}{y}\frac{dy}{dx} = x$$

so that the right-hand side is now a function of x only.

Integrating both sides with respect to x gives

$$\int \frac{1}{y}\frac{dy}{dx}\,dx = \int x\,dx$$

As $\dfrac{dy}{dx}\,dx$ can be written as dy

$$\int \frac{1}{y}\,dy = \int x\,dx$$

Both sides may now be integrated separately.

$$\ln|y| = \tfrac{1}{2}x^2 + c$$

Since you have been told $y > 0$, you may drop the modulus symbol. In this case, $|y| = y$.

❓ Explain why there is no need to put a constant of integration on both sides of the equation.

You now need to rearrange the solution above to give y in terms of x. Making both sides powers of e gives

$$e^{\ln y} = e^{\frac{1}{2}x^2+c}$$
$$\Rightarrow \quad y = e^{\frac{1}{2}x^2+c}$$
$$\Rightarrow \quad y = e^{\frac{1}{2}x^2}e^c$$

> Notice that the right-hand side is $e^{\frac{1}{2}x^2+c}$ and not $e^{\frac{1}{2}x^2} + e^c$.

This expression can be simplified by replacing e^c with a new constant A.

$$y = Ae^{\frac{1}{2}x^2}$$

Note

Usually the first part of this process is carried out in just one step.

$$\frac{dy}{dx} = xy$$

can immediately be rewritten as

$$\int \frac{1}{y}\,dy = \int x\,dx$$

This method is called *separation of variables*. It can be helpful to do this by thinking of the differential equation as though $\frac{dy}{dx}$ were a fraction and trying to rearrange the equation to obtain all the x terms on one side and all the y terms on the other. Then just insert an integration sign on each side. Remember that dy and dx must both end up in the numerator (top line).

EXAMPLE 9.6 Find the general solution of the differential equation $\frac{dy}{dx} = e^{-y}$.

SOLUTION

Separating the variables gives

$$\int \frac{1}{e^{-y}}\,dy = \int dx$$
$$\Rightarrow \quad \int e^y\,dy = \int dx$$

The right-hand side can be thought of as integrating 1 with respect to x.

$$e^y = x + c$$

Taking logarithms of both sides gives

$$y = \ln|x + c|$$

⚠ $\ln|x + c|$ is not the same as $\ln|x| + c$.

EXERCISE 9B

1. Solve the following differential equations by integration.

   (i) $\dfrac{dy}{dx} = x^2$ (ii) $\dfrac{dy}{dx} = \cos x$

   (iii) $\dfrac{dy}{dx} = e^x$ (iv) $\dfrac{dy}{dx} = \sqrt{x}$

2. Find the general solutions of the following differential equations by separating the variables.

   (i) $\dfrac{dy}{dx} = xy^2$ (ii) $\dfrac{dy}{dx} = \dfrac{x^2}{y}$

   (iii) $\dfrac{dy}{dx} = y$ (iv) $\dfrac{dy}{dx} = e^{x-y}$

   (v) $\dfrac{dy}{dx} = \dfrac{y}{x}$ (vi) $\dfrac{dy}{dx} = x\sqrt{y}$

   (vii) $\dfrac{dy}{dx} = y^2 \cos x$ (viii) $\dfrac{dy}{dx} = \dfrac{x(y^2 + 1)}{y(x^2 + 1)}$

   (ix) $\dfrac{dy}{dx} = xe^y$ (x) $\dfrac{dy}{dx} = \dfrac{x \ln x}{y^2}$

The particular solution of a differential equation

You have already seen that a differential equation has an infinite number of different solutions corresponding to different values of the constant of integration. In Example 9.4, you found that $\dfrac{dy}{dx} = 3x^2 - 2$ had a general solution of $y = x^3 - 2x + c$.

Figure 9.5 shows the curves of the solutions corresponding to some different values of c.

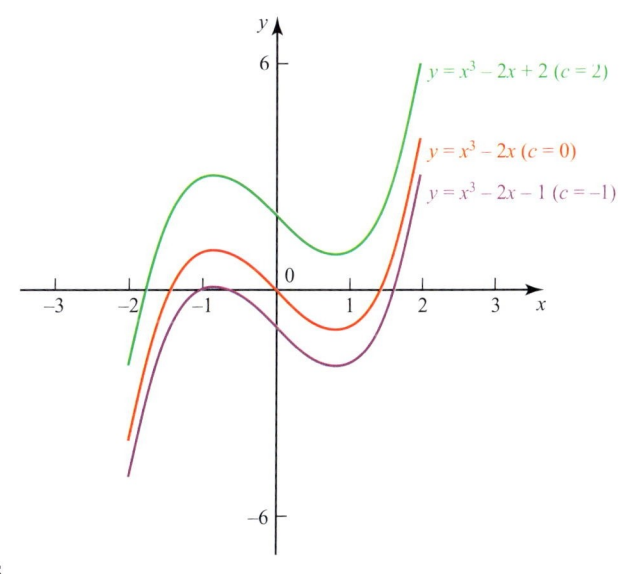

Figure 9.5

If you are given some more information, you can find out which of the possible solutions is the one that matches the situation in question. For example, you might be told that when $x = 1$, $y = 0$. This tells you that the correct solution is the one with the curve that passes through the point $(1, 0)$. You can use this information to find out the value of c for this particular solution by substituting the values $x = 1$ and $y = 0$ into the general solution.

$$y = x^3 - 2x + c$$
$$0 = 1 - 2 + c$$
$$\Rightarrow \quad c = 1$$

So the solution in this case is $y = x^3 - 2x + 1$.

This is called the *particular solution*.

EXAMPLE 9.7

(i) Find the general solution of the differential equation $\dfrac{dy}{dx} = y^2$.

(ii) Find the particular solution for which $y = 1$ when $x = 0$.

SOLUTION

(i) Separating the variables gives $\displaystyle\int \dfrac{1}{y^2}\,dy = \int dx$

$$-\dfrac{1}{y} = x + c$$

The general solution is $y = -\dfrac{1}{x + c}$.

Figure 9.6 shows a set of solution curves.

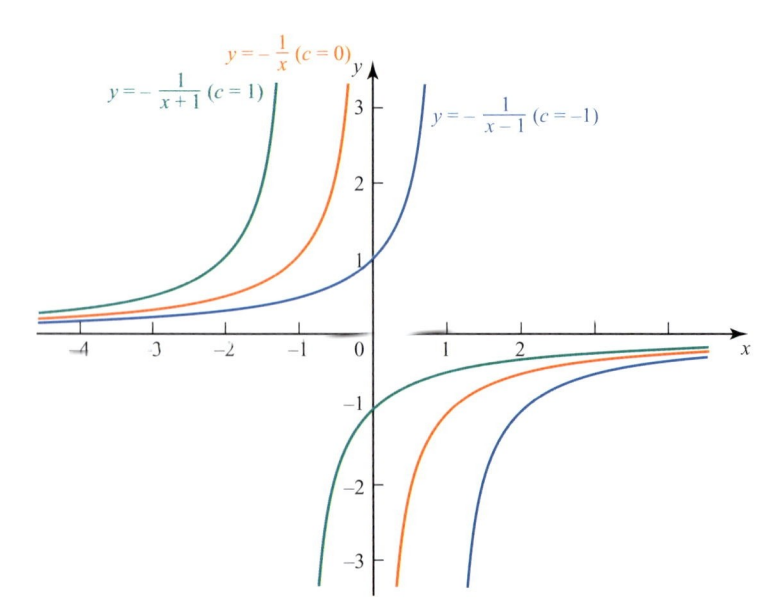

Figure 9.6

(ii) When $x = 0$, $y = 1$, which gives

$$1 = -\frac{1}{c} \implies c = -1.$$

So the particular solution is

$$y = -\frac{1}{x-1} \quad \text{or} \quad y = \frac{1}{1-x}$$

This is the blue curve illustrated in figure 9.6.

EXAMPLE 9.8

The acceleration of an object is inversely proportional to its velocity at any given time and the direction of motion is taken to be positive.
When the velocity is $1 \, \text{m s}^{-1}$, the acceleration is $3 \, \text{m s}^{-2}$.

(i) Find a differential equation to model this situation.
(ii) Find the particular solution to this differential equation for which the initial velocity is $2 \, \text{m s}^{-1}$.
(iii) In this case, how long does the object take to reach a velocity of $8 \, \text{m s}^{-1}$?

SOLUTION

(i) $\dfrac{dv}{dt} = \dfrac{k}{v}$

When $v = 1$, $\dfrac{dv}{dt} = 3$ so $k = 3$, which gives $\dfrac{dv}{dt} = \dfrac{3}{v}$.

(ii) Separating the variables:

$$\int v \, dv = \int 3 \, dt$$
$$\tfrac{1}{2} v^2 = 3t + c$$

When $t = 0$, $v = 2$ so $c = 2$, which gives

$$\tfrac{1}{2} v^2 = 3t + 2$$
$$v^2 = 6t + 4$$

Since the direction of motion is positive

$$v = \sqrt{6t + 4}$$

(iii) When $v = 8$ $\quad 64 = 6t + 4$
$\qquad\qquad\qquad\quad 60 = 6t \quad \implies \quad t = 10$

The object takes 10 seconds to reach a velocity of $8 \, \text{m s}^{-1}$.

The graph of the particular solution is shown in figure 9.7.

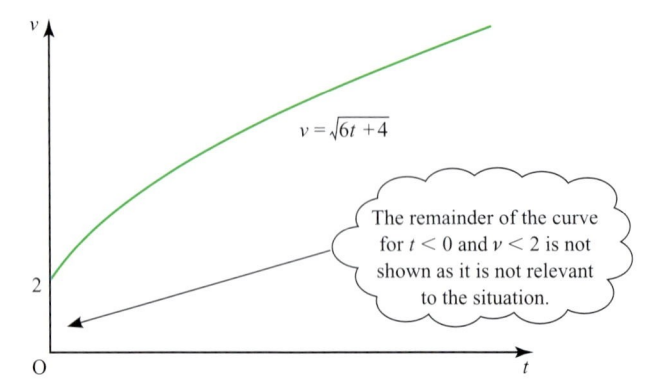

Figure 9.7

The remainder of the curve for $t < 0$ and $v < 2$ is not shown as it is not relevant to the situation.

Sometimes you will be asked to *verify* the solution of a differential equation. In that case you are expected to do two things:

- substitute the solution in the differential equation and show that it works
- show that the solution fits the conditions you have been given.

EXAMPLE 9.9

Show that $\sin y = x$ is a solution of the differential equation

$$\frac{dy}{dx} = \frac{1}{\sqrt{1-x^2}}$$

given that $y = 0$ when $x = 0$.

SOLUTION

$$\sin y = x$$

$$\Rightarrow \quad \cos y \frac{dy}{dx} = 1$$

$$\Rightarrow \quad \frac{dy}{dx} = \frac{1}{\cos y}$$

Substituting into the differential equation $\frac{dy}{dx} = \frac{1}{\sqrt{1-x^2}}$:

LHS: $\dfrac{1}{\cos y}$

RHS: $\dfrac{1}{\sqrt{1-x^2}} = \dfrac{1}{\sqrt{1-\sin^2 y}} = \dfrac{1}{\cos y}$

So the solution fits the differential equation.

Substituting $x = 0$ into the solution $\sin y = x$ gives $\sin y = 0$ and this is satisfied by $y = 0$.

So the solution also fits the particular conditions.

EXERCISE 9C

1 Find the particular solution of each of the following differential equations.

(i) $\dfrac{dy}{dx} = x^2 - 1$ $y = 2$ when $x = 3$

(ii) $\dfrac{dy}{dx} = x^2 y$ $y = 1$ when $x = 0$

(iii) $\dfrac{dy}{dx} = xe^{-y}$ $y = 0$ when $x = 0$

(iv) $\dfrac{dy}{dx} = y^2$ $y = 1$ when $x = 1$

(v) $\dfrac{dy}{dx} = x(y+1)$ $y = 0$ when $x = 1$

(vi) $\dfrac{dy}{dx} = y^2 \sin x$ $y = 1$ when $x = 0$

2 A cold liquid at temperature $\theta\,°C$, where $\theta < 20$, is standing in a warm room. The temperature of the liquid obeys the differential equation

$$\dfrac{d\theta}{dt} = 2(20 - \theta)$$

where the time t is measured in hours.

(i) Find the general solution of this differential equation.
(ii) Find the particular solution for which $\theta = 5$ when $t = 0$.
(iii) In this case, how long does the liquid take to reach a temperature of 18°C?

3 A population of rabbits increases so that the number of rabbits N (in hundreds), after t years is modelled by the differential equation

$$\dfrac{dN}{dt} = N.$$

(i) Find the general solution for N in terms of t.
(ii) Find the particular solution for which $N = 10$ when $t = 0$.
(iii) What will happen to the number of rabbits when t becomes very large? Why is this not a realistic model for an actual population of rabbits?

4 An object is moving so that its velocity $v\left(= \dfrac{ds}{dt}\right)$ is inversely proportional to its displacement s from a fixed point.
If its velocity is $1\,\text{m s}^{-1}$ when its displacement is $2\,\text{m}$, find a differential equation to model the situation.
Find the general solution of your differential equation.

5 (i) Write $\dfrac{1}{y(3-y)}$ in partial fractions.

(ii) Find $\displaystyle\int \dfrac{1}{y(3-y)}\,dy$.

(iii) Solve the differential equation

$$x\dfrac{dy}{dx} = y(3 - y)$$

where $x = 2$ when $y = 2$, giving y as a function of x.

[MEI]

6 Given that *k* is a constant, find the solution of the differential equation

$$\frac{dy}{dt} + ky = 2k$$

for which $y = 3$ when $t = 0$.

Sketch the graph of *y* against $|kt|$, making clear how it behaves for large values of $|kt|$.

[MEI]

7 A colony of bacteria which is initially of size 1500 increases at a rate proportional to its size so that, after *t* hours, its population *N* satisfies the equation $\frac{dN}{dt} = kN$.

  (i) If the size of the colony increases to 3000 in 20 hours, solve the differential equation to find *N* in terms of *t*.

  (ii) What size is the colony when $t = 80$?

  (iii) How long did it take, to the nearest minute, for the population to increase from 2000 to 3000?

[MEI]

8 (i) Show that $\frac{x^2+1}{x^2-1} = 1 + \frac{2}{x^2-1}$.

  (ii) Find the partial fractions for $\frac{2}{(x-1)(x+1)}$.

  (iii) Solve the differential equation

$$(x^2 - 1)\frac{dy}{dx} = -(x^2 + 1)y \qquad \text{(where } x > 1\text{)}$$

given that $y = 1$ when $x = 3$. Express *y* as a function of *x*.

[MEI]

9 A patch of oil pollution in the sea is approximately circular in shape. When first seen its radius was 100 m and its radius was increasing at a rate of 0.5 m per minute. At a time *t* minutes later, its radius is *r* metres. An expert believes that, if the patch is untreated, its radius will increase at a rate which is proportional to $\frac{1}{r^2}$.

  (i) Write down a differential equation for this situation, using a constant of proportionality, *k*.

  (ii) Using the initial conditions, find the value of *k*. Hence calculate the expert's prediction of the radius of the oil patch after 2 hours.

The expert thinks that if the oil patch is treated with chemicals then its radius will increase at a rate which is proportional to $\frac{1}{r^2(2+t)}$.

  (iii) Write down a differential equation for this new situation and, using the same initial conditions as before, find the value of the new constant of proportionality.

  (iv) Calculate the expert's prediction of the radius of the treated oil patch after 2 hours.

[MEI]

10 (i) Express $\dfrac{1}{(2-x)(1+x)}$ in partial fractions.

An industrial process creates a chemical C. At time t hours after the start of the process the amount of C produced is x kg. The rate at which C is produced is given by the differential equation

$$\frac{dx}{dt} = k(2-x)(1+x)e^{-t},$$

where k is a constant.

(ii) When $t = 0$, $x = 0$ and the rate of production of C is $\frac{2}{3}$ kg per hour. Calculate the value of k.

(iii) Show that $\ln\left(\dfrac{1+x}{2-x}\right) = -e^{-t} + 1 - \ln 2$, provided that $x < 2$.

(iv) Find, in hours, the time taken to produce 0.5 kg of C, giving your answer correct to 2 decimal places.

(v) Show that there is a finite limit to the amount of C which this process can produce, however long it runs, and determine the value of this limit.

[MEI]

11 (i) Use integration by parts to evaluate $\int 4x \cos 2x \, dx$.

(ii) Use part **(i)**, together with a suitable expression for $\cos^2 x$, to show that

$$\int 8x \cos^2 x \, dx = 2x^2 + 2x \sin 2x + \cos 2x + c.$$

(iii) Find the solution of the differential equation

$$\frac{dy}{dx} = \frac{8x\cos^2 x}{y}$$

which satisfies $y = \sqrt{3}$ when $x = 0$.

(iv) Show that any point (x, y) on the graph of this solution which satisfies $\sin 2x = 1$ also lies on one of the lines $y = 2x + 1$ or $y = -2x - 1$.

[MEI]

12 (i) Express $\dfrac{1-x}{(1+x)(1+x^2)}$ in the form $\dfrac{A}{1+x} + \dfrac{Bx+C}{1+x^2}$.

(ii) Hence show that the solution of the differential equation

$$\frac{dy}{dx} = \frac{y(1-x)}{(1+x)(1+x^2)},$$

given that $y = 1$ when $x = 0$, is

$$y = \frac{1+x}{\sqrt{1+x^2}}.$$

(iii) Find the first three terms of the binomial expansion of $\dfrac{1}{\sqrt{1+x^2}}$.

Hence find a polynomial approximation for $y = \dfrac{1+x}{\sqrt{1+x^2}}$ up to the term in x^5.

[MEI]

13 (i) Express $\dfrac{1}{(3x-1)x}$ in partial fractions.

A model for the way in which a population of animals in a closed environment varies with time is given, for $P > \tfrac{1}{3}$, by

$$\frac{dP}{dt} = \tfrac{1}{2}(3P^2 - P)\sin t$$

where P is the size of the population in thousands at time t.

(ii) Given that $P = \tfrac{1}{2}$ when $t = 0$, use the method of separation of variables to show that

$$\ln\left(\frac{3P-1}{P}\right) = \tfrac{1}{2}(1 - \cos t).$$

(iii) Calculate the smallest positive value of t for which $P = 1$.

(iv) Rearrange the equation at the end of part **(ii)** to show that

$$P = \frac{1}{3 - e^{\tfrac{1}{2}(1-\cos t)}}.$$

Hence find the two values between which the number of animals in the population oscillates.

[MEI]

14 (i) Use integration by parts to show that

$$\int \ln x \, dx = x \ln x - x + c.$$

(ii) Differentiate $\ln(\sin x)$ with respect to x, for $0 < x < \tfrac{\pi}{2}$.

Hence write down $\int \cot x \, dx$, for $0 < x < \tfrac{\pi}{2}$.

(iii) For $x > 0$ and $0 < y < \tfrac{\pi}{2}$, the variables y and x are connected by the differential equation

$$\frac{dy}{dx} = \frac{\ln x}{\cot y},$$

and $y = \tfrac{\pi}{6}$ when $x = e$.

Find the value of y when $x = 1$, giving your answer correct to 3 significant figures.

Use the differential equation to show that this value of y is a stationary value, and determine its nature.

[MEI]

15 (i) Using partial fractions, find

$$\int \frac{1}{y(4-y)}\,dy.$$

(ii) Given that $y = 1$ when $x = 0$, solve the differential equation

$$\frac{dy}{dx} = y(4-y),$$

obtaining an expression for y in terms of x.

(iii) State what happens to the value of y if x becomes very large and positive.

[Cambridge International AS & A Level Mathematics 9709, Paper 3 Q8 June 2005]

16 The temperature of a quantity of liquid at time t is θ. The liquid is cooling in an atmosphere whose temperature is constant and equal to A. The rate of decrease of θ is proportional to the temperature difference $(\theta - A)$. Thus θ and t satisfy the differential equation

$$\frac{d\theta}{dt} = -k(\theta - A),$$

where k is a positive constant.

(i) Find, in any form, the solution of this differential equation, given that $\theta = 4A$ when $t = 0$.

(ii) Given also that $\theta = 3A$ when $t = 1$, show that $k = \ln\frac{3}{2}$.

(iii) Find θ in terms of A when $t = 2$, expressing your answer in its simplest form.

[Cambridge International AS & A Level Mathematics 9709, Paper 32 Q9 November 2009]

17 The variables x and t are related by the differential equation

$$e^{2t}\frac{dx}{dt} = \cos^2 x,$$

where $t \geq 0$. When $t = 0$, $x = 0$.

(i) Solve the differential equation, obtaining an expression for x in terms of t.

(ii) State what happens to the value of x when t becomes very large.

(iii) Explain why x increases as t increases.

[Cambridge International AS & A Level Mathematics 9709, Paper 32 Q7 June 2010]

18 An underground storage tank is being filled with liquid as shown in the diagram (overleaf). Initially the tank is empty. At time t hours after filling begins, the volume of liquid is $V\,\text{m}^3$ and the depth of liquid is $h\,\text{m}$. It is given that $V = \frac{4}{3}h^3$.

The liquid is poured in at a rate of $20\,\text{m}^3$ per hour, but owing to leakage, liquid is lost at a rate proportional to h^2. When $h = 1$, $\frac{dh}{dt} = 4.95$.

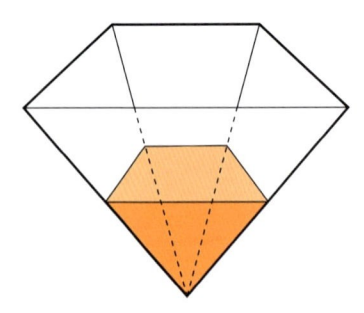

(i) Show that h satisfies the differential equation

$$\frac{dh}{dt} = \frac{5}{h^2} - \frac{1}{20}.$$

(ii) Verify that $\dfrac{20h^2}{100-h^2} \equiv -20 + \dfrac{2000}{(10-h)(10+h)}$.

(iii) Hence solve the differential equation in part (i), obtaining an expression for t in terms of h.

[Cambridge International AS & A Level Mathematics 9709, Paper 3 Q8 November 2008]

INVESTIGATION

Investigate the tea cooling problem introduced on page 208. You will need to make some assumptions about the initial temperature of the tea and the temperature of the room.

What difference would it make if you were to add some cold milk to the tea and then leave it to cool?

Would it be better to allow the tea to cool first before adding the milk?

KEY POINTS

1. A differential equation is an equation involving derivatives such as

   $\dfrac{dy}{dx}$ and $\dfrac{d^2y}{dx^2}$

2. A first-order differential equation involves a first derivative only.

3. Some first-order differential equations may be solved by separating the variables.

4. A general solution is one in which the constant of integration is left in the solution, and a particular solution is one in which additional information is used to calculate the constant of integration.

5. A general solution may be represented by a family of curves, a particular solution by a particular member of that family.

10 Vectors

> By relieving the brain of all unnecessary work, a good notation sets it free to concentrate on more advanced problems.
>
> A.N. Whitehead, 1861–1947

The vector equation of a line

Two-dimensional co-ordinate geometry involves the study of points, given as co-ordinates, and lines, given as cartesian equations. The same work may also be treated using vectors.

The co-ordinates of a point, say (3, 4), are replaced by its position vector $\begin{pmatrix} 3 \\ 4 \end{pmatrix}$ or $3\mathbf{i} + 4\mathbf{j}$. The cartesian equation of a line is replaced by its vector form, and this is introduced on page 231.

Since most two-dimensional problems are readily solved using the methods of cartesian co-ordinate geometry, as introduced in *Pure Mathematics 1*, Chapter 2, why go to the trouble of relearning it all in vectors? The answer is that vector methods are very much easier to use in many three-dimensional situations than cartesian methods are. In preparation for that, we review some familiar two-dimensional work in this section, comparing cartesian and vector methods.

The vector joining two points

In figure 10.1, start by looking at two points A(2, −1) and B(4, 3); that is the points with position vectors $\overrightarrow{OA} = \begin{pmatrix} 2 \\ -1 \end{pmatrix}$ and $\overrightarrow{OB} = \begin{pmatrix} 4 \\ 3 \end{pmatrix}$, alternatively $2\mathbf{i} - \mathbf{j}$ and $4\mathbf{i} + 3\mathbf{j}$.

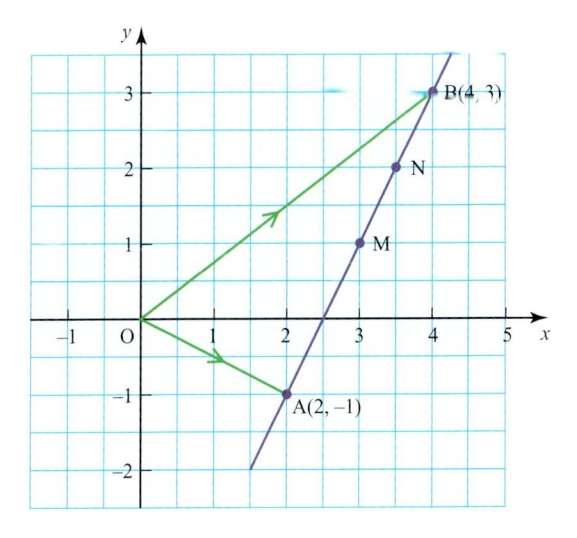

Figure 10.1

The vector joining A to B is $\overrightarrow{AB}$ and this is given by

$$\overrightarrow{AB} = \overrightarrow{AO} + \overrightarrow{OB}$$
$$= -\overrightarrow{OA} + \overrightarrow{OB}$$
$$= \overrightarrow{OB} - \overrightarrow{OA}$$
$$= \begin{pmatrix} 4 \\ 3 \end{pmatrix} - \begin{pmatrix} 2 \\ -1 \end{pmatrix} = \begin{pmatrix} 2 \\ 4 \end{pmatrix}$$

Since $\overrightarrow{AB} = \begin{pmatrix} 2 \\ 4 \end{pmatrix}$, then it follows that the length of AB is given by

$$|\overrightarrow{AB}| = \sqrt{2^2 + 4^2}$$
$$= \sqrt{20}.$$

You can find the position vectors of points along AB as follows.

The mid-point, M, has position vector $\overrightarrow{OM}$, given by

$$\overrightarrow{OM} = \overrightarrow{OA} + \tfrac{1}{2}\overrightarrow{AB}$$
$$= \begin{pmatrix} 2 \\ -1 \end{pmatrix} + \tfrac{1}{2}\begin{pmatrix} 2 \\ 4 \end{pmatrix}$$
$$= \begin{pmatrix} 3 \\ 1 \end{pmatrix}.$$

In the same way, the position vector of the point N, three-quarters of the distance from A to B, is given by

$$\overrightarrow{ON} = \begin{pmatrix} 2 \\ -1 \end{pmatrix} + \tfrac{3}{4}\begin{pmatrix} 2 \\ 4 \end{pmatrix}$$
$$= \begin{pmatrix} 3\tfrac{1}{2} \\ 2 \end{pmatrix}$$

and it is possible to find the position vector of any other point of subdivision of the line AB in the same way.

P A point P has position vector $\overrightarrow{OP} = \overrightarrow{OA} + \lambda\overrightarrow{AB}$ where λ is a fraction.

Show that this can be expressed as

$$\overrightarrow{OP} = (1 - \lambda)\overrightarrow{OA} + \lambda\overrightarrow{OB}.$$

The vector equation of a line

It is now a small step to go from finding the position vector of any point on the line AB to finding the vector form of the equation of the line AB. To take this step, you will find it helpful to carry out the following activity.

ACTIVITY 10.1 The position vectors of a set of points are given by

$$\mathbf{r} = \begin{pmatrix} 2 \\ -1 \end{pmatrix} + \lambda \begin{pmatrix} 2 \\ 4 \end{pmatrix}$$

λ is the Greek letter 'lamda'.

where λ is a parameter which may take any value.

(i) Show that $\lambda = 2$ corresponds to the point with position vector $\begin{pmatrix} 6 \\ 7 \end{pmatrix}$.

(ii) Find the position vectors of points corresponding to values of λ of $-2, -1, 0, \frac{1}{2}, \frac{3}{4}, 1, 3$.

(iii) Mark all your points on a sheet of squared paper and show that when they are joined up they give the line AB in figure 10.2.

(iv) State what values of λ correspond to the points A, B, M and N.

(v) What can you say about the position of the point if
 (a) $0 < \lambda < 1$?
 (b) $\lambda > 1$?
 (c) $\lambda < 0$?

This activity should have convinced you that

$$\mathbf{r} = \begin{pmatrix} 2 \\ -1 \end{pmatrix} + \lambda \begin{pmatrix} 2 \\ 4 \end{pmatrix}$$

The number λ is called a parameter and it can take any value. Of course, you can use other letters for the parameter such as μ, s and t.

is the equation of the line passing through $(2, -1)$ and $(4, 3)$, written in vector form.

You may find it helpful to think of this in these terms.

1 Start at the origin.
2 Move to the point A with position vector $\begin{pmatrix} 2 \\ -1 \end{pmatrix}$.
and then
3 Move λ steps of $\begin{pmatrix} 2 \\ 4 \end{pmatrix}$ (i.e. in the direction $\overrightarrow{AB}$). λ need not be a whole number and may be negative.

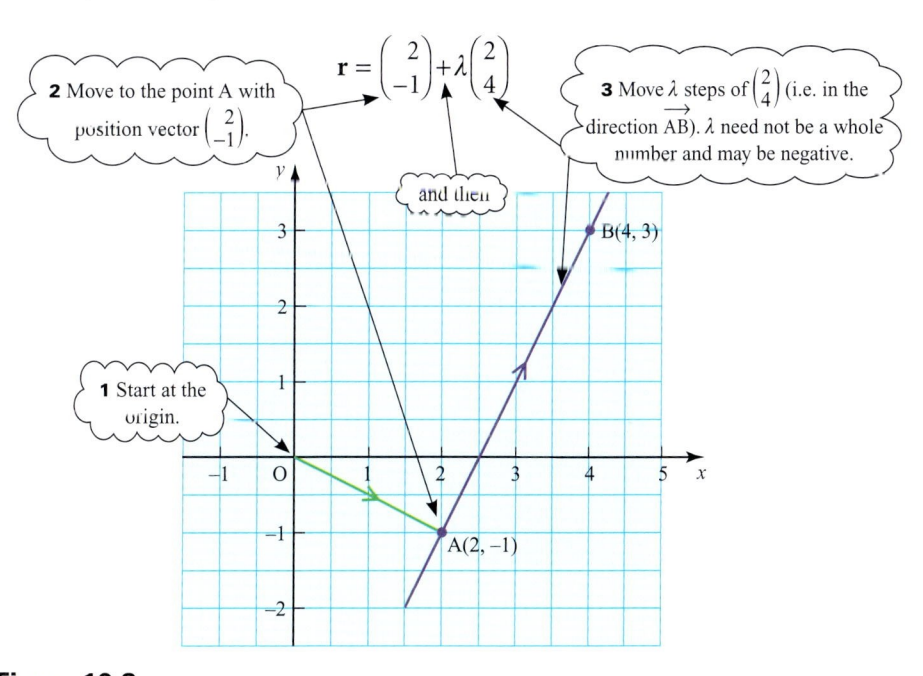

Figure 10.2

You should also have noticed that when:

$\lambda = 0$ the point corresponds to the point A
$\lambda = 1$ the point corresponds to the point B
$0 < \lambda < 1$ the point lies between A and B
$\lambda > 1$ the point lies beyond B
$\lambda < 0$ the point lies beyond A.

The vector form of the equation is not unique; there are many (in fact infinitely many) different ways in which the equation of any particular line may be expressed. There are two reasons for this: direction and location.

Direction

The direction of the line in the example is $\begin{pmatrix} 2 \\ 4 \end{pmatrix}$. That means that for every 2 units along (in the **i** direction), the line goes up 4 units (in the **j** direction). This is equivalent to stating that for every 1 unit along, the line goes up 2 units, corresponding to the equation

$$\mathbf{r} = \begin{pmatrix} 2 \\ -1 \end{pmatrix} + \lambda \begin{pmatrix} 1 \\ 2 \end{pmatrix}.$$

The only difference is that the two equations have different values of λ for particular points. In the first equation, point B, with position vector $\begin{pmatrix} 4 \\ 3 \end{pmatrix}$, corresponds to a value of λ of 1. In the second equation, the value of λ for B is 2. The direction $\begin{pmatrix} 2 \\ 4 \end{pmatrix}$ is the same as $\begin{pmatrix} 1 \\ 2 \end{pmatrix}$, or as any multiple of $\begin{pmatrix} 1 \\ 2 \end{pmatrix}$ such as $\begin{pmatrix} 3 \\ 6 \end{pmatrix}$, $\begin{pmatrix} -5 \\ -10 \end{pmatrix}$ or $\begin{pmatrix} 100.5 \\ 201 \end{pmatrix}$. Any of these could be used in the vector equation of the line.

Location

In the equation

$$\mathbf{r} = \begin{pmatrix} 2 \\ -1 \end{pmatrix} + \lambda \begin{pmatrix} 2 \\ 4 \end{pmatrix}$$

$\begin{pmatrix} 2 \\ -1 \end{pmatrix}$ is the position vector of the point A on the line, and represents the point at which the line was joined. However, this could have been any other point on the line, such as M(3, 1), B(4, 3), etc. Consequently

$$\mathbf{r} = \begin{pmatrix} 3 \\ 1 \end{pmatrix} + \lambda \begin{pmatrix} 2 \\ 4 \end{pmatrix}$$

and

$$\mathbf{r} = \begin{pmatrix} 4 \\ 3 \end{pmatrix} + \lambda \begin{pmatrix} 2 \\ 4 \end{pmatrix}$$

are also equations of the same line, and there are infinitely many other possibilities, one corresponding to each point on the line.

Notes

1. It is usual to refer to any valid vector form of the equation as *the* vector equation of the line even though it is not unique.

2. It is often a good idea to give the direction vector in its simplest integer form: for example, replacing $\begin{pmatrix} 2 \\ 4 \end{pmatrix}$ with $\begin{pmatrix} 1 \\ 2 \end{pmatrix}$.

The general vector form of the equation of a line

If A and B are points with position **a** and **b**, then the equation

$$\mathbf{r} = \overrightarrow{OA} + \lambda \overrightarrow{AB}$$

may be written as $\quad \mathbf{r} = \mathbf{a} + \lambda(\mathbf{b} - \mathbf{a})$

which implies $\quad \mathbf{r} = (1 - \lambda)\mathbf{a} + \lambda \mathbf{b}$.

This is the general vector form of the equation of the line joining two points.

ACTIVITY 10.2 Plot the following lines on the same sheet of squared paper. When you have done so, explain why certain among them are the same as each other, others are parallel to each other, and others are in different directions.

(i) $\mathbf{r} = \begin{pmatrix} 2 \\ -1 \end{pmatrix} + \lambda \begin{pmatrix} 1 \\ 2 \end{pmatrix}$
(ii) $\mathbf{r} = \begin{pmatrix} 2 \\ -1 \end{pmatrix} + \lambda \begin{pmatrix} -1 \\ 2 \end{pmatrix}$
(iii) $\mathbf{r} = \begin{pmatrix} 0 \\ 2 \end{pmatrix} + \lambda \begin{pmatrix} 1 \\ 2 \end{pmatrix}$

(iv) $\mathbf{r} = \begin{pmatrix} 1 \\ -3 \end{pmatrix} + \lambda \begin{pmatrix} 3 \\ 6 \end{pmatrix}$
(v) $\mathbf{r} = \begin{pmatrix} 4 \\ 3 \end{pmatrix} + \lambda \begin{pmatrix} 1 \\ -2 \end{pmatrix}$

The same methods can be used to find the vector equation of a line in three dimensions, as shown in this example.

EXAMPLE 10.1 The co-ordinates of A and B are (−2, 4, 1) and (2, 1, 3) respectively.

(i) Find the vector equation of the line AB.
(ii) Does the point P(6, −2, 7) lie on the line AB?
(iii) The point N lies on the line AB.

Given that $3|\overrightarrow{AN}| = |\overrightarrow{NB}|$ find the co-ordinates of N.

SOLUTION

(i) $\mathbf{a} = \overrightarrow{OA} = \begin{pmatrix} -2 \\ 4 \\ 1 \end{pmatrix}$ and $\mathbf{b} = \overrightarrow{OB} = \begin{pmatrix} 2 \\ 1 \\ 3 \end{pmatrix}$

$\overrightarrow{AB} = \mathbf{b} - \mathbf{a} = \begin{pmatrix} 2 \\ 1 \\ 3 \end{pmatrix} - \begin{pmatrix} -2 \\ 4 \\ 1 \end{pmatrix} = \begin{pmatrix} 4 \\ -3 \\ 2 \end{pmatrix}$

The vector equation of a line can be written as

$$\mathbf{r} = \overrightarrow{OA} + \lambda \overrightarrow{AB}$$

$$\Rightarrow \quad \mathbf{r} = \begin{pmatrix} -2 \\ 4 \\ 1 \end{pmatrix} + \lambda \begin{pmatrix} 4 \\ -3 \\ 2 \end{pmatrix}$$

There are other ways of writing this equation, for example

$$\mathbf{r} = \begin{pmatrix} 2 \\ 1 \\ 3 \end{pmatrix} + \lambda \begin{pmatrix} 4 \\ -3 \\ 2 \end{pmatrix} \text{ or } \mathbf{r} = \begin{pmatrix} -6 \\ 7 \\ -1 \end{pmatrix} + \lambda \begin{pmatrix} 4 \\ -3 \\ 2 \end{pmatrix}$$

but they are all equivalent to each other.

(ii) If P lies on the line AB then for some value of λ

$$\begin{pmatrix} x \\ y \\ z \end{pmatrix} = \begin{pmatrix} 6 \\ -2 \\ 7 \end{pmatrix} = \begin{pmatrix} -2 \\ 4 \\ 1 \end{pmatrix} + \lambda \begin{pmatrix} 4 \\ -3 \\ 2 \end{pmatrix}$$

Find the value of λ for the x co-ordinate.

$$x: \quad 6 = -2 + 4\lambda \quad \Rightarrow \quad \lambda = 2$$

Then check whether this value of λ gives a y co-ordinate of -2 and a z co-ordinate of 7.

$$y: \quad -2 = 4 - 3 \times 2$$
$$z: \quad 7 \neq 1 + 2 \times 2$$

So the point P(6, −2, 7) does not lie on the line.

(iii) Since $3|\overrightarrow{AN}| = |\overrightarrow{NB}|$, N must lie $\frac{1}{4}$ of the way along the line AB so the value of λ is $\frac{1}{4}$.

$$\overrightarrow{ON} = \overrightarrow{OA} + \tfrac{1}{4}\overrightarrow{AB}$$

$$\overrightarrow{ON} = \begin{pmatrix} -2 \\ 4 \\ 1 \end{pmatrix} + \frac{1}{4}\begin{pmatrix} 4 \\ -3 \\ 2 \end{pmatrix} = \begin{pmatrix} -1 \\ 3\tfrac{1}{4} \\ 1\tfrac{1}{2} \end{pmatrix}$$

So the co-ordinates of N are (−1, 3.25, 1.5).

EXERCISE 10A

1 For each of these pairs of points, A and B, write down:

 (a) the vector $\overrightarrow{AB}$

 (b) $|\overrightarrow{AB}|$

 (c) the position vector of the mid-point of AB.

 (i) A is (2, 3), B is (4, 11).
 (ii) A is (4, 3), B is (0, 0).
 (iii) A is (−2, −1), B is (4, 7).
 (iv) A is (−3, 4), B is (3, −4).
 (v) A is (−10, −8), B is (−5, 4).

2 Find the equation of each of these lines in vector form.

  (i) Joining (2, 1) to (4, 5).
  (ii) Joining (3, 5) to (0, 8).
  (iii) Joining (−6, −6) to (4, 4).
  (iv) Through (5, 3) in the same direction as **i** + **j**.
  (v) Through (2, 1) parallel to 6**i** + 3**j**.
  (vi) Through (0, 0) parallel to $\begin{pmatrix} -1 \\ 4 \end{pmatrix}$.
  (vii) Joining (0, 0) to (−2, 8).
  (viii) Joining (3, −12) to (−1, 4).

3 Find the equation of each of these lines in vector form.

  (i) Through (2, 4, −1) in the direction $\begin{pmatrix} 3 \\ 6 \\ 4 \end{pmatrix}$
  (ii) Through (1, 0, −1) in the direction $\begin{pmatrix} 1 \\ 0 \\ 0 \end{pmatrix}$
  (iii) Through (1, 0, 4) and (6, 3, −2)
  (iv) Through (0, 0, 1) and (2, 1, 4)
  (v) Through (1, 2, 3) and (−2, −4, −6)

4 Determine whether the given point P lies on the line.

  (i) P(5, 1, 4) and the line $\mathbf{r} = \begin{pmatrix} 1 \\ 3 \\ 4 \end{pmatrix} + \lambda \begin{pmatrix} 2 \\ -1 \\ 0 \end{pmatrix}$
  (ii) P(−1, 5, 1) and the line $\mathbf{r} = \begin{pmatrix} 1 \\ 3 \\ 4 \end{pmatrix} + \lambda \begin{pmatrix} 2 \\ -2 \\ 3 \end{pmatrix}$
  (iii) P(−5, 3, 12) and the line $\mathbf{r} = \begin{pmatrix} 1 \\ 0 \\ -2 \end{pmatrix} + \lambda \begin{pmatrix} -2 \\ 1 \\ 5 \end{pmatrix}$
  (iv) P(9, 0, −6) and the line $\mathbf{r} = \begin{pmatrix} 1 \\ 2 \\ 0 \end{pmatrix} + \lambda \begin{pmatrix} 4 \\ -1 \\ -2 \end{pmatrix}$
  (v) P(−9, −2, −17) and the line $\mathbf{r} = \begin{pmatrix} 1 \\ 3 \\ -2 \end{pmatrix} + \lambda \begin{pmatrix} 2 \\ 1 \\ 3 \end{pmatrix}$

5 The co-ordinates of three points are A(−1, −2, 1), B(−3, 4, −5) and C(0, −2, 4).

  (i) Find a vector equation of the line AB.
  (ii) Find the co-ordinates of the mid-point M of AB.
  (iii) The point N lies on BC.
  Given that $2|\overrightarrow{BN}| = |\overrightarrow{NC}|$, find the equation of the line MN.

The intersection of two lines

Hold a pen and a pencil to represent two distinct straight lines as follows:

- hold them to represent parallel lines;
- hold them to represent intersecting lines;
- hold them to represent lines which are not parallel and which do not intersect (even if you extend them).

In three-dimensional space two or more straight lines which are not parallel and which do not meet are known as *skew* lines. In a plane two distinct lines are either parallel or intersecting, but in three dimensions there are three possibilities: the lines may be parallel, or intersecting, or skew. The next example illustrates a method of finding whether two lines meet, and, if they do meet, the co-ordinates of the point of intersection.

EXAMPLE 10.2 Find the position vector of the point where the following lines intersect.

$$\mathbf{r} = \begin{pmatrix} 2 \\ 3 \end{pmatrix} + \lambda \begin{pmatrix} 1 \\ 2 \end{pmatrix} \quad \text{and} \quad \mathbf{r} = \begin{pmatrix} 6 \\ 1 \end{pmatrix} + \mu \begin{pmatrix} 1 \\ -3 \end{pmatrix}$$

Note here that different letters are used for the parameters in the two equations to avoid confusion.

SOLUTION

When the lines intersect, the position vector is the same for each of them.

$$\mathbf{r} = \begin{pmatrix} x \\ y \end{pmatrix} = \begin{pmatrix} 2 \\ 3 \end{pmatrix} + \lambda \begin{pmatrix} 1 \\ 2 \end{pmatrix} = \begin{pmatrix} 6 \\ 1 \end{pmatrix} + \mu \begin{pmatrix} 1 \\ -3 \end{pmatrix}$$

This gives two simultaneous equations for λ and μ.

$$x: \quad 2 + \lambda = 6 + \mu \quad \Rightarrow \quad \lambda - \mu = 4$$
$$y: \quad 3 + 2\lambda = 1 - 3\mu \quad \Rightarrow \quad 2\lambda + 3\mu = -2$$

Solving these gives $\lambda = 2$ and $\mu = -2$. Substituting in either equation gives

$$\mathbf{r} = \begin{pmatrix} 4 \\ 7 \end{pmatrix}$$

which is the position vector of the point of intersection.

EXAMPLE 10.3 Find the co-ordinates of the point of intersection of the lines joining A(1, 6) to B(4, 0), and C(1, 1) to D(5, 3).

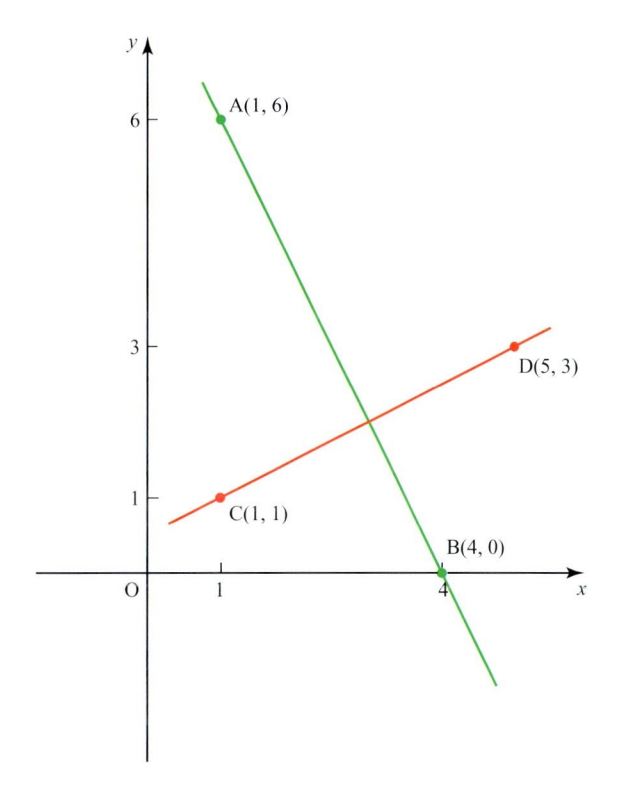

Figure 10.3

SOLUTION

$$\overrightarrow{AB} = \begin{pmatrix} 4 \\ 0 \end{pmatrix} - \begin{pmatrix} 1 \\ 6 \end{pmatrix} = \begin{pmatrix} 3 \\ -6 \end{pmatrix}$$

and so the vector equation of line AB is

$$\mathbf{r} = \overrightarrow{OA} + \lambda \overrightarrow{AB}$$

$$\mathbf{r} = \begin{pmatrix} 1 \\ 6 \end{pmatrix} + \lambda \begin{pmatrix} 3 \\ -6 \end{pmatrix}$$

$$\overrightarrow{CD} = \begin{pmatrix} 5 \\ 3 \end{pmatrix} - \begin{pmatrix} 1 \\ 1 \end{pmatrix} = \begin{pmatrix} 4 \\ 2 \end{pmatrix}$$

and so the vector equation of line CD is

$$\mathbf{r} = \overrightarrow{OC} + \mu \overrightarrow{CD}$$

$$\mathbf{r} = \begin{pmatrix} 1 \\ 1 \end{pmatrix} + \mu \begin{pmatrix} 4 \\ 2 \end{pmatrix}$$

The intersection of these lines is at

$$\mathbf{r} = \begin{pmatrix} 1 \\ 6 \end{pmatrix} + \lambda \begin{pmatrix} 3 \\ -6 \end{pmatrix} = \begin{pmatrix} 1 \\ 1 \end{pmatrix} + \mu \begin{pmatrix} 4 \\ 2 \end{pmatrix}$$

x: $1 + 3\lambda = 1 + 4\mu$ $\Rightarrow$ $3\lambda - 4\mu = 0$ ①

y: $6 - 6\lambda = 1 + 2\mu$ $\Rightarrow$ $6\lambda + 2\mu = 5$ ②

Solve ① and ② simultaneously:

①: $\quad 3\lambda - 4\mu = 0$
② × 2: $\quad 12\lambda + 4\mu = 10$
Add: $\quad 15\lambda \quad\;\; = 10$

$\Rightarrow \quad \lambda = \tfrac{2}{3}$

Substitute $\lambda = \tfrac{2}{3}$ in the equation for AB:

$\Rightarrow \mathbf{r} = \begin{pmatrix} 1 \\ 6 \end{pmatrix} + \tfrac{2}{3}\begin{pmatrix} 3 \\ -6 \end{pmatrix}$

$\Rightarrow \mathbf{r} = \begin{pmatrix} 3 \\ 2 \end{pmatrix}$

The point of intersection has co-ordinates (3, 2).

Note

Alternatively, you could have found $\mu = \tfrac{1}{2}$ and substituted in the equation for CD.

In three dimensions, lines may be parallel, they may intersect or they may be skew.

EXAMPLE 10.4 Determine whether each pair of lines are parallel, intersect or are skew.

(i) $\mathbf{r} = \begin{pmatrix} 1 \\ -2 \\ 1 \end{pmatrix} + \lambda \begin{pmatrix} -3 \\ 2 \\ 1 \end{pmatrix}$ and $\mathbf{r} = \begin{pmatrix} 1 \\ 3 \\ -2 \end{pmatrix} + \mu \begin{pmatrix} 6 \\ -4 \\ -2 \end{pmatrix}$

(ii) $\mathbf{r} = \begin{pmatrix} 1 \\ 2 \\ -1 \end{pmatrix} + \lambda \begin{pmatrix} 2 \\ -3 \\ 4 \end{pmatrix}$ and $\mathbf{r} = \begin{pmatrix} 4 \\ -2 \\ -5 \end{pmatrix} + \mu \begin{pmatrix} -1 \\ 2 \\ 1 \end{pmatrix}$

SOLUTION

(i) The vectors $\begin{pmatrix} -3 \\ 2 \\ 1 \end{pmatrix}$ and $\begin{pmatrix} 6 \\ -4 \\ -2 \end{pmatrix}$ are in the same direction as

$\begin{pmatrix} 6 \\ -4 \\ -2 \end{pmatrix} = -2 \begin{pmatrix} -3 \\ 2 \\ 1 \end{pmatrix}$

Note the lines are different as one line passes through (1, −2, 1) and the other through (1, 3, −2).

So the lines are parallel.

(ii) These lines are not parallel, so either they intersect or they are skew.

If the two lines intersect then there is a point (x, y, z) that lies on both lines.

$$\begin{pmatrix} x \\ y \\ z \end{pmatrix} = \begin{pmatrix} 1 \\ 2 \\ -1 \end{pmatrix} + \lambda \begin{pmatrix} 2 \\ -3 \\ 4 \end{pmatrix} \text{ and } \begin{pmatrix} x \\ y \\ z \end{pmatrix} = \begin{pmatrix} 4 \\ -2 \\ -5 \end{pmatrix} + \mu \begin{pmatrix} -1 \\ 2 \\ 1 \end{pmatrix}$$

This gives three simultaneous equations for λ and μ.

x: $1 + 2\lambda = 4 - \mu$ $\Rightarrow$ $2\lambda + \mu = 3$ ①

y: $2 - 3\lambda = -2 + 2\mu$ $\Rightarrow$ $3\lambda + 2\mu = 4$ ②

z: $-1 + 4\lambda = -5 + \mu$ $\Rightarrow$ $4\lambda - \mu = -4$ ③

Now solve any two of the three equations above simultaneously.

Using ① and ②:

$$\begin{cases} 2\lambda + \mu = 3 \\ 3\lambda + 2\mu = 4 \end{cases} \Rightarrow \begin{cases} 4\lambda + 2\mu = 6 \\ 3\lambda + 2\mu = 4 \end{cases} \Rightarrow \lambda = 2, \mu = -1$$

If these solutions satisfy the previously unused equation (equation ③ here) then the lines meet, and you can substitute the value of λ (or μ) into equations ①, ② and ③ to find the co-ordinates of the point of intersection.

If these solutions do not satisfy equation ③ then the lines are skew.

$$4\lambda - \mu = -4 \qquad\qquad\qquad ③$$

When $\lambda = 2$ and $\mu = -1$

$$4\lambda - \mu = 9 \neq -4$$

As there are no values for λ and μ that satisfy all three equations, the lines do not meet and so are skew; you have already seen that they are not parallel.

Note

If the equation of the second line was

$$\begin{pmatrix} x \\ y \\ z \end{pmatrix} = \begin{pmatrix} 4 \\ -2 \\ 8 \end{pmatrix} + \mu \begin{pmatrix} -1 \\ 2 \\ 1 \end{pmatrix}$$

then the values of $\lambda = 2$ and $\mu = -1$ would produce the same point for both lines:

$$\begin{pmatrix} x \\ y \\ z \end{pmatrix} = \begin{pmatrix} 1 \\ 2 \\ -1 \end{pmatrix} + 2\begin{pmatrix} 2 \\ -3 \\ 4 \end{pmatrix} = \begin{pmatrix} 5 \\ -4 \\ 7 \end{pmatrix}$$

and $\begin{pmatrix} x \\ y \\ z \end{pmatrix} = \begin{pmatrix} 4 \\ -2 \\ 8 \end{pmatrix} - 1\begin{pmatrix} -1 \\ 2 \\ 1 \end{pmatrix} = \begin{pmatrix} 5 \\ -4 \\ 7 \end{pmatrix}$.

So the lines would intersect at $(5, -4, 7)$.

EXERCISE 10B

1 Find the position vector of the point of intersection of each of these pairs of lines.

(i) $\mathbf{r} = \begin{pmatrix} 2 \\ 1 \end{pmatrix} + \lambda \begin{pmatrix} 1 \\ 0 \end{pmatrix}$ and $\mathbf{r} = \begin{pmatrix} 3 \\ 0 \end{pmatrix} + \mu \begin{pmatrix} 1 \\ 1 \end{pmatrix}$

(ii) $\mathbf{r} = \begin{pmatrix} 2 \\ -1 \end{pmatrix} + \lambda \begin{pmatrix} 1 \\ 2 \end{pmatrix}$ and $\mathbf{r} = \mu \begin{pmatrix} 1 \\ 1 \end{pmatrix}$

(iii) $\mathbf{r} = \begin{pmatrix} 0 \\ 5 \end{pmatrix} + \lambda \begin{pmatrix} -2 \\ -2 \end{pmatrix}$ and $\mathbf{r} = \begin{pmatrix} 0 \\ -7 \end{pmatrix} + \mu \begin{pmatrix} 1 \\ 2 \end{pmatrix}$

(iv) $\mathbf{r} = \begin{pmatrix} -2 \\ -3 \end{pmatrix} + \lambda \begin{pmatrix} -1 \\ 3 \end{pmatrix}$ and $\mathbf{r} = \begin{pmatrix} 1 \\ 3 \end{pmatrix} + \mu \begin{pmatrix} 2 \\ -1 \end{pmatrix}$

(v) $\mathbf{r} = \begin{pmatrix} 2 \\ 7 \end{pmatrix} + \lambda \begin{pmatrix} 1 \\ -1 \end{pmatrix}$ and $\mathbf{r} = \begin{pmatrix} 5 \\ 1 \end{pmatrix} + \mu \begin{pmatrix} 1 \\ 2 \end{pmatrix}$

2 Decide whether each of these pairs of lines intersect, are parallel or are skew. If the lines intersect, find the co-ordinates of the point of intersection.

(i) $\mathbf{r} = \begin{pmatrix} 1 \\ -6 \\ -1 \end{pmatrix} + \lambda \begin{pmatrix} 1 \\ 2 \\ 3 \end{pmatrix}$ and $\mathbf{r} = \begin{pmatrix} 9 \\ 7 \\ 2 \end{pmatrix} + \mu \begin{pmatrix} 2 \\ 3 \\ -1 \end{pmatrix}$

(ii) $\mathbf{r} = \begin{pmatrix} 1 \\ -6 \\ 0 \end{pmatrix} + \lambda \begin{pmatrix} 6 \\ -9 \\ -3 \end{pmatrix}$ and $\mathbf{r} = \begin{pmatrix} -5 \\ 3 \\ 0 \end{pmatrix} + \mu \begin{pmatrix} 2 \\ -3 \\ -1 \end{pmatrix}$

(iii) $\mathbf{r} = \begin{pmatrix} 6 \\ -4 \\ 2 \end{pmatrix} + \lambda \begin{pmatrix} 1 \\ -2 \\ 5 \end{pmatrix}$ and $\mathbf{r} = \begin{pmatrix} 1 \\ 4 \\ -17 \end{pmatrix} + \mu \begin{pmatrix} 1 \\ -1 \\ 2 \end{pmatrix}$

(iv) $\mathbf{r} = \begin{pmatrix} -1 \\ 2 \\ 4 \end{pmatrix} + \lambda \begin{pmatrix} 2 \\ 0 \\ 3 \end{pmatrix}$ and $\mathbf{r} = \begin{pmatrix} -4 \\ 4 \\ 6 \end{pmatrix} + \mu \begin{pmatrix} 5 \\ -2 \\ 1 \end{pmatrix}$

(v) $\mathbf{r} = \begin{pmatrix} 0 \\ -1 \\ 4 \end{pmatrix} + \lambda \begin{pmatrix} 5 \\ 3 \\ -3 \end{pmatrix}$ and $\mathbf{r} = \begin{pmatrix} 2 \\ 5 \\ -1 \end{pmatrix} + \mu \begin{pmatrix} 4 \\ -3 \\ 2 \end{pmatrix}$

(vi) $\mathbf{r} = \begin{pmatrix} 9 \\ 3 \\ -4 \end{pmatrix} + \lambda \begin{pmatrix} 1 \\ 2 \\ -3 \end{pmatrix}$ and $\mathbf{r} = \begin{pmatrix} 1 \\ -4 \\ 5 \end{pmatrix} + \mu \begin{pmatrix} 1 \\ -1 \\ 2 \end{pmatrix}$

(vii) $\mathbf{r} = \begin{pmatrix} 2 \\ 3 \\ 1 \end{pmatrix} + \lambda \begin{pmatrix} 1 \\ 1 \\ -2 \end{pmatrix}$ and $\mathbf{r} = \begin{pmatrix} -1 \\ -3 \\ -1 \end{pmatrix} + \mu \begin{pmatrix} 1 \\ 3 \\ 2 \end{pmatrix}$

3 In this question the origin is taken to be at a harbour and the unit vectors **i** and **j** to have lengths of 1 km in the directions E and N.

A cargo vessel leaves the harbour and its position vector t hours later is given by

$\mathbf{r}_1 = 12t\mathbf{i} + 16t\mathbf{j}$.

A fishing boat is trawling nearby and its position at time t is given by

$\mathbf{r}_2 = (10 - 3t)\mathbf{i} + (8 + 4t)\mathbf{j}$.

(i) How far apart are the two boats when the cargo vessel leaves harbour?

(ii) How fast is each boat travelling?

(iii) What happens?

4 The points A(1, 0), B(7, 2) and C(13, 7) are the vertices of a triangle.

The mid-points of the sides BC, CA and AB are L, M and N.

(i) Write down the position vectors of L, M and N.

(ii) Find the vector equations of the lines AL, BM and CN.

(iii) Find the intersections of these pairs of lines.

    (a) AL and BM   (b) BM and CN

(iv) What do you notice?

5 The line $\mathbf{r} = \begin{pmatrix} -4 \\ 4 \\ -12 \end{pmatrix} + q \begin{pmatrix} 2 \\ -10 \\ 11 \end{pmatrix}$ meets $\mathbf{r} = \begin{pmatrix} 4 \\ -15 \\ -16 \end{pmatrix} + s \begin{pmatrix} 2 \\ -3 \\ -5 \end{pmatrix}$ at A and meets

$\mathbf{r} = \begin{pmatrix} -1 \\ -29 \\ -3 \end{pmatrix} + t \begin{pmatrix} 1 \\ 1 \\ 8 \end{pmatrix}$ at B. Find the co-ordinates of A and the length of AB.

6 To support a tree damaged in a gale a tree surgeon attaches wire guys to four of the branches (see the diagram). He joins (2, 0, 3) to (−1, 2, 6) and (0, 3, 5) to (−2, −2, 4). Do the guys, assumed straight, meet?

7 Show that the three lines $\mathbf{r} = \begin{pmatrix} -7 \\ 24 \\ -4 \end{pmatrix} + q \begin{pmatrix} 4 \\ -7 \\ 4 \end{pmatrix}$, $\mathbf{r} = \begin{pmatrix} 3 \\ -10 \\ 15 \end{pmatrix} + s \begin{pmatrix} 2 \\ 2 \\ -1 \end{pmatrix}$ and

$\mathbf{r} = \begin{pmatrix} -3 \\ 6 \\ 6 \end{pmatrix} + t \begin{pmatrix} 8 \\ -3 \\ 2 \end{pmatrix}$ form a triangle and find the lengths of its sides.

8 The drawing shows an ordinary music stand, which consists of a rectangle DEFG with a vertical support OA.

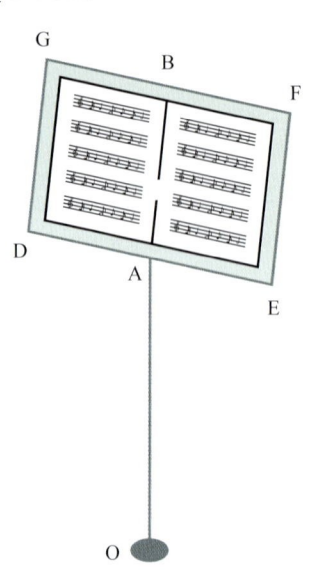

Relative to axes through the origin O, which is on the floor, the co-ordinates of various points are given (with dimensions in metres) as:

A is (0, 0, 1)   D is (−0.25, 0, 1)   F is (0.25, 0.15, 1.3).

DE and GF are horizontal, A is the mid-point of DE and B is the mid-point of GF. C is on AB so that $AC = \frac{1}{3}AB$.

(i) Write down the vector $\overrightarrow{AD}$ and show that $\overrightarrow{EF}$ is $\begin{pmatrix} 0 \\ 0.15 \\ 0.3 \end{pmatrix}$.

(ii) Calculate the co-ordinates of C.

(iii) Find the equations of the lines DE and EF in vector form.

[MEI, part]

The angle between two lines

In *Pure Mathematics 1*, Chapter 8 you learnt that the angle, θ, between two vectors $\mathbf{a} = \begin{pmatrix} a_1 \\ a_2 \\ a_3 \end{pmatrix}$ and $\mathbf{b} = \begin{pmatrix} b_1 \\ b_2 \\ b_3 \end{pmatrix}$ can be found using the formula:

$$\cos \theta = \frac{\mathbf{a} \cdot \mathbf{b}}{|\mathbf{a}||\mathbf{b}|}$$

where $\mathbf{a} \cdot \mathbf{b}$ is the scalar product and $\mathbf{a} \cdot \mathbf{b} = a_1b_1 + a_2b_2 + a_3b_3$.

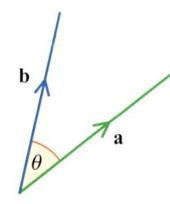

Figure 10.4

EXAMPLE 10.5

(i) Find the angle between the vectors $\begin{pmatrix} -4 \\ 3 \\ 0 \end{pmatrix}$ and $\begin{pmatrix} 2 \\ -1 \\ 3 \end{pmatrix}$.

(ii) Verify that the vectors $\begin{pmatrix} -9 \\ -2 \\ 4 \end{pmatrix}$ and $\begin{pmatrix} 2 \\ -3 \\ 3 \end{pmatrix}$ are perpendicular.

SOLUTION

(i) Let $\mathbf{a} = \begin{pmatrix} -4 \\ 3 \\ 0 \end{pmatrix} \Rightarrow |\mathbf{a}| = \sqrt{(-4)^2 + 3^2 + 0^2} = 5$

and $\mathbf{b} = \begin{pmatrix} 2 \\ -1 \\ 3 \end{pmatrix} \Rightarrow |\mathbf{b}| = \sqrt{2^2 + (-1)^2 + 3^2} = \sqrt{14}$

The scalar product $\mathbf{a}.\mathbf{b}$ is

$$\begin{pmatrix} -4 \\ 3 \\ 0 \end{pmatrix} . \begin{pmatrix} 2 \\ -1 \\ 3 \end{pmatrix} = (-4) \times 2 + 3 \times (-1) + 0 \times 3 = -11$$

Substituting into $\cos\theta = \dfrac{\mathbf{a}.\mathbf{b}}{|\mathbf{a}||\mathbf{b}|}$ gives:

$$\cos\theta = \dfrac{-11}{5\sqrt{14}}$$

$\Rightarrow \quad \theta = 126.0°$

(ii) When two vectors are perpendicular, the angle between them is 90°.
Since $\cos 90° = 0$ then $\mathbf{a}.\mathbf{b} = 0$.
So if the scalar product of two non-zero vectors is zero then the vectors are perpendicular.

$$\begin{pmatrix} -9 \\ -2 \\ 4 \end{pmatrix} . \begin{pmatrix} 2 \\ -3 \\ 3 \end{pmatrix} = (-9) \times 2 + (-2) \times (-3) + 4 \times 3$$

$= (-18) + 6 + 12$

$= 0$

Therefore, the two vectors are perpendicular.

Even if two lines do not meet, it is still possible to specify the angle between them. The lines l and m shown in figure 10.5 do not meet; they are described as *skew*.

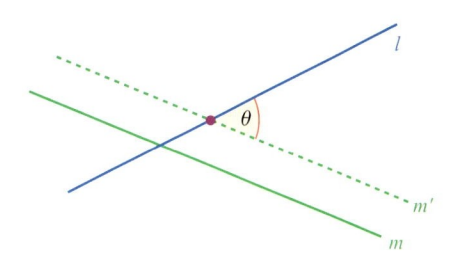

Figure 10.5

The angle between them is that between their directions; it is shown in figure 10.5 as the angle θ between the lines l and m', where m' is a translation of the line m to a position where it does intersect the line l.

EXAMPLE 10.6 Find the angle between the lines

$$\mathbf{r} = \begin{pmatrix} 1 \\ 0 \\ 4 \end{pmatrix} + \lambda \begin{pmatrix} 2 \\ -1 \\ -1 \end{pmatrix} \quad \text{and} \quad \mathbf{r} = \begin{pmatrix} 2 \\ -1 \\ 3 \end{pmatrix} + \mu \begin{pmatrix} 3 \\ 0 \\ 1 \end{pmatrix}.$$

SOLUTION

The angle between the lines is the angle between their directions $\begin{pmatrix} 2 \\ -1 \\ -1 \end{pmatrix}$ and $\begin{pmatrix} 3 \\ 0 \\ 1 \end{pmatrix}$.

Using $\cos\theta = \dfrac{\mathbf{a}\cdot\mathbf{b}}{|\mathbf{a}||\mathbf{b}|}$

$$\cos\theta = \dfrac{2\times 3 + (-1)\times 0 + (-1)\times 1}{\sqrt{2^2+(-1)^2+(-1)^2}\times\sqrt{3^2+0^2+1^2}}$$

$$\cos\theta = \dfrac{5}{\sqrt{6}\times\sqrt{10}}$$

$\Rightarrow \quad \theta = 49.8°$

EXERCISE 10C Remember $\mathbf{i} = \begin{pmatrix} 1 \\ 0 \\ 0 \end{pmatrix}, \mathbf{j} = \begin{pmatrix} 0 \\ 1 \\ 0 \end{pmatrix}$ and $\mathbf{k} = \begin{pmatrix} 0 \\ 0 \\ 1 \end{pmatrix}$.

In questions **1** to **5**, find the angle between each pair of lines.

1 $\mathbf{r} = \begin{pmatrix} 2 \\ 1 \\ 3 \end{pmatrix} + s\begin{pmatrix} 1 \\ 4 \\ 0 \end{pmatrix}$ and $\mathbf{r} = \begin{pmatrix} 6 \\ 10 \\ 4 \end{pmatrix} + t\begin{pmatrix} 2 \\ 1 \\ 1 \end{pmatrix}$

2 $\mathbf{r} = s\begin{pmatrix} 4 \\ 1 \\ 4 \end{pmatrix}$ and $\mathbf{r} = \begin{pmatrix} 7 \\ 0 \\ -3 \end{pmatrix} + t\begin{pmatrix} 1 \\ 2 \\ -1 \end{pmatrix}$

3 $\mathbf{r} = \begin{pmatrix} 4 \\ 2 \\ -1 \end{pmatrix} + s\begin{pmatrix} 3 \\ 7 \\ -4 \end{pmatrix}$ and $\mathbf{r} = \begin{pmatrix} 5 \\ 1 \\ 0 \end{pmatrix} + t\begin{pmatrix} 2 \\ 8 \\ -5 \end{pmatrix}$

4 $\mathbf{r} = 2\mathbf{i} + 3\mathbf{j} + 4\mathbf{k} + s(\mathbf{i}+\mathbf{j}-\mathbf{k})$ and $\mathbf{r} = t(\mathbf{i}-\mathbf{k})$

5 $\mathbf{r} = \mathbf{i} - 2\mathbf{j} - \mathbf{k} + s(2\mathbf{i}+3\mathbf{j}+2\mathbf{k})$ and $\mathbf{r} = 2\mathbf{i}+\mathbf{j}+t\mathbf{k}$

6 The diagram shows an extension to a house. Its base and walls are rectangular and the end of its roof, EPF, is sloping, as illustrated.

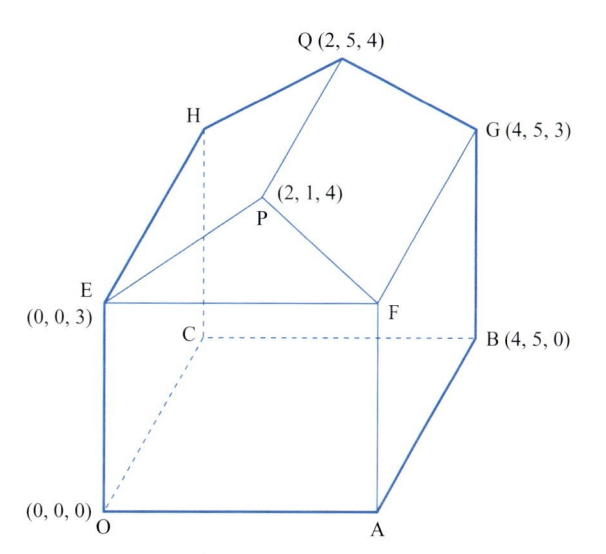

(i) Write down the co-ordinates of A and F.

(ii) Find, using vector methods, the angles FPQ and EPF.

The owner decorates the room with two streamers which are pulled taut. One goes from O to G, the other from A to H. She says that they touch each other and that they are perpendicular to each other.

(iii) Is she right?

7 The points A and B have position vectors, relative to the origin O, given by

$$\overrightarrow{OA} = \mathbf{i} + 2\mathbf{j} + 3\mathbf{k} \quad \text{and} \quad \overrightarrow{OB} = 2\mathbf{i} + \mathbf{j} + 3\mathbf{k}.$$

The line l has vector equation

$$\mathbf{r} = (1 - 2t)\mathbf{i} + (5 + t)\mathbf{j} + (2 - t)\mathbf{k}.$$

(i) Show that l does not interesect the line passing through A and B.

(ii) The point P lies on l and is such that angle PAB is equal to 60°. Given that the position vector of P is $(1 - 2t)\mathbf{i} + (5 + t)\mathbf{j} + (2 - t)\mathbf{k}$, show that $3t^2 + 7t + 2 = 0$. Hence find the only possible position vector of P.

[Cambridge International AS & A Level Mathematics 9709, Paper 3 Q10 June 2008]

The perpendicular distance from a point to a line

The scalar product is also useful when determining the distance between a point and a line.

EXAMPLE 10.7 Find the shortest distance from point P(11, −5, −3) to the line l with equation
$$\mathbf{r} = \begin{pmatrix} 1 \\ 5 \\ 0 \end{pmatrix} + \lambda \begin{pmatrix} -3 \\ 1 \\ 4 \end{pmatrix}.$$

SOLUTION

The shortest distance from P to the line l is $|\overrightarrow{NP}|$ where N is a point on the line l and PN is perpendicular to the line l.

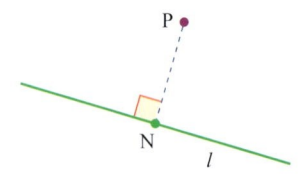

Figure 10.6

You need to find the co-ordinates of N and then you can find $|\overrightarrow{NP}|$.
N lies on the line l. Let the value of λ at N be t.
So, relative to the origin O

$$\overrightarrow{ON} = \begin{pmatrix} 1 \\ 5 \\ 0 \end{pmatrix} + t \begin{pmatrix} -3 \\ 1 \\ 4 \end{pmatrix} = \begin{pmatrix} 1-3t \\ 5+t \\ 4t \end{pmatrix}$$

and $\overrightarrow{NP} = \overrightarrow{OP} - \overrightarrow{ON}$

$$= \begin{pmatrix} 11 \\ -5 \\ -3 \end{pmatrix} - \begin{pmatrix} 1-3t \\ 5+t \\ 4t \end{pmatrix}$$

$$= \begin{pmatrix} 10+3t \\ -10-t \\ -3-4t \end{pmatrix}$$

As $\overrightarrow{NP}$ is perpendicular to the line l,

$$\overrightarrow{NP} \cdot \begin{pmatrix} -3 \\ 1 \\ 4 \end{pmatrix} = 0$$

The direction of the line l

When two vectors are perpendicular, their scalar product is 0.

$$\overrightarrow{NP} \cdot \begin{pmatrix} -3 \\ 1 \\ 4 \end{pmatrix} = \begin{pmatrix} 10+3t \\ -10-t \\ -3-4t \end{pmatrix} \cdot \begin{pmatrix} -3 \\ 1 \\ 4 \end{pmatrix}$$

$$= (10+3t) \times (-3) + (-10-t) \times 1 + (-3-4t) \times 4$$
$$= -30 - 9t - 10 - t - 12 - 16t$$
$$= -52 - 26t$$

The scalar product is 0, so

$$-52 - 26t = 0 \quad \Rightarrow \quad t = -2$$

Substituting $t = -2$ into $\overrightarrow{ON}$ and $\overrightarrow{NP}$ gives

$$\overrightarrow{ON} = \begin{pmatrix} 1 \\ 5 \\ 0 \end{pmatrix} - 2\begin{pmatrix} -3 \\ 1 \\ 4 \end{pmatrix} = \begin{pmatrix} 7 \\ 3 \\ -8 \end{pmatrix}$$

and $\overrightarrow{NP} = \begin{pmatrix} 10 + 3 \times (-2) \\ -10 - (-2) \\ -3 - 4 \times (-2) \end{pmatrix} = \begin{pmatrix} 4 \\ -8 \\ 5 \end{pmatrix}$

So $|\overrightarrow{NP}| = \sqrt{4^2 + (-8)^2 + 5^2}$
$$= \sqrt{105}$$
$$= 10.25 \text{ units}$$

EXERCISE 10D

1 For each point P and line l find

(a) the co-ordinates of the point N on the line such that PN is perpendicular to the line

(b) the distance PN.

(i) P(−2, 11, 5) and $\mathbf{r} = \begin{pmatrix} 0 \\ 2 \\ -3 \end{pmatrix} + t\begin{pmatrix} -1 \\ 2 \\ 5 \end{pmatrix}$

(ii) P(7, −1, 6) and $\mathbf{r} = \begin{pmatrix} 2 \\ 1 \\ 3 \end{pmatrix} + t\begin{pmatrix} 1 \\ -2 \\ 1 \end{pmatrix}$

(iii) P(8, 4, −1) and $\mathbf{r} = \begin{pmatrix} 1 \\ 5 \\ -3 \end{pmatrix} + t\begin{pmatrix} -1 \\ -2 \\ 0 \end{pmatrix}$

2 Find the perpendicular distance of the point P(−7, −2, 13) to the line
$$\mathbf{r} = \begin{pmatrix} 1 \\ 2 \\ 5 \end{pmatrix} + \lambda \begin{pmatrix} 1 \\ 3 \\ -4 \end{pmatrix}.$$

3 Find the distance of the point C(0, 6, 0) to the line joining the points A(−4, 2, −3) and B(−2, 0, 1).

4 The room illustrated in the diagram has rectangular walls, floor and ceiling. A string has been stretched in a straight line between the corners A and G.

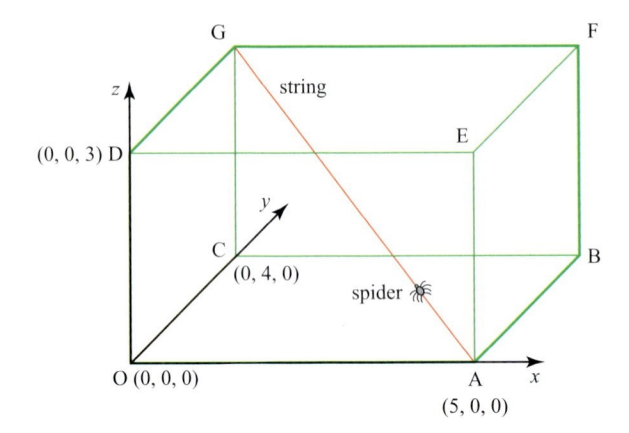

The corner O is taken as the origin. A is (5, 0, 0), C is (0, 4, 0) and D is (0, 0, 3), where the lengths are in metres.

(i) Write down the co-ordinates of G.
(ii) Find the vector $\overrightarrow{AG}$ and the length of the string $|\overrightarrow{AG}|$.
(iii) Write down the equation of the line AG in vector form.

A spider walks up the string, starting from A.

(iv) Find the position vector of the spider when it is at Q, one quarter of the way from A to G, and find the angle OQG.
(v) Show that when the spider is 1.5 m above the floor it is at its closest point to O, and find how far it is then from O.

[MEI]

5 The diagram illustrates the flight path of a helicopter H taking off from an airport.

Co-ordinate axes O*xyz* are set up with the origin O at the base of the airport control tower. The *x* axis is due east, the *y* axis due north and the *z* axis vertical.

The units of distance are kilometres throughout.

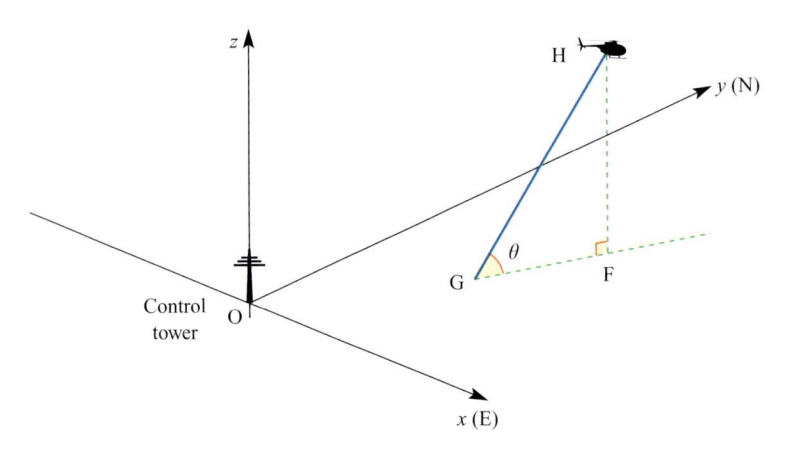

The helicopter takes off from the point G.

The position vector **r** of the helicopter t minutes after take-off is given by

$$\mathbf{r} = (1 + t)\mathbf{i} + (0.5 + 2t)\mathbf{j} + 2t\mathbf{k}.$$

(i) Write down the co-ordinates of G.

(ii) Find the angle the flight path makes with the horizontal.
(This angle is shown as θ in the diagram.)

(iii) Find the bearing of the flight path.
(This is the bearing of the line GF shown in the diagram.)

(iv) The helicopter enters a cloud at a height of 2 km.
Find the co-ordinates of the point where the helicopter enters the cloud.

(v) A mountain top is situated at M(5, 4.5, 3).
Find the value of t when HM is perpendicular to the flight path GH.
Find the distance from the helicopter to the mountain top at this time.

[MEI]

The vector equation of a plane

❓ Which balances better, a three-legged stool or a four-legged stool? Why?
What information do you need to specify a particular plane?

There are various ways of finding the equation of a plane and these are given in this book. Your choice of which one to use will depend on the information you are given.

e Finding the equation of a plane given three points on it

⚠ There are several methods used to find the equation of a plane through three given points. The shortest method involves the use of vector product which is beyond the scope of this book. The method given here develops the same ideas as were used for the equation of a line. It will help you to understand the extra concepts involved, but it is not a requirement of the Cambridge syllabus.

Vector form

To find the vector form of the equation of the plane through the points A, B and C (with position vectors $\overrightarrow{OA} = \mathbf{a}$, $\overrightarrow{OB} = \mathbf{b}$, $\overrightarrow{OC} = \mathbf{c}$), think of starting at the origin, travelling along OA to join the plane at A, and then any distance in each of the directions $\overrightarrow{AB}$ and $\overrightarrow{AC}$ to reach a general point R with position vector $\mathbf{r}$, where

$$\mathbf{r} = \overrightarrow{OA} + \lambda\overrightarrow{AB} + \mu\overrightarrow{AC}.$$

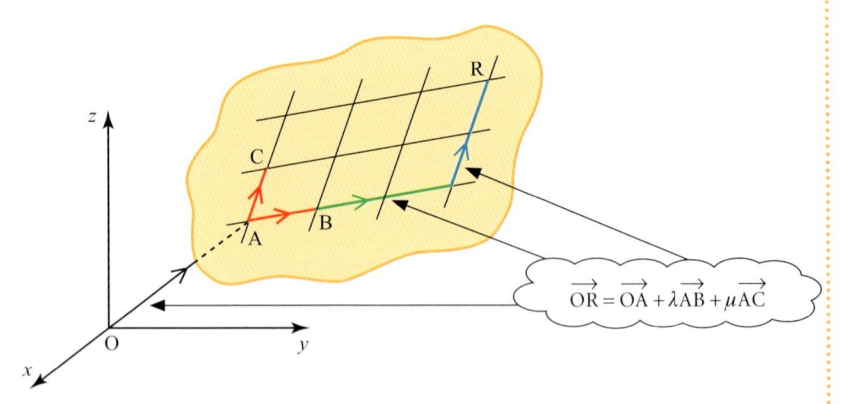

Figure 10.7

This is a vector form of the equation of the plane. Since $\overrightarrow{OA} = \mathbf{a}$, $\overrightarrow{AB} = \mathbf{b} - \mathbf{a}$ and $\overrightarrow{AC} = \mathbf{c} - \mathbf{a}$, it may also be written as

$$\mathbf{r} = \mathbf{a} + \lambda(\mathbf{b} - \mathbf{a}) + \mu(\mathbf{c} - \mathbf{a}).$$

EXAMPLE 10.8 Find the equation of the plane through A(4, 2, 0), B(3, 1, 1) and C(4, −1, 1).

SOLUTION

$$\overrightarrow{OA} = \begin{pmatrix} 4 \\ 2 \\ 0 \end{pmatrix}$$

$$\overrightarrow{AB} = \overrightarrow{OB} - \overrightarrow{OA} = \begin{pmatrix} 3 \\ 1 \\ 1 \end{pmatrix} - \begin{pmatrix} 4 \\ 2 \\ 0 \end{pmatrix} = \begin{pmatrix} -1 \\ -1 \\ 1 \end{pmatrix}$$

$$\overrightarrow{AC} = \overrightarrow{OC} - \overrightarrow{OA} = \begin{pmatrix} 4 \\ -1 \\ 1 \end{pmatrix} - \begin{pmatrix} 4 \\ 2 \\ 0 \end{pmatrix} = \begin{pmatrix} 0 \\ -3 \\ 1 \end{pmatrix}$$

So the equation $\mathbf{r} = \overrightarrow{OA} + \lambda \overrightarrow{AB} + \mu \overrightarrow{AC}$ becomes

$$\mathbf{r} = \begin{pmatrix} 4 \\ 2 \\ 0 \end{pmatrix} + \lambda \begin{pmatrix} -1 \\ -1 \\ 1 \end{pmatrix} + \mu \begin{pmatrix} 0 \\ -3 \\ 1 \end{pmatrix}.$$

This is the vector form of the equation, written using components.

Cartesian form

You can convert this equation into cartesian form by writing it as

$$\begin{pmatrix} x \\ y \\ z \end{pmatrix} = \begin{pmatrix} 4 \\ 2 \\ 0 \end{pmatrix} + \lambda \begin{pmatrix} -1 \\ -1 \\ 1 \end{pmatrix} + \mu \begin{pmatrix} 0 \\ -3 \\ 1 \end{pmatrix}$$

and eliminating λ and μ. The three equations contained in this vector equation may be simplified to give

$$\lambda = -x + 4 \qquad \qquad ①$$

$$\lambda + 3\mu = -y + 2 \qquad \qquad ②$$

$$\lambda + \mu = z \qquad \qquad ③$$

Substituting ① into ② gives

$$-x + 4 + 3\mu = -y + 2$$

$$3\mu = x - y - 2$$

$$\mu = \tfrac{1}{3}(x - y - 2)$$

Substituting this and ① into ③ gives

$$-x + 4 + \tfrac{1}{3}(x - y - 2) = z$$

$$-3x + 12 + x - y - 2 = 3z$$

$$2x + y + 3z = 10$$

and this is the cartesian equation of the plane through A, B and C.

Note

In contrast to the equation of a line, the equation of a plane is more neatly expressed in cartesian form. The general cartesian equation of a plane is often written as either

$$ax + by + cz = d \qquad \text{or} \qquad n_1 x + n_2 y + n_3 z = d.$$

Finding the equation of a plane using the direction perpendicular to it

? Lay a sheet of paper on a flat horizontal table and mark several straight lines on it. Now take a pencil and stand it upright on the sheet of paper (see figure 10.8).

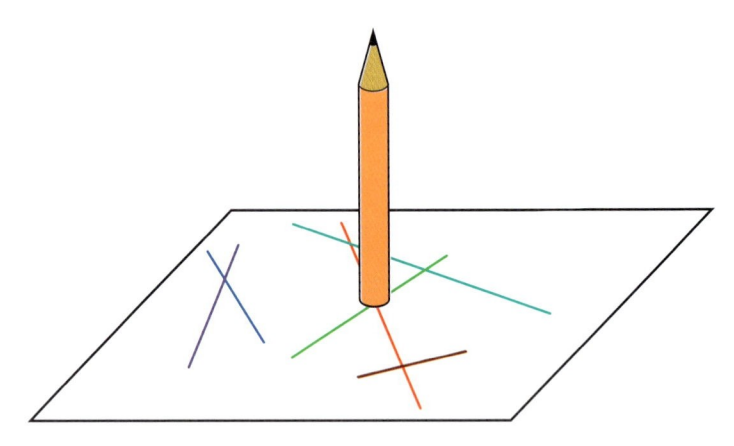

Figure 10.8

(i) What angle does the pencil make with any individual line?

(ii) Would it make any difference if the table were tilted at an angle (apart from the fact that you could no longer balance the pencil)?

The discussion above shows you that there is a direction (that of the pencil) which is at right angles to every straight line in the plane. A line in that direction is said to be perpendicular to the plane or *normal* to the plane.

This allows you to find a different vector form of the equation of a plane which you use when you know the position vector **a** of one point A in the plane and the direction $\mathbf{n} = n_1\mathbf{i} + n_2\mathbf{j} + n_3\mathbf{k}$ perpendicular to the plane.

What you want to find is an expression for the position vector **r** of a general point R in the plane (see figure 10.9). Since AR is a line in the plane, it follows that AR is at right angles to the direction **n**.

$\overrightarrow{AR} . \mathbf{n} = 0$

The point A has position vector **a**.

The point R has position vector **r**.

The vector $\overrightarrow{AR}$ is $\mathbf{r} - \mathbf{a}$.

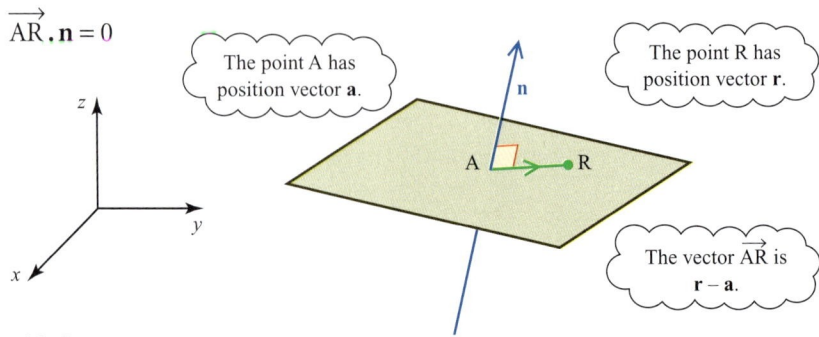

Figure 10.9

The vector $\overrightarrow{AR}$ is given by
$$\overrightarrow{AR} = \mathbf{r} - \mathbf{a}$$
and so $(\mathbf{r} - \mathbf{a}) \cdot \mathbf{n} = 0.$

> For example, the plane through A(2, 0, 0) perpendicular to $\mathbf{n} = (3\mathbf{i} - 4\mathbf{j} + \mathbf{k})$ can be written as $(\mathbf{r} - 2\mathbf{i}) \cdot (3\mathbf{i} - 4\mathbf{j} + \mathbf{k}) = 0$ which simplifies to $3x - 4y + z = 6$.

This can also be written as
$$\mathbf{r} \cdot \mathbf{n} - \mathbf{a} \cdot \mathbf{n} = 0$$

or $\begin{pmatrix} x \\ y \\ z \end{pmatrix} \cdot \begin{pmatrix} n_1 \\ n_2 \\ n_3 \end{pmatrix} - \mathbf{a} \cdot \mathbf{n} = 0$

$\Rightarrow \begin{pmatrix} x \\ y \\ z \end{pmatrix} \cdot \begin{pmatrix} n_1 \\ n_2 \\ n_3 \end{pmatrix} = \mathbf{a} \cdot \mathbf{n}$

$\Rightarrow n_1 x + n_2 y + n_3 z = d$

where $d = \mathbf{a} \cdot \mathbf{n}$.

Notice that d is a constant scalar.

EXAMPLE 10.9

Write down the equation of the plane through the point (2, 1, 3) given that the vector $\begin{pmatrix} 4 \\ 5 \\ 6 \end{pmatrix}$ is perpendicular to the plane.

SOLUTION

In this case, the position vector $\mathbf{a}$ of the point (2, 1, 3) is given by $\mathbf{a} = \begin{pmatrix} 2 \\ 1 \\ 3 \end{pmatrix}$.

The vector perpendicular to the plane is
$$\mathbf{n} = \begin{pmatrix} n_1 \\ n_2 \\ n_3 \end{pmatrix} = \begin{pmatrix} 4 \\ 5 \\ 6 \end{pmatrix}.$$

The equation of the plane is
$$n_1 x + n_2 y + n_3 z = \mathbf{a} \cdot \mathbf{n}$$
$$4x + 5y + 6z = 2 \times 4 + 1 \times 5 + 3 \times 6$$
$$4x + 5y + 6z = 31$$

Look carefully at the equation of the plane in Example 10.9. You can see at once that the vector $\begin{pmatrix} 4 \\ 5 \\ 6 \end{pmatrix}$, formed from the coefficients of x, y and z, is perpendicular to the plane.

The vector $\begin{pmatrix} n_1 \\ n_2 \\ n_3 \end{pmatrix}$ is perpendicular to all planes of the form

$$n_1 x + n_2 y + n_3 z = d$$

whatever the value of d (see figure 10.10). Consequently, all planes of that form are parallel; the coefficients of x, y and z determine the direction of the plane, the value of d its location.

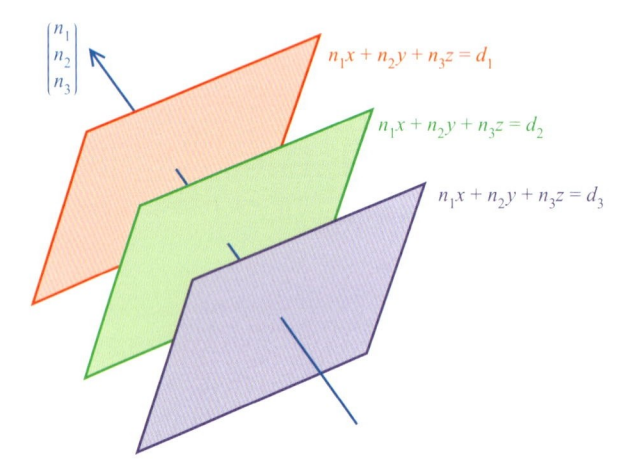

Figure 10.10

The intersection of a line and a plane

There are three possibilities for the intersection of a line and a plane.

1 The line and plane are not parallel and so they intersect in one point

2 The line and plane are parallel and so do not intersect

3 The line and plane are parallel and the line lies in the plane

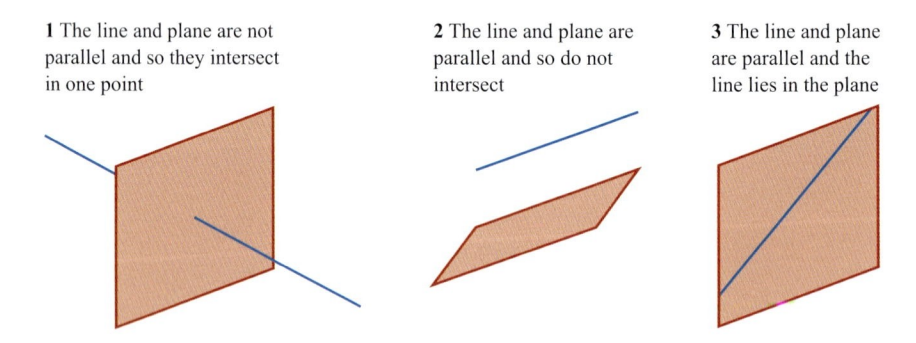

Figure 10.11

The point of intersection of a line and a plane is found by following the procedure in the next example.

EXAMPLE 10.10 Find the point of intersection of the line $\mathbf{r} = \begin{pmatrix} 2 \\ 3 \\ 4 \end{pmatrix} + \lambda \begin{pmatrix} 1 \\ 2 \\ -1 \end{pmatrix}$ with the plane $5x + y - z = 1$.

SOLUTION

The line is

$$\mathbf{r} = \begin{pmatrix} x \\ y \\ z \end{pmatrix} = \begin{pmatrix} 2 \\ 3 \\ 4 \end{pmatrix} + \lambda \begin{pmatrix} 1 \\ 2 \\ -1 \end{pmatrix}$$

and so for any point on the line

$$x = 2 + \lambda \quad y = 3 + 2\lambda \quad \text{and} \quad z = 4 - \lambda.$$

Substituting these into the equation of the plane $5x + y - z = 1$ gives

$$5(2 + \lambda) + (3 + 2\lambda) - (4 - \lambda) = 1$$
$$8\lambda = -8$$
$$\lambda = -1.$$

Substituting $\lambda = -1$ in the equation of the line gives

$$\mathbf{r} = \begin{pmatrix} x \\ y \\ z \end{pmatrix} = \begin{pmatrix} 2 \\ 3 \\ 4 \end{pmatrix} - \begin{pmatrix} 1 \\ 2 \\ -1 \end{pmatrix} = \begin{pmatrix} 1 \\ 1 \\ 5 \end{pmatrix}$$

so the point of intersection is $(1, 1, 5)$.

As a check, substitute $(1, 1, 5)$ into the equation of the plane:

$$5x + y - z = 5 + 1 - 5$$
$$= 1 \quad \text{as required.}$$

When a line is parallel to a plane, its direction vector is perpendicular to the plane's normal vector.

EXAMPLE 10.11 Show that the line $\mathbf{r} = \begin{pmatrix} 2 \\ 1 \\ 0 \end{pmatrix} + t \begin{pmatrix} 3 \\ 1 \\ -2 \end{pmatrix}$ is parallel to the plane $2x + 4y + 5z = 8$.

SOLUTION

The direction of the line is $\begin{pmatrix} 3 \\ 1 \\ -2 \end{pmatrix}$ and of the normal to the plane is $\begin{pmatrix} 2 \\ 4 \\ 5 \end{pmatrix}$.

If these two vectors are perpendicular, then the line and plane are parallel.

To prove that two vectors are perpendicular, you need to show that their scalar product is 0.

$$\begin{pmatrix} 3 \\ 1 \\ -2 \end{pmatrix} \cdot \begin{pmatrix} 2 \\ 4 \\ 5 \end{pmatrix} = 3 \times 2 + 1 \times 4 + (-2) \times 5 = 0$$

So the line and plane are parallel as required.

To prove that a line lies in a plane, you need to show the line and the plane are parallel and that any point on the line also lies in the plane.

EXAMPLE 10.12 Does the line $\mathbf{r} = \begin{pmatrix} 2 \\ 1 \\ 0 \end{pmatrix} + t \begin{pmatrix} 3 \\ 1 \\ -2 \end{pmatrix}$ lie in the plane $2x + 4y + 5z = 8$?

SOLUTION

You have already seen that this line and plane are parallel in Example 10.11.

Find a point on the line $\mathbf{r} = \begin{pmatrix} 2 \\ 1 \\ 0 \end{pmatrix} + t \begin{pmatrix} 3 \\ 1 \\ -2 \end{pmatrix}$ by setting $t = 1$.

So the point $(5, 2, -2)$ lies on the line.

Now check that this point satisfies the equation of the plane, $2x + 4y + 5z = 8$.

$$2 \times 5 + 4 \times 2 + 5(-2) = 8 \checkmark$$

The line and the plane are parallel and the point $(5, 2, -2)$ lies both on the line and in the plane. Therefore the line must lie in the plane.

> *Note*
>
> The previous two examples showed you that the line $\mathbf{r} = \begin{pmatrix} 2 \\ 1 \\ 0 \end{pmatrix} + t \begin{pmatrix} 3 \\ 1 \\ -2 \end{pmatrix}$ lies in the plane $2x + 4y + 5z = 8$. This line is parallel to all the planes in the form $2x + 4y + 5z = d$ but in the case when $d = 8$ it lies in the plane; for other values of d the line and the plane never meet.

The distance of a point from a plane

The shortest distance of a point, A, from a plane is the distance AP, where P is the point where the line through A perpendicular to the plane intersects the plane (see figure 10.12). This is usually just called the distance of the point from the plane. The process of finding this distance is shown in the next example.

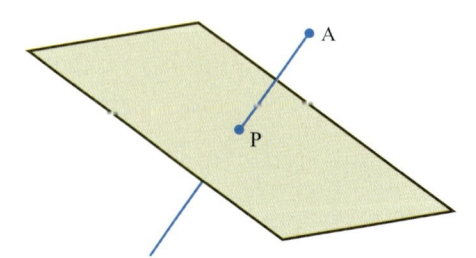

Figure 10.12

EXAMPLE 10.13 A is the point (7, 5, 3) and the plane π has the equation $3x + 2y + z = 6$. Find

(i) the equation of the line through A perpendicular to the plane π
(ii) the point of intersection, P, of this line with the plane
(iii) the distance AP.

SOLUTION

(i) The direction perpendicular to the plane $3x + 2y + z = 6$ is $\begin{pmatrix} 3 \\ 2 \\ 1 \end{pmatrix}$ so the line through (7, 5, 3) perpendicular to the plane is given by

$$\mathbf{r} = \begin{pmatrix} 7 \\ 5 \\ 3 \end{pmatrix} + \lambda \begin{pmatrix} 3 \\ 2 \\ 1 \end{pmatrix}.$$

(ii) For any point on the line

$$x = 7 + 3\lambda \qquad y = 5 + 2\lambda \qquad \text{and} \qquad z = 3 + \lambda.$$

Substituting these expressions into the equation of the plane $3x + 2y + z = 6$ gives

$$3(7 + 3\lambda) + 2(5 + 2\lambda) + (3 + \lambda) = 6$$
$$14\lambda = -28$$
$$\lambda = -2.$$

So the point P has co-ordinates (1, 1, 1).

(iii) The vector $\overrightarrow{AP}$ is given by

$$\begin{pmatrix} 1 \\ 1 \\ 1 \end{pmatrix} - \begin{pmatrix} 7 \\ 5 \\ 3 \end{pmatrix} = \begin{pmatrix} -6 \\ -4 \\ -2 \end{pmatrix}$$

and so the length AP is $\sqrt{(-6)^2 + (-4)^2 + (-2)^2} = \sqrt{56}$.

Note

In practice, you would not usually follow the procedure in Example 10.13 because there is a well-known formula for the distance of a point from a plane. You are invited to derive this in the following activity.

ACTIVITY 10.3 Generalise the work in Example 10.13 to show that the distance of the point (α, β, γ) from the plane $n_1 x + n_2 y + n_3 z = d$ is given by

$$\frac{|n_1 \alpha + n_2 \beta + n_3 \gamma - d|}{\sqrt{n_1^2 + n_2^2 + n_3^2}}.$$

The angle between a line and a plane

You can find the angle between a line and a plane by first finding the angle between the *normal* to the plane and the direction of the line. A normal to a plane is a line perpendicular to it.

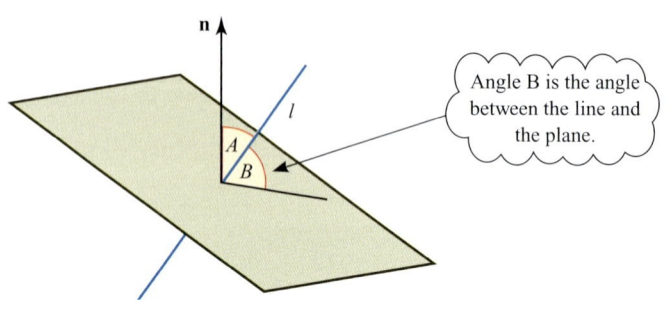

Figure 10.13

Angle B is the angle between the line and the plane.

The angle between the normal, **n**, and the plane is 90°.

Angle A is the angle between the line *l* and the normal to the plane, so the angle between the line and the plane, angle B, is 90° − A.

EXAMPLE 10.14 Find the angle between the line $\mathbf{r} = \begin{pmatrix} -1 \\ 2 \\ -3 \end{pmatrix} + t \begin{pmatrix} -1 \\ 2 \\ 5 \end{pmatrix}$ and the plane $2x + 3y + z = 4$.

SOLUTION

The normal, **n**, to the plane is $\begin{pmatrix} 2 \\ 3 \\ 1 \end{pmatrix}$. The direction, **d**, of the line is $\begin{pmatrix} -1 \\ 2 \\ 5 \end{pmatrix}$.

The angle between the normal to the plane and the direction of the line is given by:

$$\cos A = \frac{\mathbf{n} \cdot \mathbf{d}}{|\mathbf{n}||\mathbf{d}|}$$

$\mathbf{n} \cdot \mathbf{d} = 2 \times (-1) + 3 \times 2 + 1 \times 5 = 9$

$$\cos A = \frac{9}{\sqrt{14} \times \sqrt{30}}$$

$\Rightarrow \quad A = 63.95°$

$\Rightarrow \quad B = 26.05°$ Since $A + B = 90°$

So the angle between the line and the plane is 26° to the nearest degree.

EXERCISE 10E

1 Determine whether the following planes and lines are parallel.
If they are parallel, show whether the line lies in the plane.

(i) $\mathbf{r} = \begin{pmatrix} 3 \\ 1 \\ 2 \end{pmatrix} + t \begin{pmatrix} 1 \\ -1 \\ 2 \end{pmatrix}$ and $3x + y - z = 8$

(ii) $\mathbf{r} = \begin{pmatrix} 2 \\ 1 \\ -5 \end{pmatrix} + t \begin{pmatrix} 1 \\ -4 \\ 3 \end{pmatrix}$ and $x - 2y - 3z = 2$

(iii) $\mathbf{r} = \begin{pmatrix} 2 \\ 0 \\ 7 \end{pmatrix} + t \begin{pmatrix} -3 \\ 2 \\ -5 \end{pmatrix}$ and $2x - 3y + z = 5$

(iv) $\mathbf{r} = \begin{pmatrix} -2 \\ 1 \\ 4 \end{pmatrix} + t \begin{pmatrix} 3 \\ -4 \\ 0 \end{pmatrix}$ and $4x + 3y + z = -1$

(v) $\mathbf{r} = \begin{pmatrix} 2 \\ 1 \\ 0 \end{pmatrix} + t \begin{pmatrix} -5 \\ 4 \\ 7 \end{pmatrix}$ and $x + 2y - 6z = 0$

(vi) $\mathbf{r} = \begin{pmatrix} 2 \\ 3 \\ 5 \end{pmatrix} + t \begin{pmatrix} -1 \\ 2 \\ 5 \end{pmatrix}$ and $3x + 4y - z = 7$

2 The points L, M and N have co-ordinates $(0, -1, 2)$, $(2, 1, 0)$ and $(5, 1, 1)$.

(i) Write down the vectors $\overrightarrow{LM}$ and $\overrightarrow{LN}$.

(ii) Show that $\overrightarrow{LM} \cdot \begin{pmatrix} 1 \\ -4 \\ -3 \end{pmatrix} = \overrightarrow{LN} \cdot \begin{pmatrix} 1 \\ -4 \\ -3 \end{pmatrix} = 0$.

(iii) Find the equation of the plane LMN.

3 (i) Show that the points A(1, 1, 1), B(3, 0, 0) and C(2, 0, 2) all lie in the plane $2x + 3y + z = 6$.

(ii) Show that $\overrightarrow{AB} \cdot \begin{pmatrix} 2 \\ 3 \\ 1 \end{pmatrix} = \overrightarrow{AC} \cdot \begin{pmatrix} 2 \\ 3 \\ 1 \end{pmatrix} = 0$

(iii) The point D has co-ordinates (7, 6, 2). D lies on a line perpendicular to the plane through one of the points A, B or C.
Through which of these points does the line pass?

4 The lines l, $\mathbf{r} = \begin{pmatrix} 2 \\ 1 \\ 0 \end{pmatrix} + \lambda \begin{pmatrix} 1 \\ 1 \\ 1 \end{pmatrix}$, and m, $\mathbf{r} = \begin{pmatrix} 4 \\ 0 \\ 2 \end{pmatrix} + \mu \begin{pmatrix} 1 \\ 0 \\ 1 \end{pmatrix}$, lie in the same plane π.

(i) Find the co-ordinates of any two points on each of the lines.

(ii) Show that all the four points you found in part (i) lie on the plane $x - z = 2$.

(iii) Explain why you now have more than sufficient evidence to show that the plane π has equation $x - z = 2$.

(iv) Find the co-ordinates of the point where the lines l and m intersect.

5 Find the points of intersection of the following planes and lines.

(i) $x + 2y + 3z = 11$ and $\mathbf{r} = \begin{pmatrix} 1 \\ 2 \\ 4 \end{pmatrix} + \lambda \begin{pmatrix} 1 \\ 1 \\ 1 \end{pmatrix}$

(ii) $2x + 3y - 4z = 1$ and $\mathbf{r} = \begin{pmatrix} -2 \\ -3 \\ -4 \end{pmatrix} + \lambda \begin{pmatrix} 3 \\ 4 \\ 5 \end{pmatrix}$

(iii) $3x - 2y - z = 14$ and $\mathbf{r} = \begin{pmatrix} 8 \\ 4 \\ 2 \end{pmatrix} + \lambda \begin{pmatrix} 1 \\ 2 \\ 1 \end{pmatrix}$

(iv) $x + y + z = 0$ and $\mathbf{r} = \lambda \begin{pmatrix} 1 \\ 1 \\ 2 \end{pmatrix}$

(v) $5x - 4y - 7z = 49$ and $\mathbf{r} = \begin{pmatrix} 3 \\ -1 \\ 2 \end{pmatrix} + \lambda \begin{pmatrix} 2 \\ 5 \\ -3 \end{pmatrix}$

6 In each of the following examples you are given a point A and a plane π. Find

(a) the equation of the line through A perpendicular to π
(b) the point of intersection, P, of this line with π
(c) the distance AP.

(i) A is (2, 2, 3); π is $x - y + 2z = 0$
(ii) A is (2, 3, 0); π is $2x + 5y + 3z = 0$
(iii) A is (3, 1, 3); π is $x = 0$
(iv) A is (2, 1, 0); π is $3x - 4y + z = 2$
(v) A is (0, 0, 0); π is $x + y + z = 6$

7 The points U and V have co-ordinates (4, 0, 7) and (6, 4, 13).
The line UV is perpendicular to a plane and the point U lies in the plane.

(i) Find the equation of the plane in cartesian form.
(ii) The point W has co-ordinates (−1, 10, 2).
Show that $WV^2 = WU^2 + UV^2$.
(iii) What information does this give you about the position of W?
Confirm this information by a different method.

8 (i) Find the equation of the line through (13, 5, 0) parallel to the line
$\mathbf{r} = \begin{pmatrix} 2 \\ -1 \\ 4 \end{pmatrix} + \lambda \begin{pmatrix} 3 \\ 1 \\ -2 \end{pmatrix}.$

(ii) Where does this line meet the plane $3x + y - 2z = 2$?
(iii) How far is the point of intersection from (13, 5, 0)?

9 **(i)** Find the angle between the line $\mathbf{r} = \mathbf{i} + 2\mathbf{j} + t(3\mathbf{i} + 2\mathbf{j} - \mathbf{k})$ and the plane $2x - 3y - z = 1$.

(ii) Find the angle between the line $\mathbf{r} = \begin{pmatrix} -1 \\ 0 \\ 2 \end{pmatrix} + t \begin{pmatrix} 1 \\ 3 \\ -2 \end{pmatrix}$ and the plane $4x - 3z = -2$.

(iii) Find the angle between the line $\mathbf{r} = \mathbf{i} + 2\mathbf{j} + t(3\mathbf{i} + 2\mathbf{j} - \mathbf{k})$ and the plane $7x - 2y + z = 1$.

10 A is the point (1, 2, 0), B is (0, 4, 1) and C is (9, −2, 1).

(i) Show that A, B and C lie in the plane $2x + 3y - 4z = 8$.

(ii) Write down the vectors $\overrightarrow{AB}$ and $\overrightarrow{AC}$ and verify that they are at right angles to $\begin{pmatrix} 2 \\ 3 \\ -4 \end{pmatrix}$.

(iii) Find the angle BAC.

(iv) Find the area of triangle ABC (using area = $\frac{1}{2} bc \sin A$).

11 P is the point (2, −1, 3), Q is (5, −5, 3) and R is (7, 2, −3). Find

(i) the lengths of PQ and QR

(ii) the angle PQR

(iii) the area of triangle PQR

(iv) the point S such that PQRS is a parallelogram.

12 P is the point (2, 2, 4), Q is (0, 6, 8), X is (−2, −2, −3) and Y is (2, 6, 9).

(i) Write in vector form the equations of the lines PQ and XY.

(ii) Verify that the equation of the plane PQX is $2x + 5y - 4z = -2$.

(iii) Does the point Y lie in the plane PQX?

(iv) Does any point on PQ lie on XY? (That is, do the lines intersect?)

13 You are given the four points O(0, 0, 0), A(5, −12, 16), B(8, 3, 19) and C(−23, −80, 12).

(i) Show that the three points A, B and C all lie in the plane with equation $2x - y + 3z = 70$.

(ii) Write down a vector which is normal to this plane.

(iii) The line from the origin O perpendicular to this plane meets the plane at D. Find the co-ordinates of D.

(iv) Write down the equations of the two lines OA and AB in vector form.

(v) Hence find the angle OAB, correct to the nearest degree.

[MEI]

14 A pyramid in the shape of a tetrahedron has base ABC and vertex P as shown in the diagram. The vertices A, B, C, P have position vectors

$$\mathbf{a} = -4\mathbf{j} + 2\mathbf{k},$$
$$\mathbf{b} = 2\mathbf{i} + 4\mathbf{k},$$
$$\mathbf{c} = -5\mathbf{i} - 2\mathbf{j} + 6\mathbf{k},$$
$$\mathbf{p} = 3\mathbf{i} - 8\mathbf{j} + 12\mathbf{k}$$

respectively.

The equation of the plane of the base is

$$\mathbf{r}.\begin{pmatrix} 2 \\ -3 \\ 4 \end{pmatrix} = 20.$$

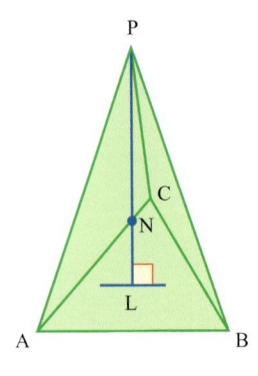

(i) Write down a vector which is normal to the base ABC.

The line through P, perpendicular to the base, cuts the base at L.

(ii) Find the equation of the line PL in vector form and use it to find the co-ordinates of L.

(iii) Find the co-ordinates of the point N on LP, such that $\overrightarrow{LN} = \frac{1}{4}\overrightarrow{LP}$.

(iv) Find the angle between PA and PL.

[MEI]

15 The position vectors of three points A, B, C on a plane ski-slope are

$$\mathbf{a} = 4\mathbf{i} + 2\mathbf{j} - \mathbf{k}, \quad \mathbf{b} = -2\mathbf{i} + 26\mathbf{j} + 11\mathbf{k}, \quad \mathbf{c} = 16\mathbf{i} + 17\mathbf{j} + 2\mathbf{k},$$

where the units are metres.

(i) Show that the vector $2\mathbf{i} - 3\mathbf{j} + 7\mathbf{k}$ is perpendicular to $\overrightarrow{AB}$ and also perpendicular to $\overrightarrow{AC}$.

Hence find the equation of the plane of the ski-slope.

The track for an overhead railway lies along DEF, where D and E have position vectors $\mathbf{d} = 130\mathbf{i} - 40\mathbf{j} + 20\mathbf{k}$ and $\mathbf{e} = 90\mathbf{i} - 20\mathbf{j} + 15\mathbf{k}$, and F is a point on the ski-slope.

(ii) Find the equation of the straight line DE.
(iii) Find the position vector of the point F.
(iv) Find the length of the track DF.

[MEI]

16 A tunnel is to be excavated through a hill. In order to define position, co-ordinates (x, y, z) are taken relative to an origin O such that x is the distance east from O, y is the distance north and z is the vertical distance upwards, with one unit equal to 100 m.

The tunnel starts at point A(2, 3, 5) and runs in the direction $\begin{pmatrix} 1 \\ 1 \\ -0.5 \end{pmatrix}$.

It meets the hillside again at B. At B the side of the hill forms a plane with equation $x + 5y + 2z = 77$.

(i) Write down the equation of the line AB in the form $\mathbf{r} = \mathbf{u} + \lambda \mathbf{t}$.

(ii) Find the co-ordinates of B.

(iii) Find the angle which AB makes with the upward vertical.

(iv) An old tunnel through the hill has equation $\mathbf{r} = \begin{pmatrix} 4 \\ 1 \\ 2 \end{pmatrix} + \mu \begin{pmatrix} 7 \\ 15 \\ 0 \end{pmatrix}$.

Show that the point P on AB where $x = 7\tfrac{1}{2}$ is directly above a point Q in the old tunnel. Find the vertical separation PQ of the tunnels at this point.

[MEI]

17 ABCD is a parallelogram. The co-ordinates of A, B and D are $(-1, 1, 2)$, $(1, 2, 0)$ and $(1, 0, 2)$ respectively.

(i) Find the co-ordinates of C.

(ii) Use a scalar product to find the size of angle BAD.

(iii) Show that the vector $\mathbf{i} + 2\mathbf{j} + 2\mathbf{k}$ is perpendicular to the plane ABCD.

(iv) The diagonals AC and BD intersect at the point E.
Find a vector equation of the straight line l through E perpendicular to the plane ABCD.

(v) A point F lies on l and is 3 units from A.
Find the co-ordinates of the two possible positions of F.

[MEI]

18 The line l has equation $\mathbf{r} = 4\mathbf{i} + 2\mathbf{j} - \mathbf{k} + t(2\mathbf{i} - \mathbf{j} - 2\mathbf{k})$. It is given that l lies in the plane with equation $2x + by + cz = 1$, where b and c are constants.

(i) Find the values of b and c.

(ii) The point P has position vector $2\mathbf{j} + 4\mathbf{k}$. Show that the perpendicular distance from P to l is $\sqrt{5}$.

[Cambridge International AS & A Level Mathematics 9709, Paper 3 Q9 June 2009]

19 With respect to the origin O, the points A and B have position vectors given by

$$\overrightarrow{OA} = 2\mathbf{i} + 2\mathbf{j} + \mathbf{k} \quad \text{and} \quad \overrightarrow{OB} = \mathbf{i} + 4\mathbf{j} + 3\mathbf{k}.$$

The line l has vector equation $\mathbf{r} = 4\mathbf{i} - 2\mathbf{j} + 2\mathbf{k} + s(\mathbf{i} + 2\mathbf{j} + \mathbf{k})$.

(i) Prove that the line l does not intersect the line through A and B.

(ii) Find the equation of the plane containing l and the point A, giving your answer in the form $ax + by + cz = d$.

[Cambridge International AS & A Level Mathematics 9709, Paper 3 Q10 June 2005]

20 The points A and B have position vectors, relative to the origin O, given by

$$\overrightarrow{OA} = \begin{pmatrix} -1 \\ 3 \\ 5 \end{pmatrix} \quad \text{and} \quad \overrightarrow{OB} = \begin{pmatrix} 3 \\ -1 \\ -4 \end{pmatrix}.$$

The line l passes through A and is parallel to OB. The point N is the foot of the perpendicular from B to l.

(i) State a vector equation for the line l.

(ii) Find the position vector of N and show that BN = 3.

(iii) Find the equation of the plane containing A, B and N, giving your answer in the form $ax + by + cz = d$.

[Cambridge International AS & A Level Mathematics 9709, Paper 3 Q10 June 2006]

21 The straight line l has equation $\mathbf{r} = \mathbf{i} + 6\mathbf{j} - 3\mathbf{k} + s(\mathbf{i} - 2\mathbf{j} + 2\mathbf{k})$. The plane p has equation $(\mathbf{r} - 3\mathbf{i}) \cdot (2\mathbf{i} - 3\mathbf{j} + 6\mathbf{k}) = 0$. The line l intersects the plane p at the point A.

(i) Find the position vector of A.

(ii) Find the acute angle between l and p.

(iii) Find a vector equation for the line which lies in p, passes through A and is perpendicular to l.

[Cambridge International AS & A Level Mathematics 9709, Paper 3 Q10 November 2007]

The intersection of two planes

If you look around you, will find objects which can be used to represent planes – walls, floors, ceilings, doors, roofs, and so on. You will see that the intersection of two planes is a straight line.

EXAMPLE 10.15 Find l, the line of intersection of the two planes

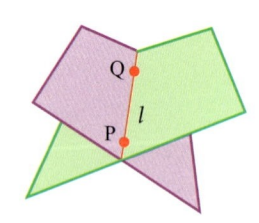

$$3x + 2y - 3z = -18 \quad \text{and} \quad x - 2y + z = 12.$$

Figure 10.14

SOLUTION 1

This solution depends on finding two points on l.

You can find one point by arbitrarily choosing to put $y = 0$ into the equations of the planes and solving simultaneously:

$$\left. \begin{matrix} 3x - 3z = -18 \\ x + z = 12 \end{matrix} \right\} \Leftrightarrow \begin{cases} x - z = -6 \\ x + z = 12 \end{cases} \Leftrightarrow x = 3, z = 9.$$

So P with co-ordinates $(3, 0, 9)$ is a point on l.

(You could run into difficulties putting $y = 0$ as it is possible that the line has no points where $y = 0$. In this case your simultaneous equations for x and z would be inconsistent; you would then choose a value for x or z instead.)

In the same way, arbitrarily choosing to put $z = 1$ into the equations gives

$$\left.\begin{array}{l}3x + 2y = -15 \\ x - 2y = 11\end{array}\right\} \Leftrightarrow \left\{\begin{array}{l}4x = -4 \\ 2y = x - 11\end{array}\right\} \Leftrightarrow x = -1, y = -6$$

so Q with co-ordinates $(-1, -6, 1)$ is a point on l.

$$\overrightarrow{PQ} = \begin{pmatrix} -1 \\ -6 \\ 1 \end{pmatrix} - \begin{pmatrix} 3 \\ 0 \\ 9 \end{pmatrix} = \begin{pmatrix} -4 \\ -6 \\ -8 \end{pmatrix} = -2\begin{pmatrix} 2 \\ 3 \\ 4 \end{pmatrix}.$$

Removing factor -2 makes the arithmetic simpler.

Use $\begin{pmatrix} 2 \\ 3 \\ 4 \end{pmatrix}$ as the direction vector for l.

The vector equation for l is $\mathbf{r} = \begin{pmatrix} -1 \\ -6 \\ 1 \end{pmatrix} + t\begin{pmatrix} 2 \\ 3 \\ 4 \end{pmatrix}$.

SOLUTION 2

In this solution the original two equations in x, y and z are solved, expressing each of x, y and z in terms of some parameter.

Put $x = \lambda$ into $\begin{cases} 3x + 2y - 3z = -18 \\ x - 2y + z = 12 \end{cases}$ and solve simultaneously for y and z:

$$\left.\begin{array}{l}2y - 3z = -18 - 3\lambda \\ -2y + z = 12 - \lambda\end{array}\right\} \Rightarrow -2z = -6 - 4\lambda \Rightarrow z = 2\lambda + 3$$

so that $2y = 3z - 18 - 3\lambda \Rightarrow 2y = 3(2\lambda + 3) - 18 - 3\lambda \Rightarrow 2y = 3\lambda - 9 \Rightarrow y = \frac{3}{2}\lambda - \frac{9}{2}$.

Thus the equations for l are

$$\begin{cases} x = \lambda \\ y = \frac{3}{2}\lambda - \frac{9}{2} \\ z = 2\lambda + 3 \end{cases} \quad \text{or} \quad \begin{pmatrix} x \\ y \\ z \end{pmatrix} = \begin{pmatrix} 0 \\ -\frac{9}{2} \\ 3 \end{pmatrix} + \lambda \begin{pmatrix} 1 \\ \frac{3}{2} \\ 2 \end{pmatrix}.$$

Note

This set of equations is different from but equivalent to the equations in Solution 1. The equivalence is most easily seen by substituting $2\mu - 1$ for λ, obtaining

$$\begin{cases} x = 2\mu - 1 \\ y = \frac{3}{2}(2\mu - 1) - \frac{9}{2} = 3\mu - 6 \\ z = 2(2\mu - 1) + 3 = 4\mu + 1 \end{cases}$$

The angle between two planes

The angle between two planes can be found by using the scalar product. As figures 10.15 and 10.16 make clear, the angle between planes π_1 and π_2 is the same as the angle between their normals, $\mathbf{n}_1$ and $\mathbf{n}_2$.

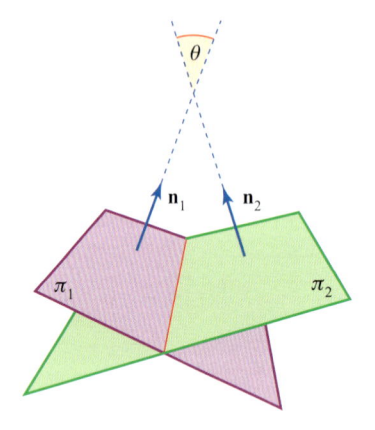

Figure 10.15

'Edge on' view

Figure 10.16

EXAMPLE 10.16 Find the acute angle between the planes π_1: $2x + 3y + 5z = 8$ and π_2: $5x + y - 4z = 12$.

SOLUTION

The planes have normals $\mathbf{n}_1 = \begin{pmatrix} 2 \\ 3 \\ 5 \end{pmatrix}$ and $\mathbf{n}_2 = \begin{pmatrix} 5 \\ 1 \\ -4 \end{pmatrix}$, so $\mathbf{n}_1 \cdot \mathbf{n}_2 = 10 + 3 - 20 = -7$.

The angle between the normals is θ, where

$$\cos\theta = \frac{\mathbf{n}_1 \cdot \mathbf{n}_2}{|\mathbf{n}_1||\mathbf{n}_2|} = \frac{-7}{\sqrt{38} \times \sqrt{42}}$$

$\Rightarrow \qquad \theta = 100.1°$ (to 1 decimal place)

Therefore the acute angle between the planes is 79.9°.

Sheaf of planes

When several planes share a common line the arrangement is known as a *sheaf of planes* (figure 10.17). The next example shows how you can find the equation of a plane which contains the line l common to two given planes, π_1 and π_2, without having to find the equation of l itself, or any points on l.

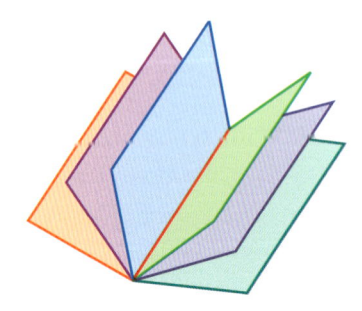

Figure 10.17

EXAMPLE 10.17 Find the equation of the plane which passes through the point $(1, 2, 3)$ and contains the common line of the planes $\pi_1: 2x + 2y + z + 3 = 0$ and $\pi_2: 2x + 3y + z + 13 = 0$.

SOLUTION

The equation

$$p(2x + 2y + z + 3) + q(2x + 3y + z + 13) = 0 \qquad ①$$

can be rearranged in the form $n_1 x + n_2 y + n_3 z = d$, where not all of a, b, c, d are zero provided p and q are not both zero. Therefore equation ① represents a plane. Further, any point (x, y, z) which satisfies both π_1 and π_2 will also satisfy equation ①. Thus equation ① represents a plane containing the common line of planes π_1 and π_2. Substituting $(1, 2, 3)$ into ① gives

$$12p + 24q = 0 \quad \Leftrightarrow \quad p = -2q.$$

The required equation is

$$-2q(2x + 2y + z + 3) + q(2x + 3y + z + 13) = 0$$
$$\Leftrightarrow \qquad -q(2x + y + z - 7) = 0$$

so that the required plane has equation $2x + y + z = 7$.

❓ Planes π_1 and π_2 have equations $a_1 x + b_1 y + c_1 z - d_1 = 0$ and $a_2 x + b_2 y + c_2 z - d_2 = 0$ respectively. Plane π_3 has equation

$$p(a_1 x + b_1 y + c_1 z - d_1) + q(a_2 x + b_2 y + c_2 z - d_2) = 0.$$

How is π_3 related to π_1 and π_2 if π_1 and π_2 are parallel?

EXERCISE 10F

1 Find the vector equation of the line of intersection of each of these pairs of planes.
   (i) $x + y - 6z = 4$, $5x - 2y - 3z = 13$
   (ii) $5x - y + z = 8$, $x + 3y + z = -4$
   (iii) $3x + 2y - 6z = 4$, $x + 5y - 7z = 2$
   (iv) $5x + 2y - 3z = -2$, $3x - 3y - z = 2$

2 Find the acute angle between each pair of planes in question **1**.

3 Find the vector equation of the line which passes through the given point and which is parallel to the line of intersection of the two planes.
   (i) $(-2, 3, 5)$, $4x - y + 3z = 5$, $3x - y + 2z = 7$
   (ii) $(4, -3, 2)$, $2x + 3y + 2z = 6$, $4x - 3y + z = 11$

4 Find the equation of the plane which goes through $(3, 2, -2)$ and which contains the common line of $x + 7y - 2z = 3$ and $2x - 3y + 2z = 1$.

5 Find the equation of the plane which contains the point $(1, -2, 3)$ and which is perpendicular to the common line of $5x - 3y - 4z = 2$ and $2x + y + 5z = 7$.

6 Find the equation of the line which goes through $(4, -2, -7)$ and which is parallel to both $2x - 5y - 2z = 8$ and $x + 3y - 3z = 12$.

7 The diagram shows the co-ordinates of the corners of parts of the roof of a warehouse.

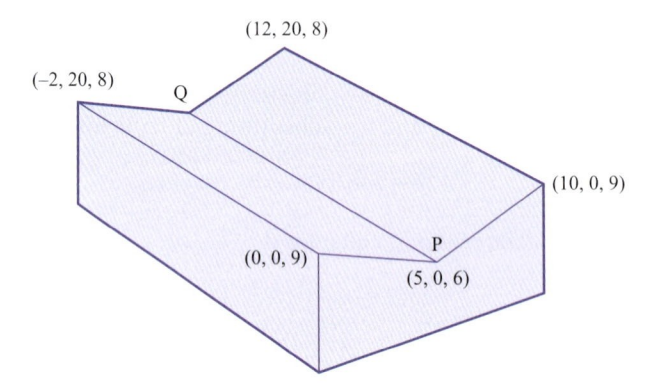

Find the equations of both roof sections, and the vector equation of the line PQ. Assuming that the z axis is vertical, what angle does PQ make with the horizontal?

8 Test drilling in the Namibian desert has shown the existence of gold deposits at $(400, 0, -400)$, $(-50, 500, -250)$, $(-200, -100, -200)$, where the units are in metres, the x axis points east, the y axis points north, and the z axis points up. Assume that these deposits are part of the same seam, contained in plane π.

(i) Find the equation of plane π.
(ii) Find the angle at which π is tilted to the horizontal.

The drilling positions $(400, 0, 3)$, $(-50, 500, 7)$, $(-200, -100, 5)$ are on the desert floor. Take the desert floor as a plane, Π.

(iii) Find the equation of Π.
(iv) Find the equation of the line where the plane containing the gold seam intersects the desert floor.
(v) How far south of the origin does the line found in part (iv) pass?

9 The diagram shows an arrow embedded in a target. The line of the arrow passes through the point A(2, 3, 5) and has direction vector $3\mathbf{i} + \mathbf{j} - 2\mathbf{k}$. The arrow intersects the target at the point B. The plane of the target has equation $x + 2y - 3z = 4$. The units are metres.

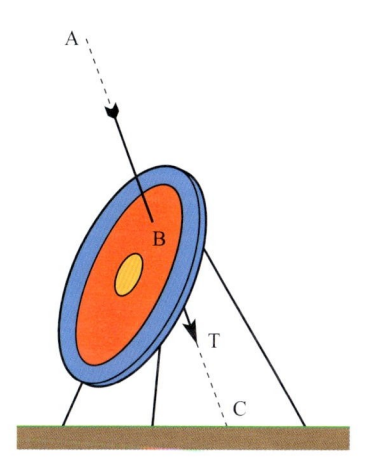

(i) Write down the vector equation of the line of the arrow in the form
   $\mathbf{r} = \mathbf{p} + \lambda \mathbf{q}$.

(ii) Find the value of λ which corresponds to B. Hence write down the co-ordinates of B.

(iii) The point C is where the line of the arrow meets the ground, which is the plane $z = 0$. Find the co-ordinates of C.

(iv) The tip, T, of the arrow is one-third of the way from B to C. Find the co-ordinates of T and the length of BT.

(v) Write down a normal vector to the plane of the target. Find the acute angle between the arrow and this normal.

[MEI]

10 A plane π has equation $ax + by + z = d$.

(i) Write down, in terms of a and b, a vector which is perpendicular to π.

Points A(2, −1, 2), B(4, −4, 2), C(5, −6, 3) lie on π.

(ii) Write down the vectors $\overrightarrow{AB}$ and $\overrightarrow{AC}$.

(iii) Use scalar products to obtain two equations for a and b.

(iv) Find the equation of the plane π.

(v) Find the angle which the plane π makes with the plane $x = 0$.

(vi) Point D is the mid-point of AC. Point E is on the line between D and B such that DE : EB = 1 : 2. Find the co-ordinates of E.

[MEI]

11 The diagram, which is not to scale, illustrates part of the roof of a building. Lines OA and OD are horizontal and at right angles. Lines BC and BE are also horizontal and at right angles. Line BC is parallel to OA and BE is parallel to OD.

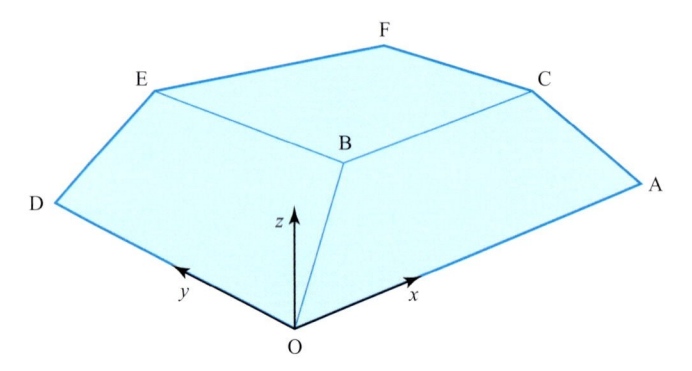

Axes are taken with O as origin, the x axis along OA, the y axis along OD and the z axis vertically upwards. The units are metres.

Point A has the co-ordinates (50, 0, 0) and point D has the co-ordinates (0, 20, 0).

The equation of line OB is $\begin{pmatrix} x \\ y \\ z \end{pmatrix} = \lambda \begin{pmatrix} 4 \\ 3 \\ 2 \end{pmatrix}$. The equation of plane CBEF is $z = 3$.

(i) Find the co-ordinates of B.
(ii) Verify that the equation of plane AOBC is $2y - 3z = 0$.
(iii) Find the equation of plane DOBE.
(iv) Write down normal vectors for planes AOBC and DOBE. Find the angle between these normal vectors. Hence write down the internal angle between the two roof surfaces AOBC and DOBE.

[MEI, *adapted*]

12 The plane p has equation $3x + 2y + 4z = 13$. A second plane q is perpendicular to p and has equation $ax + y + z = 4$, where a is a constant.

(i) Find the value of a.
(ii) The line with equation $\mathbf{r} = \mathbf{j} - \mathbf{k} + \lambda(\mathbf{i} + 2\mathbf{j} + 2\mathbf{k})$ meets the plane p at the point A and the plane q at the point B. Find the length of AB.

[Cambridge International AS & A Level Mathematics 9709, Paper 32 Q9 June 2010]

13 The diagram shows a set of rectangular axes Ox, Oy and Oz, and three points A, B and C with position vectors $\overrightarrow{OA} = \begin{pmatrix} 2 \\ 0 \\ 0 \end{pmatrix}$, $\overrightarrow{OB} = \begin{pmatrix} 1 \\ 2 \\ 0 \end{pmatrix}$ and $\overrightarrow{OC} = \begin{pmatrix} 1 \\ 1 \\ 2 \end{pmatrix}$.

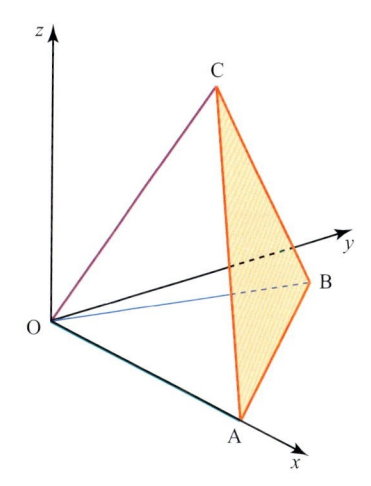

(i) Find the equation of the plane ABC, giving your answer in the form $ax + by + cz = d$.

(ii) Calculate the acute angle between the planes ABC and OAB.

[Cambridge International AS & A Level Mathematics 9709, Paper 3 Q9 June 2007]

14 Two planes have equations $2x - y - 3z = 7$ and $x + 2y + 2z = 0$.

(i) Find the acute angle between the planes.

(ii) Find a vector equation for their line of intersection.

[Cambridge International AS & A Level Mathematics 9709, Paper 3 Q7 November 2008]

15 The plane p has equation $2x - 3y + 6z = 16$. The plane q is parallel to p and contains the point with position vector $\mathbf{i} + 4\mathbf{j} + 2\mathbf{k}$.

(i) Find the equation of q, giving your answer in the form $ax + by + cz = d$.

(ii) Calculate the perpendicular distance between p and q.

(iii) The line l is parallel to the plane p and also parallel to the plane with equation $x - 2y + 2z = 5$. Given that l passes through the origin, find a vector equation for l.

[Cambridge International AS & A Level Mathematics 9709, Paper 32 Q10 November 2009]

KEY POINTS

1 The position vector $\overrightarrow{OP}$ of a point P is the vector joining the origin to P.

2 The vector $\overrightarrow{AB}$ is $\mathbf{b} - \mathbf{a}$, where $\mathbf{a}$ and $\mathbf{b}$ are the position vectors of A and B.

3 The vector $\mathbf{r}$ often denotes the position vector of a general point.

4 The vector equation of the line through A with direction vector $\mathbf{u}$ is given by

$\mathbf{r} = \mathbf{a} + \lambda \mathbf{u}$.

5 The vector equation of the line through points A and B is given by

$$\mathbf{r} = \overrightarrow{OA} + \lambda \overrightarrow{AB}$$
$$= \mathbf{a} + \lambda(\mathbf{b} - \mathbf{a})$$
$$= (1 - \lambda)\mathbf{a} + \lambda\mathbf{b}.$$

6 The vector equation of the line through (a_1, a_2, a_3) in the direction $\begin{pmatrix} u_1 \\ u_2 \\ u_3 \end{pmatrix}$ is

$$\mathbf{r} = \begin{pmatrix} a_1 \\ a_2 \\ a_3 \end{pmatrix} + \lambda \begin{pmatrix} u_1 \\ u_2 \\ u_3 \end{pmatrix}.$$

7 The angle between two vectors, **a** and **b**, is given by θ in

$$\cos\theta = \frac{\mathbf{a} \cdot \mathbf{b}}{|\mathbf{a}||\mathbf{b}|}$$

where $\mathbf{a} \cdot \mathbf{b} = a_1 b_1 + a_2 b_2$ (in two dimensions)
$= a_1 b_1 + a_2 b_2 + a_3 b_3$ (in three dimensions).

8 The cartesian equation of a plane perpendicular to the vector $\mathbf{n} = \begin{pmatrix} n_1 \\ n_2 \\ n_3 \end{pmatrix}$ is

$$n_1 x + n_2 y + n_3 z = d.$$

e 9 The vector equation of the plane through the points A, B and C is

$$\mathbf{r} = \overrightarrow{OA} + \lambda \overrightarrow{AB} + \mu \overrightarrow{AC}.$$

10 The equation of the plane through the point with position vector **a**, and perpendicular to **n**, is given by $(\mathbf{r} - \mathbf{a}) \cdot \mathbf{n} = 0$.

11 The distance of the point (α, β, γ) from the plane $n_1 x + n_2 y + n_3 z = d$ is

$$\frac{|n_1\alpha + n_2\beta + n_3\gamma - d|}{\sqrt{n_1^2 + n_2^2 + n_3^2}}.$$

If the plane is written $ax + by + cz = d$, the formula for the distance is

$$\frac{|a\alpha + b\beta + c\gamma - d|}{\sqrt{a^2 + b^2 + c^2}}$$

12 The angle between a line and a plane is found by first considering the angle between the line and a normal to the plane.

13 To find the equation of l, the line of intersection of the planes

$$a_1 x + b_1 y + c_1 z = d_1 \quad \text{and} \quad a_2 x + b_2 y + c_2 z = d_2$$

- find a point P on l by choosing a value for one of x, y, or z, substituting this into both equations, and then solving simultaneously to find the other two variables;
- then write down the vector equation of l.

14 The angle between two planes is the same as the angle between their normals.

11 Complex numbers

> ... that wonder of analysis, that portent of the ideal world, that amphibian between being and not-being, which we call the imaginary root of negative unity.
>
> *Leibniz, 1702*

The growth of the number system

The number system we use today has taken thousands of years to develop. In primitive societies all that are needed are the *counting numbers*, 1, 2, 3, ... (or even just the first few of these).

The concept of a *fraction* was first recorded in a systematic way in an Egyptian papyrus of about 1650 BC. By 500 BC the Greeks had developed ways of calculating with whole numbers and their ratios (which accounts for calling fractions *rational numbers*). The followers of Pythagoras believed that everything in geometry and in applications of mathematics could be explained in terms of rational numbers.

It came as a great shock, therefore, when one of them proved that $\sqrt{2}$ was not a rational number. However, Greek thinkers gradually came to terms with the existence of such *irrational numbers*, and by 370 BC Eudoxus had devised a very careful theory of proportion which included both rational and irrational numbers.

It took about another thousand years for the next major development, when the Hindu mathematician Brahmagupta (in about AD 630) described *negative numbers* and gave the rules for dealing with negative signs. Surprisingly, the first use of a symbol for zero came even later, in AD 876. This was the final element needed to complete the set of real numbers, consisting of positive and negative rational and irrational numbers and zero.

Figure 11.1 (overleaf) shows the relationships between the different types of numbers.

Complex numbers

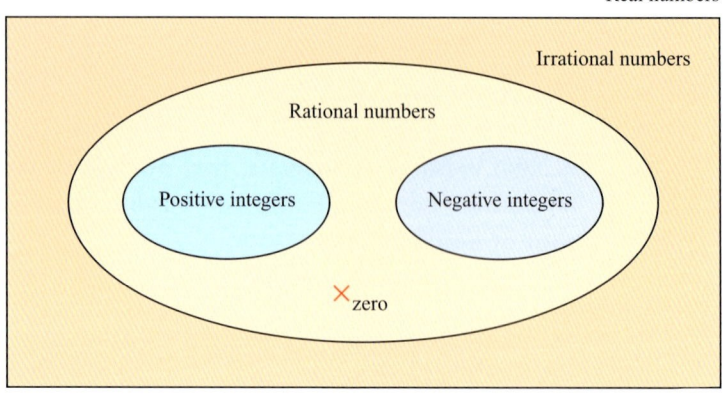

Figure 11.1

ACTIVITY 11.1 Copy figure 11.1 and write the following numbers in the correct positions.

$$3 \quad \pi \quad \frac{355}{113} \quad -1 \quad -1.4142 \quad -\sqrt{2}$$

Draw also a real number line and mark the same numbers on it.

The number system expanded in this way because people wanted to increase the range of problems they could tackle. This can be illustrated in terms of the sorts of equation that can be solved at each stage, although of course the standard algebraic way of writing these is relatively modern.

ACTIVITY 11.2 For each of these equations, make up a simple problem that would lead to the equation and say what sort of number is needed to solve the equation.

(i) $x + 7 = 10$
(ii) $7x = 10$
(iii) $x^2 = 10$
(iv) $x + 10 = 7$
(v) $x^2 + 7x = 0$
(vi) $x^2 + 10 = 0$

You will have hit a snag with equation **(vi)**. Since the square of every real number is positive or zero, there is no real number with a square of −10. This is a simple example of a quadratic equation with no real roots. The existence of such equations was recognised and accepted for hundreds of years, just as the Greeks had accepted that $x + 10 = 7$ had no solution.

Then two 16th century Italians, Tartaglia and Cardano, found methods of solving cubic and quartic (fourth degree) equations which forced mathematicians to take seriously the square roots of negative numbers. This required a further extension of the number system, to produce what are called *complex numbers*.

Complex numbers were regarded with great suspicion for many years. Descartes called them 'imaginary', Newton called them 'impossible', and Leibniz's mystification has already been quoted. But complex numbers turned out to be very useful, and had become accepted as an essential tool by the time Gauss first gave them a firm logical basis in 1831.

Working with complex numbers

Faced with the problem of wanting the square root of a negative number, we make the following Bold Hypothesis.

> The real number system can be extended by including a new number, denoted by i, which combines with itself and the real numbers according to the usual laws of algebra, but which has the additional property that $i^2 = -1$.

The original notation for i was ι, the Greek letter iota. The letter j is also commonly used instead of i.

The first thing to note is that we do not need further symbols for other square roots. For example, since $-196 = 196 \times (-1) = 14^2 \times i^2$, we see that -196 has two square roots, $\pm 14i$. The following example uses this idea to solve a quadratic equation with no real roots.

EXAMPLE 11.1 Solve the equation $z^2 - 6z + 58 = 0$, and check the roots.

(We use the letter z for the variable here because we want to keep x and y to stand for *real* numbers.)

SOLUTION

Using the quadratic formula:

$$z = \frac{6 \pm \sqrt{6^2 - 4 \times 58}}{2}$$

$$= \frac{6 \pm \sqrt{-196}}{2}$$

$$= \frac{6 \pm 14i}{2}$$

$$= 3 \pm 7i$$

To check:

$$z = 3 + 7i \Rightarrow z^2 - 6z + 58 = (3 + 7i)^2 - 6(3 + 7i) + 58$$
$$= 9 + 42i + 49i^2 - 18 - 42i + 58$$
$$= 9 + 42i - 49 - 18 - 42i + 58 \quad \leftarrow i^2 = -1$$
$$= 0$$

Notice that here 0 means $0 + 0i$.

ACTIVITY 11.3 Check the other root, $z = 3 - 7i$.

A number z of the form $x + iy$, where x and y are real, is called a *complex number*. x is called the real part of the complex number, denoted by Re(z), and y is called the imaginary part, denoted by Im(z). So if, for example, $z = 3 - 7i$ then Re(z) = 3 and Im(z) = -7. Notice in particular that the imaginary part is real!

In Example 11.1 you did some simple calculations with complex numbers. The general methods for addition, subtraction and multiplication are similarly straightforward.

Addition: add the real parts and add the imaginary parts.

$$(x+iy) + (u+iv) = (x+u) + i(y+v)$$

Subtraction: subtract the real parts and subtract the imaginary parts.

$$(x+iy) - (u+iv) = (x-u) + i(y-v)$$

Multiplication: multiply out the brackets in the usual way and simplify, remembering that $i^2 = -1$.

$$(x+iy)(u+iv) = xu + ixv + iyu + i^2yv$$
$$= (xu - yv) + i(xv + yu)$$

Division of complex numbers is dealt with later in the chapter.

❓ What are the values of i^3, i^4, i^5?

Explain how you would work out the value of i^n for any positive integer value of n.

Complex conjugates

The complex number $x - iy$ is called the *complex conjugate*, or just the *conjugate*, of $x + iy$. Simarly $x + iy$ is the complex conjugate of $x - iy$. $x + iy$ and $x - iy$ are a conjugate pair. The complex conjugate of z is denoted by z^*. If a polynomial equation, such as a quadratic, has real coefficients, then any complex roots will be conjugate pairs. This is the case in Example 11.1. If, however, the coefficients are not all real, this is no longer the case.

You can solve quadratic equations with complex coefficients in the same way as an ordinary quadratic, either by completing the square or by using the quadratic formula. This is shown in the next example.

EXAMPLE 11.2

Solve $z^2 - 4iz - 13 = 0$.

SOLUTION

Substitute $a = 1$, $b = -4i$ and $c = -13$ into the quadratic formula.

$$z = \frac{-b \pm \sqrt{b^2 - 4ac}}{2a}$$

$$= \frac{4i \pm \sqrt{(-4i)^2 - 4 \times 1 \times (-13)}}{2}$$

$$= \frac{4i \pm \sqrt{-16 + 52}}{2}$$

$$= \frac{4i \pm \sqrt{36}}{2}$$

$$= \frac{4i \pm 6}{2}$$

$$= 2i \pm 3$$

So the roots are $3 + 2i$ and $-3 + 2i$.

ACTIVITY 11.4

(i) Let $z = 3 + 5i$ and $w = 1 - 2i$.
Find the following.

(a) $z + z^*$ (b) $w + w^*$ (c) zz^* (d) ww^*

What do you notice about your answers?

(ii) Let $z = x + iy$.
Show that $z + z^*$ and zz^* are real for any values of x and y.

EXERCISE 11A

1 Express the following in the form $x + iy$.

(i) $(8 + 6i) + (6 + 4i)$
(ii) $(9 - 3i) + (-4 + 5i)$
(iii) $(2 + 7i) - (5 + 3i)$
(iv) $(5 - i) - (6 - 2i)$
(v) $3(4 + 6i) + 9(1 - 2i)$
(vi) $3i(7 - 4i)$
(vii) $(9 + 2i)(1 + 3i)$
(viii) $(4 - i)(3 + 2i)$
(ix) $(7 + 3i)^2$
(x) $(8 + 6i)(8 - 6i)$
(xi) $(1 + 2i)(3 - 4i)(5 + 6i)$
(xii) $(3 + 2i)^3$

2 Solve each of the following equations, and check the roots in each case.

(i) $z^2 + 2z + 2 = 0$
(ii) $z^2 - 2z + 5 = 0$
(iii) $z^2 - 4z + 13 = 0$
(iv) $z^2 + 6z + 34 = 0$
(v) $4z^2 - 4z + 17 = 0$
(vi) $z^2 + 4z + 6 = 0$

3 Solve each of the following equations.

(i) $z^2 - 4iz - 4 = 0$
(ii) $z^2 - 2iz + 15 = 0$
(iii) $z^2 - 2iz - 2 = 0$
(iv) $z^2 + 6iz - 13 = 0$
(v) $z^2 + 8iz - 17 = 0$
(vi) $z^2 + iz + 6 = 0$

4 Given that $z = 2 + 3i$ and $w = 6 - 4i$, find the following.

(i) $\text{Re}(z)$ (ii) $\text{Im}(w)$
(iii) z^* (iv) w^*
(v) $z^* + w^*$ (vi) $z^* - w^*$
(vii) $\text{Im}(z + z^*)$ (viii) $\text{Re}(w - w^*)$
(ix) $zz^* - ww^*$ (x) $(z^3)^*$
(xi) $(z^*)^3$ (xii) $zw^* - z^*w$

5 Let $z = x + iy$.
Show that $(z^*)^* = z$.

6 Let $z_1 = x_1 + iy_1$ and $z_2 = x_2 + iy_2$.
Show that $(z_1 + z_2)^* = z_1^* + z_2^*$.

Division of complex numbers

Before tackling the slightly complicated problem of dividing by a complex number, you need to know what is meant by equality of complex numbers.

Two complex numbers $z = x + iy$ and $w = u + iv$ are equal if both $x = u$ and $y = v$. If $u \neq x$ or $v \neq y$, or both, then z and w are not equal.

You may feel that this is making a fuss about something which is obvious. However, think about the similar question of the equality of rational numbers. The rational numbers $\frac{x}{y}$ and $\frac{u}{v}$ are equal if $x = u$ and $y = v$.

❓ Is it possible for the rational numbers $\frac{x}{y}$ and $\frac{u}{v}$ to be equal if $u \neq x$ and $v \neq y$?

So for two complex numbers to be equal, the real parts must be equal and the imaginary parts must be equal. When we use this result we say that we are *equating real and imaginary parts*.

Equating real and imaginary parts is a very useful method which often yields 'two for the price of one' when working with complex numbers. The following example illustrates this.

EXAMPLE 11.3   Find real numbers p and q such that $p + qi = \dfrac{1}{3 + 5i}$.

SOLUTION

You need to find real numbers p and q such that
$$(p + iq)(3 + 5i) = 1.$$

Expanding gives
$$3p - 5q + i(5p + 3q) = 1.$$

Equating real and imaginary parts gives

Real: $3p - 5q = 1$
Imaginary: $5p + 3q = 0$

These simultaneous equations give $p = \frac{3}{34}$, $q = -\frac{5}{34}$ and so

$$\frac{1}{3+5i} = \frac{3}{34} - \frac{5}{34}i$$

ACTIVITY 11.5 By writing $\frac{1}{x+iy} = p + iq$, show that $\frac{1}{x+iy} = \frac{x-iy}{x^2+y^2}$.

This result shows that there is an easier way to find the reciprocal of a complex number. First, notice that

$$(x+iy)(x-iy) = x^2 - i^2y^2$$
$$= x^2 + y^2$$

which is real.

So to find the reciprocal of a complex number you multiply numerator and denominator by the complex conjugate of the denominator.

EXAMPLE 11.4 Find the real and imaginary parts of $\frac{1}{5+2i}$.

SOLUTION

Multiply numerator and denominator by $5 - 2i$. ◀ *5 − 2i is the conjugate of the denominator, 5 + 2i.*

$$\frac{1}{5+2i} = \frac{5-2i}{(5+2i)(5-2i)}$$
$$= \frac{5-2i}{25+4}$$
$$= \frac{5-2i}{29}$$

so the real part is $\frac{5}{29}$ and the imaginary part is $-\frac{2}{29}$.

Note

You may have noticed that this process is very similar to the process of rationalising a denominator. To make the denominator of $\frac{1}{3+\sqrt{2}}$ rational you have to multiply the numerator and denominator by $3 - \sqrt{2}$.

Similarly, division of complex numbers is carried out by multiplying both numerator and denominator by the conjugate of the denominator, as in the next example.

EXAMPLE 11.5 Express $\dfrac{9-4i}{2+3i}$ as a complex number in the form $x + iy$.

SOLUTION

$$\dfrac{9-4i}{2+3i} = \dfrac{9-4i}{2+3i} \times \dfrac{2-3i}{2-3i}$$

$$= \dfrac{18 - 27i - 8i + 12i^2}{2^2 + 3^2}$$

$$= \dfrac{6 - 35i}{13}$$

$$= \dfrac{6}{13} - \dfrac{35}{13}i$$

The square root of a complex number

The next example shows you how to find the square root of a complex number.

EXAMPLE 11.6 Find the two square roots of $8 + 6i$.

SOLUTION

Let $(x + iy)^2 = 8 + 6i$ ⎛ $+i^2y^2 = -y^2$ ⎞

$\Rightarrow \quad x^2 + 2ixy - y^2 = 8 + 6i$

Equating the real and imaginary parts gives:

    Real: $x^2 - y^2 = 8$ ①
    Imaginary: $2xy = 6$ ②

Rearranging ② gives

$$y = \dfrac{3}{x} \qquad ③$$

Substituting ③ into ① gives

$$x^2 - \dfrac{9}{x^2} = 8$$

$$x^4 - 9 = 8x^2$$

$$x^4 - 8x^2 - 9 = 0 \quad \longleftarrow \text{This is a quadratic in } x^2.$$

$$(x^2 - 9)(x^2 + 1) = 0$$

$\Rightarrow x^2 = -1$ which has no real roots

or $x^2 = 9 \Rightarrow x = \pm 3$.

⎛ Remember that x and y are both real numbers. ⎞

When $x = 3$, $y = 1$

When $x = -3$, $y = -1$

So the square roots of $8 + 6i$ are $3 + i$ and $-3 - i$.

> What are the values of $\frac{1}{i}, \frac{1}{i^2}$ and $\frac{1}{i^3}$?
>
> Explain how you would work out the value of $\frac{1}{i^n}$ for any positive integer value of n.

⚠ **e** **The collapse of a Bold Hypothesis**

You have just avoided a mathematical inconvenience (that −1 has no real square root) by introducing a new mathematical object, i, which has the property that you want: $i^2 = -1$.

What happens if you try the same approach to get rid of the equally inconvenient ban on dividing by zero? The problem here is that there is no real number equal to $1 \div 0$. So try making the Bold Hypothesis that you can introduce a new mathematical object which equals $1 \div 0$ but otherwise behaves like a real number. Denote this new object by ∞.

Then $1 \div 0 = \infty$, and so $1 = 0 \times \infty$.

But then you soon meet a contradiction:

$$2 \times 0 = 3 \times 0$$
$$\Rightarrow (2 \times 0) \times \infty = (3 \times 0) \times \infty$$
$$\Rightarrow 2 \times (0 \times \infty) = 3 \times (0 \times \infty)$$
$$\Rightarrow 2 \times 1 = 3 \times 1$$
$$\Rightarrow 2 = 3 \qquad \text{which is impossible.}$$

So this Bold Hypothesis quickly leads to trouble. How can you be sure that the same will never happen with complex numbers? For the moment you will just have to take on trust that there is an answer, and that all is well.

EXERCISE 11B

1 Express these complex numbers in the form $x + iy$.

(i) $\dfrac{1}{3+i}$ (ii) $\dfrac{1}{6-i}$ (iii) $\dfrac{5i}{6-2i}$

(iv) $\dfrac{7+5i}{6-2i}$ (v) $\dfrac{3+2i}{1+i}$ (vi) $\dfrac{47-23i}{6+i}$

(vii) $\dfrac{2-3i}{3+2i}$ (viii) $\dfrac{5-3i}{4+3i}$ (ix) $\dfrac{6+i}{2-5i}$

(x) $\dfrac{12-8i}{(2+2i)^2}$

2 Find real numbers a and b with $a > 0$ such that

(i) $(a+ib)^2 = 21 + 20i$ (ii) $(a+ib)^2 = -40 - 42i$
(iii) $(a+ib)^2 = -5 - 12i$ (iv) $(a+ib)^2 = -9 + 40i$
(v) $(a+ib)^2 = 1 - 1.875i$ (vi) $(a+ib)^2 = i$.

3 Find real numbers a and b such that

$$\frac{a}{3+i} + \frac{b}{1+2i} = 1 - i.$$

4 Solve these equations.

   (i) $(1+i)z = 3+i$

   (ii) $(3-4i)(z-1) = 10-5i$

   (iii) $(2+i)(z-7+3i) = 15-10i$

   (iv) $(3+5i)(z+2-5i) = 6+3i$

5 Find all the complex numbers z for which $z^2 = 2z^*$.

6 For $z = x + iy$, find $\frac{1}{z} + \frac{1}{z^*}$ in terms of x and y.

7 Show that

   (i) $\text{Re}(z) = \frac{z + z^*}{2}$

   (ii) $\text{Im}(z) = \frac{z + z^*}{2i}$.

8 (i) Expand and simplify $(a + ib)^3$.

   (ii) Deduce that if $(a + ib)^3$ is real then either $b = 0$ or $b^2 = 3a^2$.

   (iii) Hence find all the complex numbers z for which $z^3 = 1$.

9 (i) Expand and simplify $(z - \alpha)(z - \beta)$.
   Deduce that the quadratic equation with roots α and β is

   $$z^2 - (\alpha + \beta)z + \alpha\beta = 0,$$

   that is:

   $$z^2 - (\text{sum of roots})z + \text{product of roots} = 0.$$

   (ii) Using the result from part (i), find quadratic equations in the form $az^2 + bz + c = 0$ with the following roots.

   (a) $7 + 4i$, $7 - 4i$

   (b) $\frac{5i}{3}, -\frac{5i}{3}$

   (c) $-2 + \sqrt{8}i, -2 - \sqrt{8}i$

   (d) $2 + i, 3 + 2i$

10 Find the two square roots of each of these.

   (i) -9

   (ii) $3 + 4i$

   (iii) $-16 + 30i$

   (iv) $-7 - 24i$

   (v) $21 - 20i$

   (vi) $-5 - 12i$

Representing complex numbers geometrically

Since each complex number $x + iy$ can be defined by the ordered pair of real numbers (x, y), it is natural to represent $x + iy$ by the point with cartesian co-ordinates (x, y).

For example, in figure 11.2,

$2 + 3i$ is represented by $(2, 3)$
$-5 - 4i$ is represented by $(-5, -4)$
$2i$ is represented by $(0, 2)$
7 is represented by $(7, 0)$.

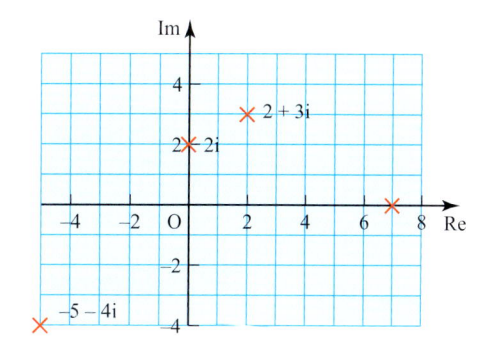

Figure 11.2

All real numbers are represented by points on the x axis, which is therefore called the *real axis*. Pure imaginary numbers (of the form $0 + iy$) give points on the y axis, which is called the *imaginary axis*. It is useful to label these Re and Im respectively. This geometrical illustration of complex numbers is called the *complex plane* or the *Argand diagram* after Jean-Robert Argand (1768–1822), a self-taught Swiss book-keeper who published an account of it in 1806.

ACTIVITY 11.6

(i) Copy figure 11.2.
For each of the four given points z mark also the point $-z$.
Describe the geometrical transformation which maps the point representing z to the point representing $-z$.

(ii) For each of the points z mark the point z^*, the complex conjugate of z.
Describe the geometrical transformation which maps the point representing z to the point representing z^*.

You will have seen in this activity that the points representing z and $-z$ have half-turn symmetry about the origin, and that the points representing z and z^* are reflections of each other in the real axis.

? How would you describe points that are reflections of each other in the imaginary axis?

Representing the sum and difference of complex numbers

Several mathematicians before Argand had used the complex plane representation. In particular, a Norwegian surveyor, Caspar Wessel (1745–1818), wrote a paper in 1797 (largely ignored until it was republished in French a century later) in which the complex number $x + iy$ is represented by the position vector $\begin{pmatrix} x \\ y \end{pmatrix}$, as shown in figure 11.3 (overleaf).

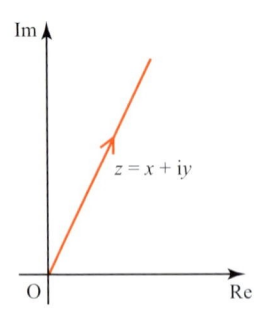

Figure 11.3

The advantage of this is that the addition of complex numbers can then be shown by the addition of the corresponding vectors.

$$\begin{pmatrix} x_1 \\ y_1 \end{pmatrix} + \begin{pmatrix} x_2 \\ y_2 \end{pmatrix} = \begin{pmatrix} x_1 + x_2 \\ y_1 + y_2 \end{pmatrix}$$

In an Argand diagram the position vectors representing z_1 and z_2 form two sides of a parallelogram, the diagonal of which is the vector $z_1 + z_2$ (see figure 11.4).

Figure 11.4

You can also represent z by any other directed line segment with components $\begin{pmatrix} x \\ y \end{pmatrix}$, not anchored at the origin as a position vector. Then addition can be shown as a triangle of vectors (see figure 11.5).

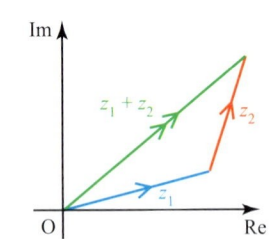

Figure 11.5

If you draw the other diagonal of the parallelogram, and let it represent the complex number w (see figure 11.6), then

$$z_2 + w = z_1 \Rightarrow w = z_1 - z_2.$$

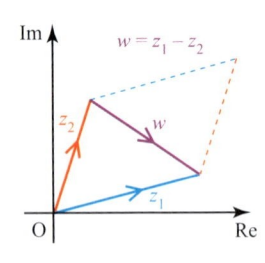

Figure 11.6

This gives a useful illustration of subtraction: the complex number $z_1 - z_2$ is represented by the vector from the point representing z_2 to the point representing z_1, as shown in figure 11.7. Notice the order of the points: the vector $z_1 - z_2$ starts at the point z_2 and goes to the point z_1.

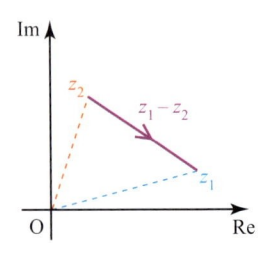

Figure 11.7

ACTIVITY 11.7 (i) Draw a diagram to illustrate $z_2 - z_1$.

(ii) Draw a diagram to illustrate that $z_1 - z_2 = z_1 + (-z_2)$.
Show that $z_1 + (-z_2)$ gives the same vector, $z_1 - z_2$ as before, but represented by a line segment in a different place.

The modulus of a complex number

Figure 11.8 shows the point representing $z = x + iy$ on an Argand diagram.

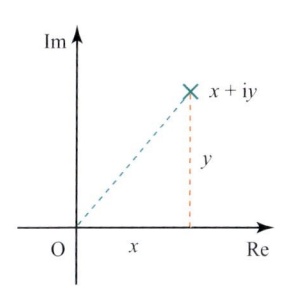

Figure 11.8

Using Pythagoras' theorem, you can see that the distance of this point from the origin is $\sqrt{x^2 + y^2}$. This distance is called the modulus of z, and is denoted by $|z|$.

So for the complex number $z = x + iy$, $|z| = \sqrt{x^2 + y^2}$.

If z is real, $z = x$ say, then $|z| = \sqrt{x^2}$, which is the absolute value of x, i.e. $|x|$. So the use of the modulus sign with complex numbers fits with its previous meaning for real numbers.

EXERCISE 11C

1 Represent each of the following complex numbers on a single Argand diagram, and find the modulus of each complex number.

(i) $3 + 2i$ (ii) $4i$ (iii) $-5 + i$
(iv) -2 (v) $-6 - 5i$ (vi) $4 - 3i$

2 Given that $z = 2 - 4i$, represent the following by points on a single Argand diagram.

(i) z (ii) $-z$ (iii) z^*
(iv) $-z^*$ (v) iz (vi) $-iz$
(vii) iz^* (viii) $(iz)^*$

3 Given that $z = 10 + 5i$ and $w = 1 + 2i$, represent the following complex numbers on an Argand diagram.

(i) z (ii) w (iii) $z + w$
(iv) $z - w$ (v) $w - z$

4 Given that $z = 3 + 4i$ and $w = 5 - 12i$, find the following.

  (i) $|z|$ (ii) $|w|$ (iii) $|zw|$

  (iv) $\left|\dfrac{z}{w}\right|$ (v) $\left|\dfrac{w}{z}\right|$

  What do you notice?

5 Let $z = 1 + i$.

  (i) Find z^n and $|z^n|$ for $n = -1, 0, 1, 2, 3, 4, 5$.
  (ii) Plot each of the points z^n from part (i) on a single Argand diagram. Join each point to its predecessor and to the origin.
  (iii) What do you notice?

6 Give a geometrical proof that $(-z)^* = -(z^*)$.

Sets of points in an Argand diagram

? In the last section, you saw that $|z|$ is the distance of the point representing z from the origin in the Argand diagram.

What do you think that $|z_2 - z_1|$ represents?

If $z_1 = x_1 + iy_1$ and $z_2 = x_2 + iy_2$, then $z_2 - z_1 = x_2 - x_1 + i(y_2 - y_1)$.

So $|z_2 - z_1| = \sqrt{(x_2 - x_1)^2 + (y_2 - y_1)^2}$.

Figure 11.9 shows an Argand diagram with the points representing the complex numbers $z_1 = x_1 + iy_1$ and $z_2 = x_2 + iy_2$ marked.

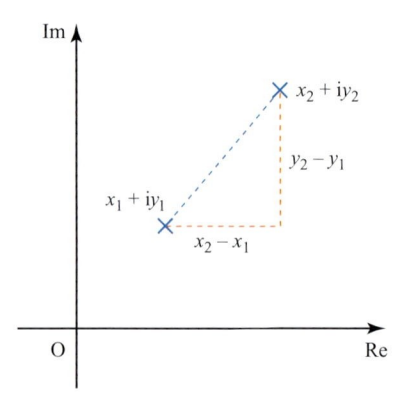

Figure 11.9

Using Pythagoras' theorem, you can see that the distance between z_1 and z_2 is given by $\sqrt{(x_2 - x_1)^2 + (y_2 - y_1)^2}$.

So $|z_2 - z_1|$ is the distance between the points z_1 and z_2.

This is the key to solving many questions about sets of points in an Argand diagram, as in the following examples.

EXAMPLE 11.7 Draw an Argand diagram showing the set of points z for which $|z - 3 - 4i| = 5$.

SOLUTION

$|z - 3 - 4i|$ can be written as $|z - (3 + 4i)|$, and this is the distance from the point $3 + 4i$ to the point z.

This equals 5 if the point z lies on the circle with centre $3 + 4i$ and radius 5 (see figure 11.10).

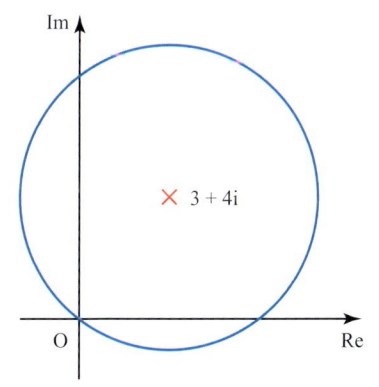

Figure 11.10

> How would you show the sets of points for which
>
> (i) $|z - 3 - 4i| \leq 5$
> (ii) $|z - 3 - 4i| < 5$
> (iii) $|z - 3 - 4i| \geq 5$?

EXAMPLE 11.8 Draw an Argand diagram showing the set of points z for which $|z - 3 - 4i| \leq |z + 1 - 2i|$.

SOLUTION

The condition can be written as $|z - (3 + 4i)| \leq |z - (-1 + 2i)|$.

$|z-(3+4i)|$ is the distance of point z from the point $3+4i$, point A in figure 11.11, and $|z-(-1+2i)|$ is the distance of point z from the point $-1+2i$, point B in figure 11.11.

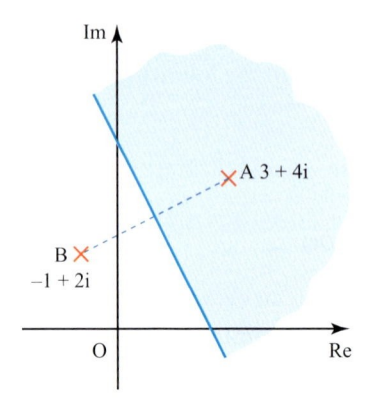

Figure 11.11

These distances are equal if z is on the perpendicular bisector of AB.

So the given condition holds if z is on this bisector or in the half plane on the side of it containing A, shown shaded in figure 11.11.

? How would you show the sets of points for which

(i) $|z-3-4i|=|z+1-2i|$

(ii) $|z-3-4i|<|z+1-2i|$

(iii) $|z-3-4i|>|z+1-2i|$?

EXERCISE 11D

1 For each of parts **(i)** to **(viii)**, draw an Argand diagram showing the set of points z for which the given condition is true.

(i) $|z|=2$
(ii) $|z-4|\leq 3$
(iii) $|z-5i|=6$
(iv) $|z+3-4i|<5$
(v) $|6-i-z|\geq 2$
(vi) $|z+2+4i|=0$
(vii) $2\leq|z-1+i|\leq 3$
(viii) $\text{Re}(z)=-2$

2 Draw an Argand diagram showing the set of points z for which $|z-12+5i|\leq 7$. Use the diagram to prove that, for these z, $6\leq|z|\leq 20$.

3 (i) On an Argand diagram, show the region R for which $|z-5+4i|\leq 3$.
   (ii) Find the greatest and least values of $|z+3-2i|$ in the region R.

4 By using an Argand diagram see if it is possible to find values of z for which $|z-2+i|\geq 10$ and $|z+4+2i|\leq 2$ simultaneously.

5 For each of parts **(i)** to **(iv)**, draw an Argand diagram showing the set of points z for which the given condition is true.

(i) $|z| = |z - 4|$

(ii) $|z| \geq |z - 2i|$

(iii) $|z + 1 - i| = |z - 1 + i|$

(iv) $|z + 5 + 7i| \leq |z - 2 - 6i|$

The modulus–argument form of complex numbers

The position of the point z in an Argand diagram can be described by means of the length of the line connecting this point to the origin, and the angle which this line makes with the positive real axis (see figure 11.12).

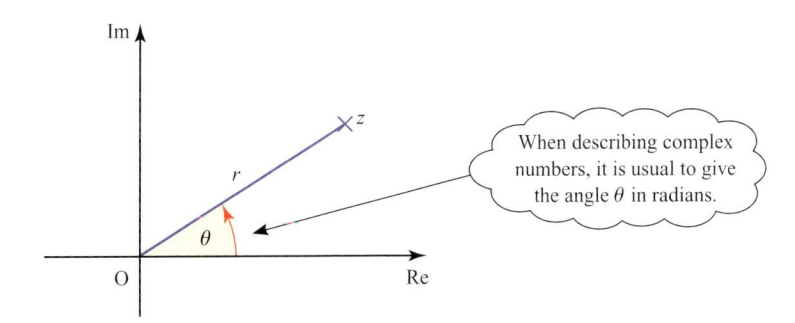

When describing complex numbers, it is usual to give the angle θ in radians.

Figure 11.12

The distance r is of course $|z|$, the modulus of z as defined on page 283.

The angle θ is slightly more complicated: it is measured anticlockwise from the positive real axis, normally in radians. However, it is not uniquely defined since adding any multiple of 2π to θ gives the same direction. To avoid confusion, it is usual to choose that value of θ for which $-\pi < \theta \leq \pi$. This is called the *principal argument* of z, denoted by arg z. Then every complex number except zero has a unique principal argument. The argument of zero is undefined.

For example, with reference to figure 11.13,

$\arg(-4) = \pi$

$\arg(-2i) = -\dfrac{\pi}{2}$

$\arg(1.5) = 0$

$\arg(-3 + 3i) = \dfrac{3\pi}{4}$

Remember that π radians = 180°.

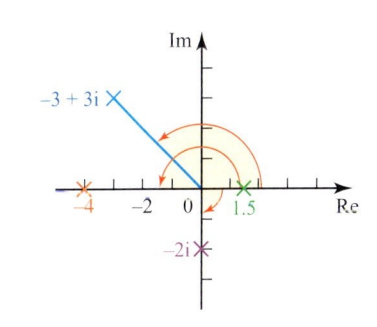

Figure 11.13

? Without using your calculator, state the values of the following.

(i) arg i (ii) arg(−4 − 4i) (iii) arg(2 − 2i)

You can see from figure 11.14 that

$$x = r\cos\theta \qquad y = r\sin\theta$$
$$r = \sqrt{x^2 + y^2} \qquad \tan\theta = \frac{y}{x}$$

and the same relations hold in the other quadrants too.

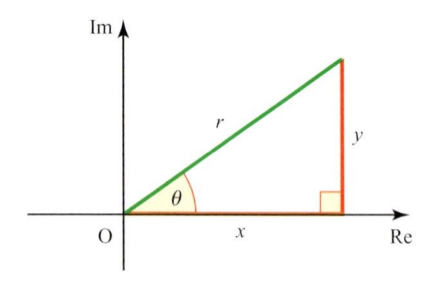

Figure 11.14

Since $x = r\cos\theta$ and $y = r\sin\theta$, we can write the complex number $z = x + iy$ in the form

$$z = r(\cos\theta + i\sin\theta).$$

This is called the *modulus–argument* or *polar* form.

ACTIVITY 11.8

(i) Set your calculator to degrees and use it to find the following.

(a) $\tan^{-1} 1$ (b) $\tan^{-1} 2$ (c) $\tan^{-1} 100$
(d) $\tan^{-1}(-2)$ (e) $\tan^{-1}(-50)$ (f) $\tan^{-1}(-200)$

What are the largest and smallest possible values, in degrees, of $\tan^{-1} x$?

(ii) Now set your calculator to radians.
Find $\tan^{-1} x$ for some different values of x.
What are the largest and smallest possible values, in radians, of $\tan^{-1} x$?

If you know the modulus and argument of a complex number, it is easy to use the relations $x = r\cos\theta$ and $y = r\sin\theta$ to find the real and imaginary parts of the complex number.

Similarly, if you know the real and imaginary parts, you can find the modulus and argument of the complex number using the relations $r = \sqrt{x^2 + y^2}$ and $\tan\theta = \frac{y}{x}$, but you do have to be quite careful in finding the argument. It is tempting to say that $\theta = \tan^{-1}\left(\frac{y}{x}\right)$, but, as you saw in the last activity, this gives a value between $-\frac{\pi}{2}$ and $\frac{\pi}{2}$, which is correct only if z is in the first or fourth quadrants.

For example, suppose that the point $z_1 = 2 - 3i$ has argument θ_1, and the point $z_2 = -2 + 3i$ has argument θ_2. It is true to say that $\tan\theta_1 = \tan\theta_2 = -\frac{3}{2}$. In the case of z_1, which is in the fourth quadrant, θ_1 is correctly given by $\tan^{-1}\left(-\frac{3}{2}\right) \approx -0.98$ rad ($\approx -56°$). However, in the case of z_2, which is in the second quadrant, θ_2 is given by $\left(-\frac{3}{2}\right) + \pi \approx 2.16$ rad ($\approx 124°$). These two points are illustrated in figure 11.15.

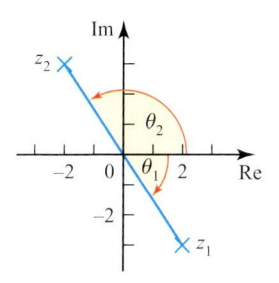

Figure 11.15

Figure 11.16 shows the values of the argument in each quadrant. It is wise always to draw a sketch diagram when finding the argument of a complex number.

$$\arg z = \tan^{-1}\left(\frac{y}{x}\right) + \pi \qquad \arg z = \tan^{-1}\left(\frac{y}{x}\right)$$

$$\arg z = \tan^{-1}\left(\frac{y}{x}\right) - \pi \qquad \arg z = \tan^{-1}\left(\frac{y}{x}\right)$$

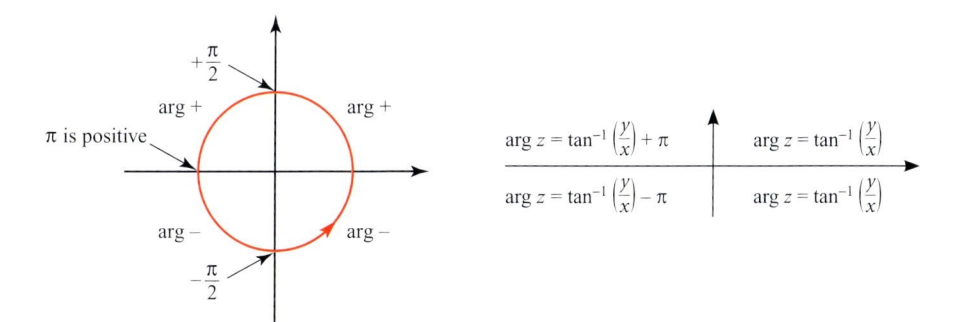

Figure 11.16

ACTIVITY 11.9 Mark the points $1 + i$, $1 - i$, $-1 + i$, $-1 - i$ on an Argand diagram. Find arg z for each of these, and check that your answers are consistent with figure 11.16.

Note

The modulus–argument form of a complex number is also called the *polar* form, as the modulus of a complex number is its distance from the origin, sometimes called the *pole*.

ACTIVITY 11.10 Most calculators can convert from (x, y) to (r, θ) (called *rectangular* to *polar*, and often shown as R → P) and from (r, θ) to (x, y) (polar to rectangular, P → R). Find out how to use these facilities on your calculator, and compare with other available types of calculator.

Does your calculator always give the correct θ, or do you sometimes have to add or subtract π (or 180°)?

A complex number in polar form must be given in the form $z = r(\cos\theta + i\sin\theta)$, not, for example, in the form $z = r(\cos\theta - i\sin\theta)$. The value of r must also be positive. So, for example, the complex number $-2(\cos\alpha + i\sin\alpha)$ is not in polar form. However, by using some of the relationships

$$\cos(\pi - \alpha) = -\cos\alpha \qquad \sin(\pi - \alpha) = \sin\alpha$$
$$\cos(\alpha - \pi) = -\cos\alpha \qquad \sin(\alpha - \pi) = -\sin\alpha$$
$$\cos(-\alpha) = \cos\alpha \qquad \sin(-\alpha) = -\sin\alpha$$

you can rewrite the complex number, for example

$$-2(\cos\alpha + i\sin\alpha) = 2(-\cos\alpha - i\sin\alpha)$$
$$= 2(\cos(\alpha - \pi) + i\sin(\alpha - \pi)).$$

This is now written correctly in polar form. The modulus is 2 and the argument is $\alpha - \pi$.

❓ How would you rewrite the following in polar form?

(i) $-2(\cos\alpha - i\sin\alpha)$ (ii) $2(\cos\alpha - i\sin\alpha)$

When you use the polar form of a complex number, remember to give the argument in radians, and to use a simple rational multiple of π where possible.

ACTIVITY 11.11 Copy and complete this table.

Give your answers in terms of $\sqrt{2}$ or $\sqrt{3}$ where appropriate, rather than as decimals. You may find figure 11.17 helpful.

	$\frac{\pi}{4}$	$\frac{\pi}{6}$	$\frac{\pi}{3}$
tan			
sin			
cos			

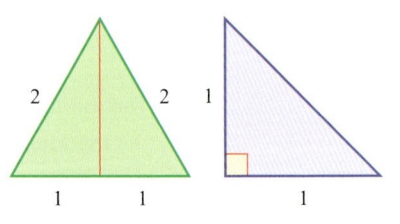

Figure 11.17

EXAMPLE 11.9 Write the following complex numbers in polar form.

(i) $4 + 3i$ 　　(ii) $-1 + i$ 　　(iii) $-1 - \sqrt{3}i$

SOLUTION

(i) $x = 4, y = 3$

Modulus $= \sqrt{3^2 + 4^2} = 5$

Since $4 + 3i$ lies in the first quadrant, the argument $= \tan^{-1}\frac{3}{4}$.

$4 + 3i = 5(\cos\alpha + i\sin\alpha)$, where $\alpha = \tan^{-1}\frac{3}{4} \approx 0.644$ radians

(ii) $x = -1, y = 1$

Modulus $= \sqrt{1^2 + 1^2} = \sqrt{2}$

Since $-1 + i$ lies in the second quadrant,

$$\text{argument} = \tan^{-1}(-1) + \pi$$
$$= -\frac{\pi}{4} + \pi = \frac{3\pi}{4}$$

$$-1 + i = \sqrt{2}\left(\cos\frac{3\pi}{4} + i\sin\frac{3\pi}{4}\right)$$

(iii) $x = -1, y = -\sqrt{3}$

Modulus $= \sqrt{1 + 3} = 2$

Since $-1 - \sqrt{3}i$ lies in the third quadrant,

$$\text{argument} = \tan^{-1}\sqrt{3} - \pi$$
$$= \frac{\pi}{3} - \pi = -\frac{2\pi}{3}$$

$$-1 - \sqrt{3}i = 2\left(\cos\left(-\frac{2\pi}{3}\right) + i\sin\left(-\frac{2\pi}{3}\right)\right)$$

EXERCISE 11E

1 Write down the values of the modulus and the principal argument of each of these complex numbers.

(i) $8\left(\cos\frac{\pi}{5} + i\sin\frac{\pi}{5}\right)$ 　　(ii) $\frac{\cos 2.3 + i\sin 2.3}{4}$

(iii) $4\left(\cos\frac{\pi}{3} - i\sin\frac{\pi}{3}\right)$ 　　(iv) $-3(\cos(-3) + i\sin(-3))$

2 For each complex number, find the modulus and principal argument, and hence write the complex number in polar form.
   Give the argument in radians, either as a simple rational multiple of π or correct to 3 decimal places.

   (i)   1
   (ii) -2
   (iii) $3i$
   (iv) $-4i$
   (v) $1 + i$
   (vi) $-5 - 5i$
   (vii) $1 - \sqrt{3}i$
   (viii) $6\sqrt{3} + 6i$
   (ix) $3 - 4i$
   (x) $-12 + 5i$
   (xi) $4 + 7i$
   (xii) $-58 - 93i$

3 Write each complex number with the given modulus and argument in the form $x + iy$, giving surds in your answer where appropriate.

   (i) $|z| = 2$, $\arg z = \dfrac{\pi}{2}$
   (ii) $|z| = 3$, $\arg z = \dfrac{\pi}{3}$
   (iii) $|z| = 7$, $\arg z = \dfrac{5\pi}{6}$
   (iv) $|z| = 1$, $\arg z = -\dfrac{\pi}{4}$
   (v) $|z| = 5$, $\arg z = -\dfrac{2\pi}{3}$
   (vi) $|z| = 6$, $\arg z = -2$

4 Given that $\arg(5 + 2i) = \alpha$, find the principal argument of each of the following in terms of α.

   (i) $-5 - 2i$
   (ii) $5 - 2i$
   (iii) $-5 + 2i$
   (iv) $2 + 5i$
   (v) $-2 + 5i$

5 The variable complex number z is given by
   $$z = 1 + \cos 2\theta + i \sin 2\theta,$$
   where θ takes all values in the interval $-\tfrac{1}{2}\pi < \theta < \tfrac{1}{2}\pi$.

   (i) Show that the modulus of z is $2 \cos \theta$ and the argument of z is θ.
   (ii) Prove that the real part of $\dfrac{1}{z}$ is constant.

   [Cambridge International AS & A Level Mathematics 9709, Paper 32 Q8 June 2010]

6 The variable complex number z is given by
   $$z = 2\cos\theta + i(1 - 2\sin\theta),$$
   where θ takes all values in the interval $-\pi < \theta \leqslant \pi$.

   (i) Show that $|z - i| = 2$, for all values of θ. Hence sketch, in an Argand diagram, the locus of the point representing z.
   (ii) Prove that the real part of $\dfrac{1}{z + 2 - i}$ is constant for $-\pi < \theta < \pi$.

   [Cambridge International AS & A Level Mathematics 9709, Paper 3 Q5 June 2008]

7 (i) The complex number z is given by $z = \dfrac{4 - 3i}{1 - 2i}$.

   (a) Express z in the form $x + iy$, where x and y are real.

   (b) Find the modulus and argument of z.

(ii) Find the two square roots of the complex number $5 - 12i$, giving your answers in the form $x + iy$, where x and y are real.

[Cambridge International AS & A Level Mathematics 9709, Paper 3 Q8 November 2007]

8 The complex number $2 + i$ is denoted by u. Its complex conjugate is denoted by u^*.

(i) Show, on a sketch of an Argand diagram with origin O, the points A, B and C representing the complex numbers u, u^* and $u + u^*$ respectively. Describe in geometrical terms the relationship between the four points O, A, B and C.

(ii) Express $\dfrac{u}{u^*}$ in the form $x + iy$, where x and y are real.

(iii) By considering the argument of $\dfrac{u}{u^*}$, or otherwise, prove that

$$\tan^{-1}\left(\frac{4}{3}\right) = 2\tan^{-1}\left(\frac{1}{2}\right).$$

[Cambridge International AS & A Level Mathematics 9709, Paper 3 Q7 June 2006]

Sets of points using the polar form

> ❓ You already know that $\arg z$ gives the angle between the line connecting the point z with the origin and the real axis.
>
> What do you think $\arg(z_2 - z_1)$ represents?

If $z_1 = x_1 + iy_1$ and $z_2 = x_2 + iy_2$, then $z_2 - z_1 = x_2 - x_1 + i(y_2 - y_1)$.

$$\arg(z_2 - z_1) = \tan^{-1}\frac{y_2 - y_1}{x_2 - x_1}$$

Figure 11.18 shows an Argand diagram with the points representing the complex numbers $z_1 = x_1 + iy_1$ and $z_2 = x_2 + iy_2$ marked.

The angle between the line joining z_1 and z_2 and a line parallel to the real axis is given by

$$\tan^{-1}\frac{y_2 - y_1}{x_2 - x_1}.$$

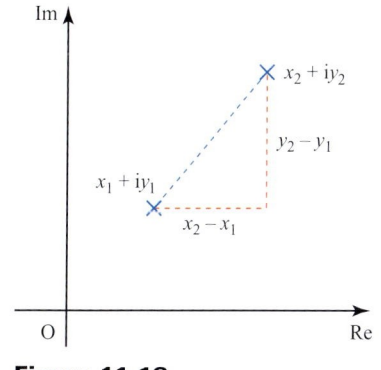

Figure 11.18

So $\arg(z_1 - z_2)$ is the angle between the line joining z_1 and z_2 and a line parallel to the real axis.

EXAMPLE 11.10 Draw Argand diagrams showing the sets of points z for which

(i) $\arg z = \dfrac{\pi}{4}$

(ii) $\arg(z - i) = \dfrac{\pi}{4}$

(iii) $0 \leqslant \arg(z - i) \leqslant \dfrac{\pi}{4}$.

SOLUTION

(i) $\arg z = \dfrac{\pi}{4}$

$\Leftrightarrow$ the line joining the origin to the point z has direction $\dfrac{\pi}{4}$

$\Leftrightarrow$ z lies on the half-line from the origin in the $\dfrac{\pi}{4}$ direction, see figure 11.19.

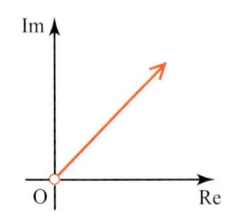

Figure 11.19

(Note that the origin is not included, since $\arg 0$ is undefined.)

(ii) $\arg(z - i) = \dfrac{\pi}{4}$

$\Leftrightarrow$ the line joining the point i to the point z has direction $\dfrac{\pi}{4}$

$\Leftrightarrow$ z lies on the half-line from the point i in the $\dfrac{\pi}{4}$ direction, see figure 11.20.

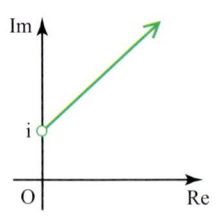

Figure 11.20

(iii) $0 \leqslant \arg(z - i) \leqslant \dfrac{\pi}{4}$

$\Leftrightarrow$ the line joining the point i to the point z has direction between 0 and $\dfrac{\pi}{4}$ (inclusive)

$\Leftrightarrow$ z lies in the one-eighth plane shown in figure 11.21.

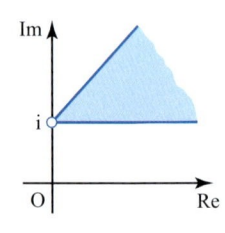

Figure 11.21

EXERCISE 11F

1 For each of parts **(i)** to **(vi)** draw an Argand diagram showing the set of points z for which the given condition is true.

(i) $\arg z = -\dfrac{\pi}{3}$

(ii) $\arg(z - 4i) = 0$

(iii) $\arg(z + 3) \geqslant \dfrac{\pi}{2}$

(iv) $\arg(z + 1 + 2i) = \dfrac{3\pi}{4}$

(v) $\arg(z - 3 + i) \leqslant -\dfrac{\pi}{6}$

(vi) $-\dfrac{\pi}{4} \leqslant \arg(z + 5 - 3i) \leqslant \dfrac{\pi}{3}$

2 Find the least and greatest possible values of $\arg z$ if $|z - 8i| \leqslant 4$.

3 You are given the complex number $w = -\sqrt{3} + 3i$.

   (i) Find $\arg w$ and $|w - 2i|$.

   (ii) On an Argand diagram, shade the region representing complex numbers z which satisfy both of these inequalities.
   $$|z - 2i| \leq 2 \quad \text{and} \quad \tfrac{1}{2}\pi \leq \arg z \leq \tfrac{2}{3}\pi.$$

   Indicate the point on your diagram which corresponds to w.

   (iii) Given that z satisfies both the inequalities in part **(ii)**, find the greatest possible value of $|z - w|$.

   [MEI, *part*]

4 The complex number w is given by $w = -\tfrac{1}{2} + i\tfrac{\sqrt{3}}{2}$.

   (i) Find the modulus and argument of w.

   (ii) The complex number z has modulus R and argument θ, where $-\tfrac{1}{3}\pi < \theta < \tfrac{1}{3}\pi$. State the modulus and argument of wz and the modulus and argument of $\tfrac{z}{w}$.

   (iii) Hence explain why, in an Argand diagram, the points representing z, wz and $\tfrac{z}{w}$ are the vertices of an equilateral triangle.

   (iv) In an Argand diagram, the vertices of an equilateral triangle lie on a circle with centre at the origin. One of the vertices represents the complex number $4 + 2i$. Find the complex numbers represented by the other two vertices. Give your answers in the form $x + iy$, where x and y are real and exact.

   [Cambridge International AS & A Level Mathematics 9709, Paper 3 Q10 November 2008]

5 (i) Solve the equation $z^2 - 2iz - 5 = 0$, giving your answers in the form $x + iy$ where x and y are real.

   (ii) Find the modulus and argument of each root.

   (iii) Sketch an Argand diagram showing the points representing the roots.

   [Cambridge International AS & A Level Mathematics 9709, Paper 3 Q3 June 2005]

6 (i) Solve the equation $z^2 + (2\sqrt{3})iz - 4 = 0$, giving your answers in the form $x + iy$, where x and y are real.

   (ii) Sketch an Argand diagram showing the points representing the roots.

   (iii) Find the modulus and argument of each root.

   (iv) Show that the origin and the points representing the roots are the vertices of an equilateral triangle.

   [Cambridge International AS & A Level Mathematics 9709, Paper 3 Q7 June 2009]

7 The complex numbers $-2 + i$ and $3 + i$ are denoted by u and v respectively.

   (i) Find, in the form $x + iy$, the complex numbers

   (a) $u + v$,

   (b) $\tfrac{u}{v}$, showing all your working.

(ii) State the argument of $\frac{u}{v}$.

In an Argand diagram with origin O, the points A, B and C represent the complex numbers u, v and $u + v$ respectively.

(iii) Prove that angle AOB $= \frac{3}{4}\pi$.

(iv) State fully the geometrical relationship between the line segments OA and BC.

[Cambridge International AS & A Level Mathematics 9709, Paper 32 Q7 November 2009]

8 The complex number $\frac{2}{-1+i}$ is denoted by u.

(i) Find the modulus and argument of u and u^2.

(ii) Sketch an Argand diagram showing the points representing the complex numbers u and u^2. Shade the region whose points represent the complex numbers z which satisfy both the inequalities $|z| < 2$ and $|z - u^2| < |z - u|$.

[Cambridge International AS & A Level Mathematics 9709, Paper 3 Q8 June 2007]

Working with complex numbers in polar form

The polar form quickly leads to an elegant geometrical interpretation of the multiplication of complex numbers. For if

$$z_1 = r_1(\cos\theta_1 + i\sin\theta_1) \text{ and } z_2 = r_2(\cos\theta_2 + i\sin\theta_2)$$

then
$$z_1 z_2 = r_1 r_2 (\cos\theta_1 + i\sin\theta_1)(\cos\theta_2 + i\sin\theta_2)$$
$$= r_1 r_2 [\cos\theta_1 \cos\theta_2 - \sin\theta_1 \sin\theta_2 + i(\sin\theta_1 \cos\theta_2 + \cos\theta_1 \sin\theta_2)].$$

Using the compound-angle formulae gives

$$z_1 z_2 = r_1 r_2 [\cos(\theta_1 + \theta_2) + i\sin(\theta_1 + \theta_2)].$$

This is the complex number with modulus $r_1 r_2$ and argument $(\theta_1 + \theta_2)$, so we have the beautiful result that

$$|z_1 z_2| = |z_1||z_2|$$

and

$\arg(z_1 z_2) = \arg z_1 + \arg z_2$ ($\pm 2\pi$ if necessary, to give the principal argument).

So to multiply complex numbers in polar form you *multiply* their moduli and *add* their arguments.

ACTIVITY 11.12 Using this interpretation, investigate

(i) multiplication by i

(ii) multiplication by -1.

The corresponding results for division are easily obtained by letting $\dfrac{z_1}{z_2} = w$. Then $z_1 = wz_2$ so that

$$|z_1| = |w||z_2| \text{ and arg } z_1 = \text{arg } w + \text{arg } z_2 \ (\pm 2\pi \text{ if necessary}).$$

Therefore $|w| = \left|\dfrac{z_1}{z_2}\right| = \dfrac{|z_1|}{|z_2|}$

and $\quad \text{arg } w = \text{arg } \dfrac{z_1}{z_2}$

$\qquad\qquad = \text{arg } z_1 - \text{arg } z_2 \ (\pm 2\pi \text{ if necessary, to give the principal argument}).$

So to divide complex numbers in polar form you *divide* their moduli and *subtract* their arguments.

This gives the following simple geometrical interpretation of multiplication and division.

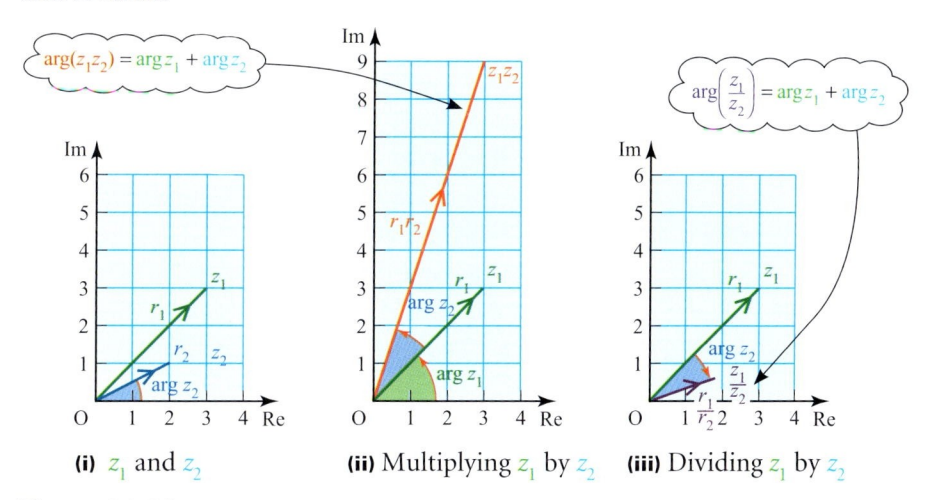

(i) z_1 and z_2 　　**(ii)** Multiplying z_1 by z_2 　　**(iii)** Dividing z_1 by z_2

Figure 11.22

To obtain the vector $z_1 z_2$ enlarge the vector z_1 by the scale factor $|z_2|$ and rotate it through arg z_2 anticlockwise about O (see Figure 11.22 **(ii)**).

To obtain the vector $\dfrac{z_1}{z_2}$ enlarge the vector z_1 by scale factor $\dfrac{1}{|z_2|}$ and rotate it clockwise through arg z_2 about O (see Figure 11.22 **(iii)**).

This combination of an enlargement followed by a rotation is called a *spiral dilatation*.

In summary:

$$|z_1 z_2| = |z_1||z_2| \qquad \text{arg }(z_1 z_2) = \text{arg } z_1 + \text{arg } z_2$$

and $\left|\dfrac{z_1}{z_2}\right| = \dfrac{|z_1|}{|z_2|} \qquad \text{arg}\left(\dfrac{z_1}{z_2}\right) = \text{arg } z_1 - \text{arg } z_2$

ACTIVITY 11.13 Check this by accurate drawing and measurement for the case $z_1 = 2 + i$, $z_2 = 3 + 4i$. Then do the same with z_1 and z_2 interchanged.

EXAMPLE 11.11 Find

(i) $6\left(\cos\dfrac{\pi}{2} + i\sin\dfrac{\pi}{2}\right) \times 2\left(\cos\dfrac{\pi}{4} + i\sin\dfrac{\pi}{4}\right)$

(ii) $6\left(\cos\dfrac{\pi}{2} + i\sin\dfrac{\pi}{2}\right) \div 2\left(\cos\dfrac{\pi}{4} + i\sin\dfrac{\pi}{4}\right)$

SOLUTION

(i) Remember that

$$r_1(\cos\theta_1 + i\sin\theta_1) \times r_2(\cos\theta_2 + i\sin\theta_2) = r_1 r_2(\cos(\theta_1 + \theta_2) + i\sin(\theta_1 + \theta_2))$$

So to multiply complex numbers you

- multiply the moduli
- add the arguments.

$6 \times 2 = 12;\ \dfrac{\pi}{2} + \dfrac{\pi}{4} = \dfrac{3\pi}{4}$

$$6\left(\cos\dfrac{\pi}{2} + i\sin\dfrac{\pi}{2}\right) \times 2\left(\cos\dfrac{\pi}{4} + i\sin\dfrac{\pi}{4}\right) = 12\left(\cos\dfrac{3\pi}{4} + i\sin\dfrac{3\pi}{4}\right)$$

(ii) To divide complex numbers you

- divide the moduli
- subtract the arguments.

$6 \div 2 = 3;\ \dfrac{\pi}{2} - \dfrac{\pi}{4} = \dfrac{\pi}{4}$

$$6\left(\cos\dfrac{\pi}{2} + i\sin\dfrac{\pi}{2}\right) \div 2\left(\cos\dfrac{\pi}{4} + i\sin\dfrac{\pi}{4}\right) = 3\left(\cos\dfrac{\pi}{4} + i\sin\dfrac{\pi}{4}\right)$$

This leads to an alternative method of finding the square root of a complex number.

Writing $8 + 6i$ in polar form gives

$$r = \sqrt{8^2 + 6^2} = 10$$
$$\tan\theta = \tfrac{6}{8} \Rightarrow \theta = 0.6435 \text{ radians}$$

So $8 + 6i = 10(\cos 0.6435 + i\sin 0.6435)$

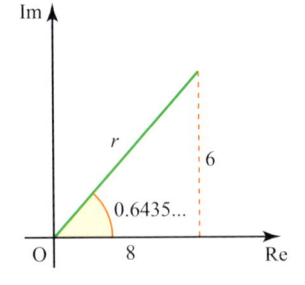

Figure 11.23

Notice that if you add 2π to the argument you will end up with exactly the same complex number on the Argand diagram (as you have just rotated through one full turn).

So $8 + 6i$ is also the same as $10[\cos(0.6435 + 2\pi) + i\sin(0.6435 + 2\pi)]$.

Let $r(\cos\theta + i\sin\theta)$ be the square root of $8 + 6i$ so that

$$r(\cos\theta + i\sin\theta) \times r(\cos\theta + i\sin\theta) = r^2(\cos 2\theta + i\sin 2\theta) = 8 + 6i$$

$\Rightarrow \qquad r(\cos 2\theta + i\sin 2\theta) = 10(\cos 0.6435 + i\sin 0.6435)$ ①

and $\qquad r(\cos 2\theta + i\sin 2\theta) = 10[\cos(0.6435 + 2\pi) + i\sin(0.6435 + 2\pi)]$ ②

From ① or ②	$r^2 = 10$	$\Rightarrow$	$r = \sqrt{10}$	*The square root of the modulus of 8 + 6i.*
From ①	$2\theta = 0.6435$	$\Rightarrow$	$\theta = 0.32175$	
From ②	$2\theta = 0.6435 + 2\pi$	$\Rightarrow$	$\theta = 0.32175 + \pi$	

So one square root of 8 + 6i is *Half of the argument of 8 + 6i.*

$$\sqrt{10}(\cos 0.32175 + i\sin 0.32175) = 3 + i$$

and the other square root is *Half of the argument of 8 + 6i, plus π.*

$$\sqrt{10}[\cos(0.32175 + \pi) + i\sin(0.32175 + \pi)] = -3 - i$$

Compare this method with that used on page 278 to find the square roots of 8 + 6i.

❓ What are the square roots of $10[\cos(0.6435 + 2n\pi) + i\sin(0.6435 + 2n\pi)]$, where n is an integer?

Complex exponents

When multiplying complex numbers in polar form you add the arguments, and when multiplying powers of the same base you add the exponents. This suggests that there may be a link between the familiar expression $\cos\theta + i\sin\theta$ and the seemingly remote territory of the exponential function. This was first noticed in 1714 by the young Englishman Roger Cotes two years before his death at the age of 28 (when Newton remarked 'If Cotes had lived we might have known something'), and made widely known through an influential book published by Euler in 1748.

Let $z = \cos\theta + i\sin\theta$. Since i behaves like any other constant in algebraic manipulation, to differentiate z with respect to θ you simply differentiate the real and imaginary parts separately. This gives

$$\begin{aligned}\frac{dz}{d\theta} &= -\sin\theta + i\cos\theta \\ &= i^2\sin\theta + i\cos\theta \\ &= i(\cos\theta + i\sin\theta) \\ &= iz\end{aligned}$$

So $z = \cos\theta + i\sin\theta$ is a solution of the differential equation $\frac{dz}{d\theta} = iz$.

If i continues to behave like any other constant when it is used as an index, then the general solution of $\frac{dz}{d\theta} = iz$ is $z = e^{i\theta + c}$, where c is a constant, just as $x = e^{kt+c}$ is the general solution of $\frac{dx}{dt} = kx$.

Therefore $\cos\theta + i\sin\theta = e^{i\theta+c}$.

Putting $\theta = 0$ gives

$$\cos 0 + i\sin 0 = e^{0+c}$$
$$\Rightarrow \quad 1 = e^c$$
$$\Rightarrow \quad c = 0$$

and it follows that

$$\cos\theta + i\sin\theta = e^{i\theta}.$$

The problem with this is that you have no way of knowing how i behaves as an index. But this does not matter. Since no meaning has yet been given to e^z when z is complex, the following *definition* can be made, suggested by this work with differential equations but not dependent on it:

$$e^{i\theta} = \cos\theta + i\sin\theta.$$

Note

The particular case when $\theta = \pi$ gives $e^{i\pi} = \cos\pi + i\sin\pi = -1$, so that

$$e^{i\pi} + 1 = 0.$$

This remarkable statement, linking the five fundamental numbers 0, 1, i, e and π, the three fundamental operations of addition, multiplication and exponentiation, and the fundamental relation of equality, has been described as a 'mathematical poem'.

EXAMPLE 11.12

Find

(i) (a) $4e^{5i} \times 3e^{2i}$

(b) $6e^{9i} \div 3e^{2i}$

(ii) Write these results as complex numbers in polar form.

SOLUTION

(i) (a) $4e^{5i} \times 3e^{2i} = 12e^{7i}$ ⟵ $4 \times 3 = 12;\ 5i + 2i = 7i$

(b) $6e^{9i} \div 3e^{2i} = 2e^{7i}$ ⟵ $6 \div 3 = 2;\ 9i - 2i = 7i$

(ii) (a) $4(\cos 5 + i\sin 5) \times 3(\cos 2 + i\sin 2) = 12(\cos 7 + i\sin 7)$

(b) $6(\cos 9 + i\sin 9) \div 3(\cos 2 + i\sin 2) = 2(\cos 7 + i\sin 7)$

EXERCISE 11G

1 Find the following.

(i) $8(\cos 0.2 + i\sin 0.2) \times 4(\cos 0.4 + i\sin 0.4)$

(ii) $8(\cos 0.2 + i\sin 0.2) \div 4(\cos 0.4 + i\sin 0.4)$

(iii) $6\left(\cos\frac{\pi}{3} + i\sin\frac{\pi}{3}\right) \times 2\left(\cos\frac{\pi}{6} + i\sin\frac{\pi}{6}\right)$

(iv) $6\left(\cos\frac{\pi}{3} + i\sin\frac{\pi}{3}\right) \div 2\left(\cos\frac{\pi}{6} + i\sin\frac{\pi}{6}\right)$

(v) $12(\cos\pi + i\sin\pi) \times 2\left(\cos\frac{\pi}{4} + i\sin\frac{\pi}{4}\right)$

(vi) $12(\cos\pi + i\sin\pi) \div 2\left(\cos\frac{\pi}{4} + i\sin\frac{\pi}{4}\right)$

2 Given that $z = 2\left(\cos\frac{\pi}{4} + i\sin\frac{\pi}{4}\right)$ and $w = 3\left(\cos\frac{\pi}{3} + i\sin\frac{\pi}{3}\right)$, find the following in polar form.

(i) wz (ii) $\frac{w}{z}$ (iii) $\frac{z}{w}$

(iv) $\frac{1}{z}$ (v) w^2 (vi) z^5

(vii) $w^3 z^4$ (viii) $5iz$ (ix) $(1+i)w$

3 Prove that, in general, $\arg\frac{1}{z} = -\arg z$, and deal with the exceptions.

4 Given the points 1 and z on a Argand diagram, explain how to find the following points by geometrical construction.

(i) $3z$ (ii) $2iz$ (iii) $(3+2i)z$

(iv) z^* (v) $|z|$ (vi) z^2

5 Find the real and imaginary parts of $\dfrac{-1+i}{1+\sqrt{3}\,i}$.

Express $-1+i$ and $1+\sqrt{3}\,i$ in polar form.

Hence show that $\cos\dfrac{5\pi}{12} = \dfrac{\sqrt{3}-1}{2\sqrt{2}}$, and find an exact expression for $\sin\dfrac{5\pi}{12}$.

6 The complex numbers α and β are given by $\frac{\alpha + 4}{\alpha} = 2 - i$ and $\beta = -\sqrt{6} + \sqrt{2}i$.

   (i) Show that $\alpha = 2 + 2i$.
   (ii) Show that $|\alpha| = |\beta|$. Find $\arg \alpha$ and $\arg \beta$.
   (iii) Find the modulus and argument of $\alpha\beta$. Illustrate the complex numbers α, β and $\alpha\beta$ on an Argand diagram.
   (iv) Describe the locus of points in the Argand diagram representing complex numbers z for which $|z - \alpha| = |z - \beta|$. Draw this locus on your diagram
   (v) Show that $z = \alpha + \beta$ satisfies $|z - \alpha| = |z - \beta|$. Mark the point representing $\alpha + \beta$ on your diagram, and find the exact value of $\arg(\alpha + \beta)$.
   [MEI]

7 Express e^z in the form $x + iy$ where z is the given complex number.

   (i) $-i\pi$ (ii) $\frac{i\pi}{4}$ (iii) $\frac{2 + 5i\pi}{6}$ (iv) $3 - 4i$

8 Find the following.

   (i) (a) $2e^{3i} \times 5e^{-2i}$ (b) $8e^{5i} \div 2e^{5i}$ (c) $3e^{7i} \times 2e^{i}$
       (d) $12e^{5i} \div 4e^{4i}$ (e) $3e^{2i} \times e^{i}$ (f) $8e^{3i} \div 2e^{4i}$

   (ii) Write these results as complex numbers in polar form.

Complex numbers and equations

The reason for inventing complex numbers was to provide solutions for quadratic equations which have no real roots, i.e. to solve $az^2 + bz + c = 0$ when the discriminant $b^2 - 4ac$ is negative. This is straightforward since if $b^2 - 4ac = -k^2$ (where k is real) then the formula for solving quadratic equations gives $z = \frac{-b \pm ik}{2a}$. These are the two complex roots of the equation. Notice that when the coefficients of the quadratic equation are real, these roots are a pair of *conjugate* complex numbers.

It would be natural to think that to solve cubic equations would require a further extension of the number system to give some sort of 'super-complex' numbers, with ever more extensions to deal with higher degree equations. But luckily things are much simpler. It turns out that *all* polynomial equations (even those with complex coefficients) can be solved by means of complex numbers. This was realised as early as 1629 by Albert Girard, who stated that an nth degree polynomial equation has precisely n roots, including complex roots and taking into account repeated roots. (For example, the fifth degree equation $(z - 2)(z - 4)^2(z^2 + 9) = 0$ has five roots: 2, 4 (twice), 3i and −3i.) Many great mathematicians tried to prove this. The chief difficulty is to show that every polynomial equation must have at least *one* root: this is called the *Fundamental Theorem of Algebra* and was first proved by Gauss (again!) in 1799.

The Fundamental Theorem, which is too difficult to prove here, is an example of an *existence theorem*: it tells us that a solution exists, but does not say what it is. To find the solution of a particular equation you may be able to use an exact method, such as the formula for the roots of a quadratic equation. (There are much more complicated formulae for solving cubic or quartic equations, but not in general for equations of degree five or more.) Alternatively, there are good approximate methods for finding roots to any required accuracy, and your calculator probably has this facility.

ACTIVITY 11.14 Find out how to use your calculator to solve polynomial equations.

You have already noted that the complex roots of a quadratic equation with real coefficients occur as a conjugate pair. The same is true of the complex roots of any polynomial equation with real coefficients. This is very useful in solving polynomial equations with complex roots, as shown in the following examples.

EXAMPLE 11.13 Given that $1 + 2i$ is a root of $4z^3 - 11z^2 + 26z - 15 = 0$, find the other roots.

SOLUTION

Since the coefficients are real, the conjugate $1 - 2i$ is also a root.

Therefore $[z - (1 + 2i)]$ and $[z - (1 - 2i)]$ are both factors of $4z^3 - 11z^2 + 26z - 15 = 0$.

This means that $(z - 1 - 2i)(z - 1 + 2i)$ is a factor of $4z^3 - 11z^2 + 26z - 15 = 0$.

$$(z - 1 - 2i)(z - 1 + 2i) = [(z - 1) - 2i][(z - 1) + 2i]$$
$$= (z - 1)^2 + 4$$
$$= z^2 - 2z + 5$$

By looking at the coefficient of z^3 and the constant term, you can see that the remaining factor is $4z - 3$.

$$4z^3 - 11z^2 + 26z - 15 = (z^2 - 2z + 5)(4z - 3)$$

The third root is therefore $\frac{3}{4}$.

EXAMPLE 11.14 Given that $-2 + i$ is a root of the equation $z^4 + az^3 + bz^2 + 10z + 25 = 0$, find the values of a and b, and solve the equation.

SOLUTION

$z = -2 + i$

$z^2 = (-2 + i)^2 = 4 - 4i + (i)^2 = 4 - 4i - 1 = 3 - 4i$
$z^3 = (-2 + i)z^2 = (-2 + i)(3 - 4i) = -6 + 11i + 4 = -2 + 11i$
$z^4 = (-2 + i)z^3 = (-2 + i)(-2 + 11i) = 4 - 24i - 11 = -7 - 24i$

Now substitute these into the equation.

$$-7 - 24i + a(-2 + 11i) + b(3 - 4i) + 10(-2 + i) + 25 = 0$$
$$(-7 - 2a + 3b - 20 + 25) + (-24 + 11a - 4b + 10)i = 0$$

Equating real and imaginary parts gives

$$-2a + 3b - 2 = 0$$
$$11a - 4b - 14 = 0$$

Solving these equations simultaneously gives $a = 2$, $b = 2$.

The equation is $z^4 + 2z^3 + 2z^2 + 10z + 25 = 0$.

Since $-2 + i$ is one root, $-2 - i$ is another root.

So $(z + 2 - i)(z + 2 + i) = (z + 2)^2 + 1$
$$= z^2 + 4z + 5$$

is a factor.

Using polynomial division or by inspection

$$z^4 + 2z^3 + 2z^2 + 10z + 25 = (z^2 + 4z + 5)(z^2 - 2z + 5).$$

The other two roots are the solutions of the quadratic equation $z^2 - 2z + 5 = 0$.

Using the quadratic formula

$$z = \frac{2 \pm \sqrt{4 - 4 \times 5}}{2}$$
$$= \frac{2 \pm \sqrt{-16}}{2}$$
$$= \frac{2 \pm 4i}{2}$$
$$= 1 \pm 2i$$

The roots of the equation are $-2 \pm i$ and $1 \pm 2i$.

EXERCISE 11H

1. Check that $2 + i$ is a root of $z^3 - z^2 - 7z + 15 = 0$, and find the other roots.

2. One root of $z^3 - 15z^2 + 76z - 140 = 0$ is an integer.
   Solve the equation.

3. Given that $1 - i$ is a root of $z^3 + pz^2 + qz + 12 = 0$, find the real numbers p and q, and the other roots.

4. One root of $z^4 - 10z^3 + 42z^2 - 82z + 65 = 0$ is $3 + 2i$.
   Solve the equation.

5. The equation $z^4 - 8z^3 + 20z^2 - 72z + 99 = 0$ has a pure imaginary root.
   Solve the equation.

6 You are given the complex number $w = 1 - i$.

  (i) Express w^2, w^3 and w^4 in the form $a + bi$.

  (ii) Given that $w^4 + 3w^3 + pw^2 + qw + 8 = 0$, where p and q are real numbers, find the values of p and q.

  (iii) Write down two roots of the equation $z^4 + 3z^3 + pz^2 + qz + 8 = 0$, where p and q are the real numbers found in part **(ii)**.

  [MEI, *part*]

7 (i) Given that $\alpha = -1 + 2i$, express α^2 and α^3 in the form $a + bi$.
  Hence show that α is a root of the cubic equation

  $$z^3 + 7z^2 + 15z + 25 = 0.$$

  (ii) Find the other two roots of this cubic equation.

  (iii) Illustrate the three roots of the cubic equation on an Argand diagram, and find the modulus and argument of each root.

  (iv) L is the locus of points in the Argand diagram representing complex numbers z for which $\left|z + \frac{5}{2}\right| = \frac{5}{2}$. Show that all three roots of the cubic equation lie on L, and draw the locus L on your diagram.

  [MEI]

8 The cubic equation $z^3 + 6z^2 + 12z + 16 = 0$ has one real root α and two complex roots β, γ.

  (i) Verify that $\alpha = -4$, and find β and γ in the form $a + bi$.
  (Take β to be the root with positive imaginary part.)

  (ii) Find $\frac{1}{\beta}$ and $\frac{1}{\gamma}$ in the form $a + bi$.

  (iii) Find the modulus and argument of each of α, β and γ.

  (iv) Illustrate the six complex numbers $\alpha, \beta, \gamma, \frac{1}{\alpha}, \frac{1}{\beta}, \frac{1}{\gamma}$ on an Argand diagram, making clear any geometrical relationships between the points.

  [MEI, *part*]

9 You are given that the complex number $\alpha = 1 + 4i$ satisfies the cubic equation

  $$z^3 + 5z^2 + kz + m = 0,$$

where k and m are real constants.

  (i) Find α^2 and α^3 in the form $a + bi$.

  (ii) Find the value of k and show that $m = 119$.

  (iii) Find the other two roots of the cubic equation.
  Give the arguments of all three roots.

  (iv) Verify that there is a constant c such that all three roots of the cubic equation satisfy

  $$|z + 2| = c.$$

  Draw an Argand diagram showing the locus of points representing all complex numbers z for which $|z + 2| = c$.
  Mark the points corresponding to the three roots of the cubic equation.

  [MEI]

10 In this question, α is the complex number $-1 + 3i$.

(i) Find α^2 and α^3.

It is given that λ and μ are real numbers such that $\lambda\alpha^3 + 8\alpha^2 + 34\alpha + \mu = 0$.

(ii) Show that $\lambda = 3$, and find the value of μ.

(iii) Solve the equation $\lambda z^3 + 8z^2 + 34z + \mu = 0$, where λ and μ are as in part (ii). Find the modulus and argument of each root, and illustrate the three roots on an Argand diagram.

[MEI, *part*]

11 The cubic equation $z^3 + z^2 + 4z - 48 = 0$ has one real root α and two complex roots β and γ.

(i) Verify that $\alpha = 3$ and find β and γ in the form $a + bi$.
Take β to be the root with positive imaginary part, and give your answers in an exact form.

(ii) Find the modulus and argument of each of the numbers $\alpha, \beta, \gamma, \dfrac{\beta}{\gamma}$, giving the arguments in radians between $-\pi$ and π.
Illustrate these four numbers on an Argand diagram.

(iii) On your Argand diagram, draw the locus of points representing complex numbers z such that

$$\arg(z - \alpha) = \arg\beta.$$

[MEI, *part*]

12 The equation $2x^3 + x^2 + 25 = 0$ has one real root and two complex roots.

(i) Verify that $1 + 2i$ is one of the complex roots.

(ii) Write down the other complex root of the equation.

(iii) Sketch an Argand diagram showing the point representing the complex number $1 + 2i$. Show on the same diagram the set of points representing the complex numbers z which satisfy

$$|z| = |z - 1 - 2i|.$$

[Cambridge International AS & A Level Mathematics 9709, Paper 3 Q7 November 2005]

KEY POINTS

1. Complex numbers can be written in the form $z = x + iy$ with $i^2 = -1$.
   x is called the real part, Re(z), and y is called the imaginary part, Im(z).

2. The conjugate of z is $z^* = x - iy$.

3. To add or subtract complex numbers, add or subtract the real and imaginary parts separately.

$$(x_1 + iy_1) \pm (x_2 + iy_2) = (x_1 \pm x_2) + i(y_1 \pm y_2)$$

4 To multiply complex numbers, multiply out the brackets and simplify.
$$(x_1 + iy_1)(x_2 + iy_2) = (x_1x_2 - y_1y_2) + i(x_1y_2 + x_2y_1)$$

5 To divide complex numbers, multiply top and bottom by the conjugate of the bottom.
$$\frac{x_1 + iy_1}{x_2 + iy_2} = \frac{(x_1x_2 + y_1y_2) + i(x_2y_1 - x_1y_2)}{x_2^2 + y_2^2}$$

6 The complex number z can be represented geometrically as the point (x, y). This is known as an Argand diagram.

7 The modulus of $z = x + iy$ is $|z| = \sqrt{x^2 + y^2}$.
This is the distance of the point z from the origin.

8 The distance between the points z_1 and z_2 in an Argand diagram is $|z_2 - z_1|$.

9 The principal argument of z, arg z, is the angle θ, $-\pi < \theta \leq \pi$, between the line connecting the origin and the point z and the positive real axis.

10 The modulus–argument or polar form of z is $z = r(\cos\theta + i\sin\theta)$, where $r = |z|$ and $\theta = \arg z$.

11 $x = r\cos\theta$ $y = r\sin\theta$
    $r = \sqrt{x^2 + y^2}$ $\tan\theta = \frac{y}{x}$

12 To multiply complex numbers in polar form, multiply the moduli and add the arguments.
$$z_1z_2 = r_1r_2[\cos(\theta_1 + \theta_2) + i\sin(\theta_1 + \theta_2)]$$

13 To divide complex numbers in polar form, divide the moduli and subtract the arguments.
$$\frac{z_1}{z_2} = \frac{r_1}{r_2}[\cos(\theta_1 - \theta_2) + i\sin(\theta_1 - \theta_2)]$$

14 $e^{i\theta} = \cos\theta + i\sin\theta$, $e^{-i\theta} = \cos\theta - i\sin\theta$

15 A polynomial equation of degree n has n roots, taking into account complex roots and repeated roots. In the case of polynomial equations with real coefficients, complex roots always occur in conjugate pairs.

Answers

Neither University of Cambridge International Examinations nor OCR bear any responsibility for the example answers to questions taken from their past question papers which are contained in this publication.

Chapter 1

❓ (Page 2)

For the green curve you can try $y = f(x)$ where $f(x) = kx(x-1)(x-2)$. This passes through $(0, 0)$, $(0, 1)$ and $(2, 0)$ but its maximum is not quite when $x = \frac{1}{2}$. A value of k of 0.208 gives a maximum value of 0.08. The blue curve is then $y = -f(x)$. A better fit can be obtained by taking a two part function

$$f(x) = 0.32x(1-x) \text{ for } 0 \leq x \leq 1$$
and $f(x) = 0.32(x-1)(x-2)$
for $1 \leq x \leq 2$.

Exercise 1A (Page 7)

1 (i) 3   (ii) 12   (iii) 7
2   $2x^3 - 4$
3   $x^4 + 4x^3 + 6x^2 + 4x + 1$
4   $x^3 + 2x^2 + 5x + 7$
5   $-x^2 + 15x + 18$
6   $2x^4 + 8$
7   $x^4 + 4x^3 + 6x^2 + 4x + 1$
8   $x^4 - 5x^2 + 4$
9   $x^4 - 10x^2 + 9$
10   $x^{11} - 1$
11   $2x - 2$
12   $10x^2$
13   4
14   $2x^2 - 2x$
15   $-8x^3 - 8x$
16   $x^2 - 2x - 3$
17   $x^2 + 3x$
18   $2x^2 - 5x + 5$
19   $x^3 + x^2 + 2x + 2$
20   $2x^2 + 3$
21   $x^3 + 2x^2 + 5$
22   $x^3 + 2x^2 + x$
23   $2x^3 + 3x^2 + x + 4$
24   $2x^2 + 2x + 3$
25   $x^2 + 3x + 1$
26   $x^2 + 4$
27   $x^2 - 2x - 2$
28   $x^2 + 2x - 2$

❓ (Page 13)

Its order will be less than n.

Exercise 1B (Page 14)

1 (i) 0, 0, −8, −18, −24, −20, 0
(ii) $(x+3)(x+2)(x-3)$
(iii) −3, −2 or 3
(iv)

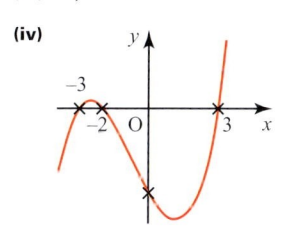

2 (i) −15, 0, 3, 0, −3, 0, 15
(ii) $x(x+2)(x-2)$
(iii) −2, 0 or 2
(iv)

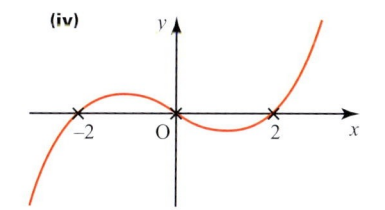

3 (i) 30, 0, $(x-3)$
(ii) $p = 2$, $q = -15$
(iii) −5, 2 or 3
(iv)

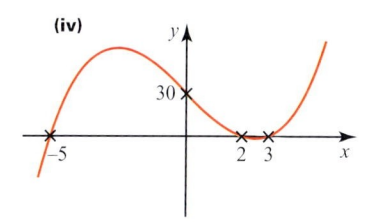

4 (ii) −2, 3 or 4
(iii)

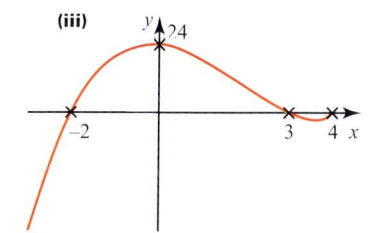

5 (i) 0
(ii) $-1 \pm \sqrt{2}$
(iii)

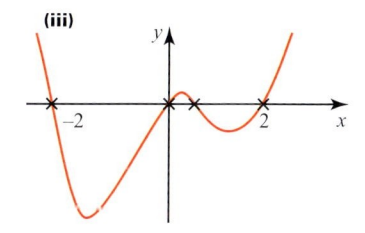

6 (i) −4
(ii) $(x-1)^2$
(iii)

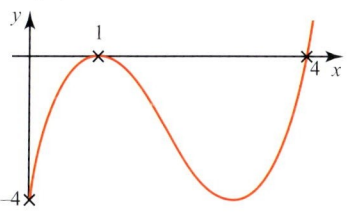

7 (i) $a = 2$, $b = 1$, $c = 2$
(ii) 0, $\sqrt{3}$ or $-\sqrt{3}$

8 (i) $(x^2-4)(x^2-1)$

(ii) $(x+2)(x-2)(x^2+1)$

(iii) Two real roots: -2 and 2

9 (ii) $2x^2+9x+11$ remainder 19

10 (i) $\pm 6, \pm 3, \pm 2, \pm 1,$

(ii) $-1, 2$ or 3

11 (i) $(x-1)(x-2)(x+2)$

(ii) $(x+1)(x^2-x+1)$

(iii) $(x-2)(x^2+2x+5)$

(iv) $(x+2)(x^2-x+3)$

12 (ii) (a) x^3-2x^2+2x+2 remainder -6

(b) x^3-3x^2+6x-6 remainder -2

(c) x^2-2x+4 remainder $-2x-4$

13 -12

14 2 or -5

15 $-5; 4; 4$

16 $-1; -7; 1, -2$ or $\frac{3}{2}$

17 (i) $a=2, b=3$

(ii) $2x+1$

18 $a=2, b=-3$

19 (i) -13

(ii) $(x+2)(2x+1)(x-3)$

20 (i) $a=-4, b=1$

(ii) $(x-3)$ and $(x+1)$

21 (i) 4

(ii) x^2-2x+2

? (Page 17)

$g(3)=3, g(-3)=3$

$|3+3|=6, |3-3|=0,$
$|3|+|3|=6, |3|+|-3|=6$

? (Page 19)

$|x|<2$ and $x\geqslant 0 \Rightarrow 0\leqslant x<2$

$|x|<2$ and $x<0 \Rightarrow -2<x<0$

Exercise 1C (Page 21)

1 (i) $x=-9$ or $x=1$

(ii) $x=-7$ or $x=1$

(iii) $x=-1$ or $x=7$

(iv) $x=-\frac{3}{2}$ or $x=2$

(v) $x=-3$ or $x=2$

(vi) $x=1$ or $x=7$

(vii) $x=-2$ or $x=4$

(viii) $x=-\frac{8}{3}$ or $x=2$

(ix) $x=-1$ or $x=\frac{3}{2}$

2 (i) $-8<x<2$

(ii) $0\leqslant x\leqslant 4$

(iii) $x<-1$ or $x>11$

(iv) $x\leqslant -3$ or $x\geqslant 1$

(v) $-2<x<5$

(vi) $-\frac{2}{3}\leqslant x\leqslant 2$

3 (i) $|x-1|<2$

(ii) $|x-5|<3$

(iii) $|x-1|<3$

(iv) $|x-2.5|<3.5$

(v) $|x-10|<0.1$

(vi) $|x-4|<3.5$

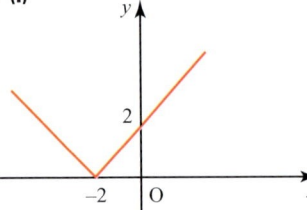

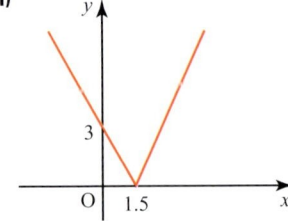

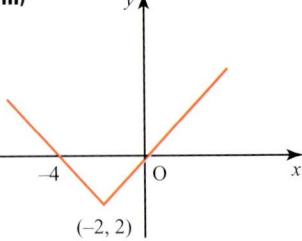

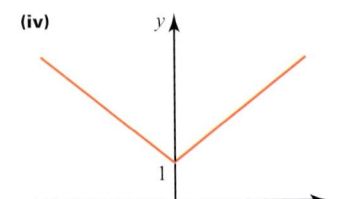

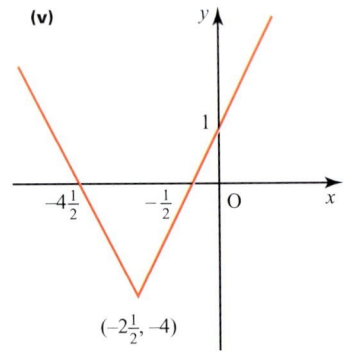

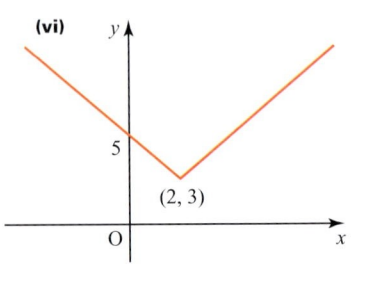

4 (i) [graph: V-shape with vertex at $(-2, 0)$, y-intercept 2]

(ii) [graph: V-shape with vertex at $(1.5, 0)$, y-intercept 3]

(iii) [graph: V-shape with vertex at $(-2, 2)$, x-intercept -4]

(iv) [graph: V-shape with vertex at $(0, 1)$]

(v) [graph: V-shape with vertex at $(-2\frac{1}{2}, -4)$, x-intercepts $-4\frac{1}{2}$ and $-\frac{1}{2}$]

(vi) [graph: V-shape with vertex at $(2, 3)$, y-intercept 5]

5 (i) $x<\frac{1}{2}$

(ii) $x<\frac{7}{2}$

(iii) $x\geqslant -\frac{1}{2}$

(iv) $-1\leqslant x\leqslant 3$

(v) $x<-1$ or $x>3$

(vi) $x\leqslant -6$ or $x\geqslant -\frac{4}{3}$

6 $\frac{1}{2}<x<1$

7 $x>\frac{1}{3}$

8 $x<2a$

Chapter 2

❓ (Page 23)

Without using logarithms, you would probably use trial and improvement to find x where $x^3 = 500$.

Investigation (Page 24)

(i) 10

(ii) $\frac{1}{2}$

(iii) $10^{8.6/85} = 1.26$

Activity 2.1 (Page 27)

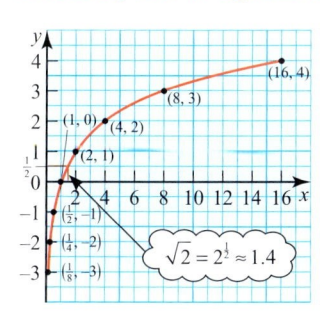

❓ (Page 28)

- $a^0 = 1$, $\log_a 1 = 0$
- $a^m = x > 0$ (for $a > 0$) so $\log_a(x) = m$ is defined only for $x > 0$
- Putting $x = \frac{1}{y}$ in $\log\left(\frac{1}{y}\right) = -\log y$
  $\Rightarrow \log x = -\log\left(\frac{1}{x}\right)$: as $x \to 0$,
  $-\log\left(\frac{1}{x}\right) \to -\infty$
- There is no limit to m in $a^m = x$ and $\log_a x = m$; think, for example, of base 2, i.e. $a = 2$. Then $x = 2^y$. When $y = 1, 2, 3, 4, \ldots$ then $x = 2, 4, 8, 16, \ldots$. So increases in y are accompanied by ever larger increases in x and so a decreasing gradient. This is the case not just for $a = 2$ but for any value of a greater than 1.
- $\log_a a = 1$

Exercise 2A (Page 29)

1. (i) $x = \log_3 9$, 2
   (ii) $x = \log_4 64$, 3
   (iii) $x = \log_2 \frac{1}{4}$, -2
   (iv) $x = \log_5 \frac{1}{5}$, -1
   (v) $x = \log_7 1$, 0
   (vi) $x = \log_{16} 2$, $\frac{1}{4}$

2. (i) $3^y = 9$, 2
   (ii) $5^y = 125$, 3
   (iii) $2^y = 16$, 4
   (iv) $6^y = 1$, 0
   (v) $64^y = 8$, $\frac{1}{2}$
   (vi) $5^y = \frac{1}{25}$, -2

3. (i) 4
   (ii) -4
   (iii) $\frac{1}{2}$
   (iv) 0
   (v) 4
   (vi) -4
   (vii) $\frac{3}{2}$
   (viii) $\frac{1}{4}$
   (ix) $\frac{1}{2}$
   (x) -3

4. (i) $\log 10$
   (ii) $\log 2$
   (iii) $\log 36$
   (iv) $\log \frac{1}{7}$
   (v) $\log 3$
   (vi) $\log 4$
   (vii) $\log 4$
   (viii) $\log \frac{1}{3}$
   (ix) $\log \frac{1}{2}$
   (x) $\log 12$

5. (i) $2 \log x$
   (ii) $3 \log x$
   (iii) $\frac{1}{2} \log x$
   (iv) $\frac{11}{6} \log x$
   (v) $6 \log x$
   (vi) $\frac{5}{2} \log x$

6. (i) $x < 7$
   (ii) $x \geqslant 3$
   (iii) $x \geqslant 3$
   (iv) $x > 0.437$
   (v) $x \leqslant 1$
   (vi) $x \geqslant 0.322$
   (vii) $0.431 \leqslant x < 1.29$
   (viii) $0 \leqslant x < 0.827$
   (ix) $1 < x < 2.58$
   (x) $0.68 < x < 1.49$

7. $\log_{10} \frac{x^2}{7}$, $x = 21$

8. (i) $x = 19.93$
   (ii) $x = -9.97$
   (iii) $x = 9.01$
   (iv) $x = 48.32$
   (v) $x = 1375$

9. 9

10. (i) 25
    (ii) 17

11. (i) $4 < y < 6$
    (ii) $1.26 < x < 1.63$

12. $y = \dfrac{\log 4 - \log x}{\log 3}$

13. $x = 0.802$

Exercise 2B (Page 35)

Some of the questions in this exercise involve drawing a line of best fit by eye. Consequently your answers may reasonably vary a little from those given.

311

P2

Answers

1 (i) If the relationship is of the form $R = kT^n$, the graph of log R against log T will be a straight line.

(ii) Values of log R: 5.46, 5.58, 5.72, 6.09, 6.55

Values of log T: 0.28, 0.43, 0.65, 1.20, 1.90

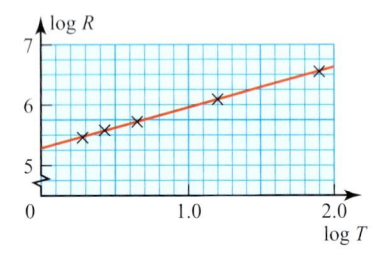

(iii) $k = 1.8 \times 10^5$, $n = 0.690$

(iv) 0.7 days

2 (ii) Plotting log A against t will test the model: if it is a straight line the model fits the data.

(iii) $b = 1.4$, $k = 0.89$

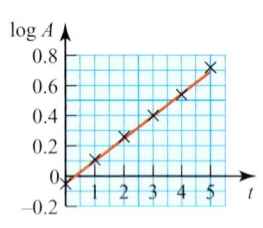

(iv) (a) $t = 2.4$ days

(b) $A = 3.0$ cm^2

(v) Exponential growth

3 (ii) $k = 3.2 \times 10^6$, $a = 0.98$

The constant k is the original number of trees.

4 (ii) $k = 1100$, $n = 1.6$

(iii) $s = 2500$ m

(iv) The train would not continue to accelerate like this throughout its journey. After 10 minutes it would probably be travelling at constant speed, or possibly even slowing down.

5 (ii) $b = 1.37$, $k = 1.58$

6 Taking logs of both sides,
log y = log A + B log x.

Plotting log y against log x gives a straight line of gradient B and intercept log A.
This gives $A = 1.5$, $B = 0.78$.
The value of y that is wrong is 6.21. If x is 5.07, y should be 5.32 according to the equation.

7 log y = B log x + log A.
He should plot log y against log x. If this gives a straight line, there is a relationship of the form $y = Ax^B$. If there is no such relationship, the points will not be in a straight line.
The value of log A is given by the intercept on the log y axis.
The value of B is the gradient of the line.

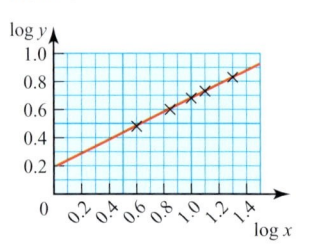

From the graph, $A = 1.5$, $B = 0.5$.
The formula is therefore
$y = 1.5x^{0.5}$.

8 (i)

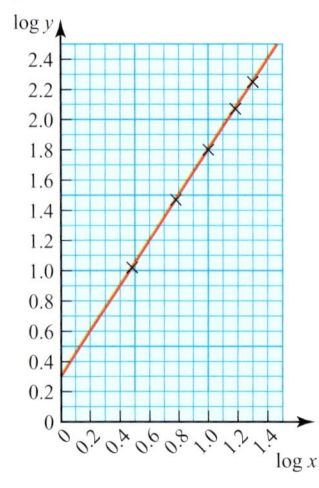

(ii) The graph is a straight line.

(iii) $A = 2.0$, $n = 1.5$

9 (ii)

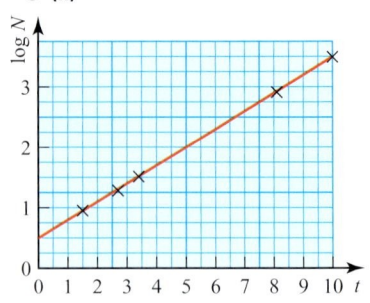

(iii) $a \approx 3$, $b \approx 2$

(v) Just over 3 million.

10 (ii)

$\log_{10} d$	$\log_{10} z$
2.89	0.32
2.91	0.41
2.94	0.51
2.97	0.60
3.00	0.68
3.02	0.75
3.05	0.77
3.07	0.79

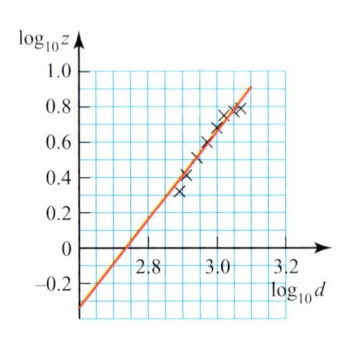

(iii) $D \approx 1050$

(iv) $n = 3$

(v) $d = 840$ (nearest 10)

11 (i) $\dfrac{\log 4}{\log 3}$ **(ii)** 3.42

? (Page 40)

$\dfrac{1}{x} = x^{-1}$. This means that $n = -1$ and so $n + 1 = 0$. You cannot divide by zero.

Investigation (Page 40)

(i) 1.099

(ii) 0.693

(iii) 1.792

$$\int_1^3 \frac{1}{x}dx + \int_1^2 \frac{1}{x}dx = \int_1^6 \frac{1}{x}dx$$

Activity 2.2 (Page 41)

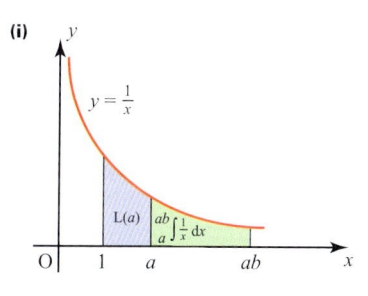

(i)

(ii) $x = az \Rightarrow dx = a\,dz$

Converting the limits:
$x = a \Rightarrow z = 1$
$x = ab \Rightarrow z = b$

$$\int_a^{ab} \frac{1}{x}dx = \int_1^b \frac{1}{az} \times a\,dz$$

$$= \int_1^b \frac{1}{z}dz$$

$$\int_1^b \frac{1}{z}dz = \int_1^b \frac{1}{x}dx = L(b)$$

(iii) $L(a) + \int_a^{ab} \frac{1}{x}dx = L(ab)$

$\Rightarrow L(a) + L(b) = L(ab)$

Activity 2.3 (Page 41)

(i) $L(1) = \int_1^1 \frac{1}{x}dx = 0$

(ii) $L(a) - L(b) = \int_1^a \frac{1}{x}dx - \int_1^b \frac{1}{x}dx$

$$= \int_b^a \frac{1}{x}dx$$

Let $x = bz$

$$\int_b^a \frac{1}{x}dx = \int_1^{a/b} \frac{1}{z}dz$$

$$= L\left(\frac{a}{b}\right)$$

(iii) $L(a^n) = \int_1^{a^n} \frac{1}{x}dx$

Let $x = z^n$ then $dx = nz^{n-1}dz$.

$$\int_1^{a^n} \frac{1}{x}dx = \int_1^a \frac{1}{z^n} \times nz^{n-1}dz$$

$$= n\int_1^a \frac{1}{z}dz$$

$$= nL(a)$$

Activity 2.4 (Page 42)

$e = 2.72$ (2 d.p.)

Exercise 2C (Page 47)

1 $x = x_0 e^{kt}$

2 $t = \frac{1}{k}\ln\left(\frac{s_0}{s}\right)$

3 $p = 25e^{-0.02t}$

4 $x = \ln\left(\frac{y-5}{y_0-5}\right)$

5 (i) $x = 0.0540$

(ii) $x = 0.0339$

(iii) $x = 0.238$

(iv) $x = 0.693$

(v) $x = 1.386$

(vi) $x = 1.099$

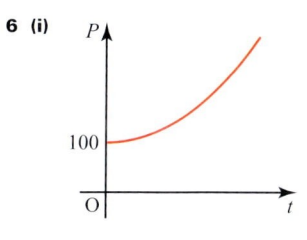

6 (i)

(ii) 100

(iii) 1218

(iv) 184 years

7 (i) 1 m

(ii) 4.61 m, 6.09 years

(iii) $a = e^{-2} = 0.135$, $b = 2.5$

(iv) 11 years

8 $y = \dfrac{5}{x^2 - 1}$

9 4.11

10 $x = 0.481$

11 $A = 3.67$, $b = 1.28$

12 $x = -1.68$

13 $A = 2.01$, $n = 0.25$

Chapter 3

? (Page 51)

Possible answers are:

Bridge: wavelength 50–100 m; amplitute 15–30 m

Ripple: wavelength 0.02–0.05 m; amplitude 0.005–0.01 m

Bridge: $a = 15$–30; $b = \dfrac{\pi}{50} - \dfrac{\pi}{25}$ (about 0.06–0.13)

Ripple: $a = 0.005$–0.01; $b = 125$–300

Exercise 3A (Page 54)

1 (i) $x = 90°$

(ii) $x = 60°, 300°$

(iii) $x = 14.0°, 194.0°$

(iv) $x = 109.5°, 250.5°$

(v) $x = 135°, 315°$

(vi) $x = 210°, 330°$

2 (i) -1

(ii) $\dfrac{-2}{\sqrt{3}}$

(iii) $\dfrac{-2}{\sqrt{3}}$

(iv) $\dfrac{-2}{\sqrt{3}}$

(v) 0

(vi) $-\sqrt{2}$

3 (i) $B = 60°$, $C = 30°$

(ii) $\sqrt{3}$

4 (i) $L = 45°$, $N = 45°$

(ii) $\sqrt{2}, \sqrt{2}, 1$

5 (ii) 14.0°

6 (i) $0 \leqslant \alpha \leqslant 90°$

(ii) No, for each of the second, third and fourth quadrants a different function is positive.

(iii) No, the graphs of the three functions do not intersect at a single point.

7 (i) $x = 0°, 180°, 360°$

(ii) $x = 45°, 225°$

(iii) $x = 60°, 300°$

(iv) $x = 54.7, 125.3°, 234.7°, 305.3°$

(v) $x = 18.4°, 71.6°, 198.4°, 251.6°$

(vi) $x = 45°, 135°, 225°, 315°$

Activity 3.1 (Page 55)

$y = \sin(\theta + 60°)$ is obtained from $y = \sin\theta$ by a translation $\begin{pmatrix} -60° \\ 0 \end{pmatrix}$.

$y = \cos(\theta - 60°)$ is obtained from $y = \sin\theta$ by a translation $\begin{pmatrix} 60° \\ 0 \end{pmatrix}$.

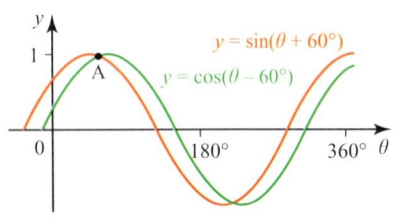

[Graph showing $y = \sin(\theta + 60°)$ and $y = \cos(\theta - 60°)$ with point A between them, axes at 180°, 360°]

It appears that the θ co-ordinate of A is midway between the two maxima (30°, 1) and (60°, 1).

Checking:
$\theta = 45° \Rightarrow \sin(\theta + 60°) = 0.966$
$\cos(\theta - 60°) = 0.966$.

If 60° is replaced by 35°, using the trace function on a graphic calculator would enable the solutions to be found.

ⓟ (Page 56)

Area of a triangle = $\frac{1}{2}$ base × height.
The definitions of sine and cosine in a right-angled triangle.

Activity 3.2 (Page 57)

(i) $\sin(\theta + \phi)$
$= \sin\theta\cos\phi + \cos\theta\sin\phi$
$\Rightarrow \sin[(90° - \theta) + \phi]$
$= \sin(90° - \theta)\cos\phi + \cos(90° - \theta)\sin\phi$
$\Rightarrow \sin[90° - (\theta - \phi)]$
$= \cos\theta\cos\phi + \sin\theta\sin\phi$
$\Rightarrow \cos(\theta - \phi)$
$= \cos\theta\cos\phi + \sin\theta\sin\phi$

(ii) $\Rightarrow \cos[\theta - (-\phi)]$
$= \cos\theta\cos(-\phi) + \sin\theta\sin(-\phi)$
$\Rightarrow \cos(\theta + \phi)$
$= \cos\theta\cos\phi - \sin\theta\sin\phi$

(iii) $\tan(\theta + \phi) = \dfrac{\sin(\theta + \phi)}{\cos(\theta + \phi)}$

$= \dfrac{\sin\theta\cos\phi + \cos\theta\sin\phi}{\cos\theta\cos\phi - \sin\theta\sin\phi}$

$= \dfrac{\dfrac{\sin\theta\cos\phi}{\cos\theta\cos\phi} + \dfrac{\cos\theta\sin\phi}{\cos\theta\cos\phi}}{\dfrac{\cos\theta\cos\phi}{\cos\theta\cos\phi} - \dfrac{\sin\theta\sin\phi}{\cos\theta\cos\phi}}$

$= \dfrac{\tan\theta + \tan\phi}{1 - \tan\theta\tan\phi}$

(iv) $\tan[\theta + (-\phi)] = \dfrac{\tan\theta + \tan(-\phi)}{1 - \tan\theta\tan(-\phi)}$

$\tan(\theta - \phi) = \dfrac{\tan\theta - \tan\phi}{1 + \tan\theta\tan\phi}$

ⓟ (Page 57)

No. In part (iii) you get

$\tan 90° = \dfrac{\sqrt{3} + \dfrac{1}{\sqrt{3}}}{1 - \sqrt{3} \times \dfrac{1}{\sqrt{3}}}$

Neither $\tan 90°$ nor $\dfrac{1}{1-1}$ is defined. For the result to be valid you must exclude the case when $\theta + \phi = 90°$ (or 270°, 450°, …).

Similarly in part (iv) you must exclude $\theta - \phi = 90°, 270°$, etc.

Exercise 3B (Page 59)

1 (i) $\dfrac{\sqrt{3}}{2\sqrt{2}} + \dfrac{1}{2\sqrt{2}}$

(ii) $-\dfrac{1}{\sqrt{2}}$

(iii) $\dfrac{\sqrt{3}-1}{\sqrt{3}+1}$

(iv) $\dfrac{\sqrt{3}+1}{\sqrt{3}-1}$

2 (i) $\dfrac{1}{\sqrt{2}}(\sin\theta + \cos\theta)$

(ii) $\tfrac{1}{2}(\sqrt{3}\cos\theta + \sin\theta)$

(iii) $\tfrac{1}{2}(\sqrt{3}\cos\theta - \sin\theta)$

(iv) $\dfrac{1}{\sqrt{2}}(\cos 2\theta - \sin 2\theta)$

(v) $\dfrac{\tan\theta + 1}{1 - \tan\theta}$

(vi) $\dfrac{\tan\theta - 1}{1 + \tan\theta}$

3 (i) $\sin\theta$

(ii) $\cos 8\phi$

(iii) 0

(iv) $\cos 2\theta$

4 (i) $\theta = 15°$

(ii) $\theta = 157.5°$

(iii) $\theta = 0°$ or $180°$

(iv) $\theta = 111.7°$

(v) $\theta = 165°$

5 (i) $\theta = \dfrac{\pi}{8}$

(ii) $\theta = 2.79$ radians

6 (i) $\dfrac{1}{\sqrt{5}}$

(ii) $\sin\beta = \tfrac{3}{5}, \cos\beta = \tfrac{4}{5}$

7 (ii) $x = 10.9°, -169.1°$

8 (ii) $x = 22.5°, 112.5°$

9 $\alpha = 26.6°$ and $\beta = 45°$ or $\alpha = 135°$ and $\beta = 116.6°$

10 (ii) $\theta = 24.7°, 95.3°$

ⓟ (Page 61)

For $\sin 2\theta$ and $\cos 2\theta$, substituting $\theta = 45°$ is helpful.
You know that $\sin 45° = \cos 45° = \dfrac{1}{\sqrt{2}}$
and that $\sin 90° = 1$ and $\cos 90° = 0$.

For $\tan 2\theta$ you cannot use $\theta = 45°$.
Take $\theta = 30°$ instead; $\tan 30° = \dfrac{1}{\sqrt{3}}$
and $\tan 60° = \sqrt{3}$.

No, checking like this is not the same as proof.

Exercise 3C (Page 65)

1 (i) $\theta = 14.5°, 90°, 165.5°, 270°$

(ii) $\theta = 0°, 35.3°, 144.7°, 180°, 215.3°, 324.7°, 360°$

(iii) $\theta = 90°, 210°, 330°$

(iv) $\theta = 30°, 150°, 210°, 330°$

(v) $\theta = 0°, 138.6°, 221.4°, 360°$

2 (i) $\theta = -\pi, 0, \pi$

  (ii) $\theta = -\pi, 0, \pi$

  (iii) $\theta = \frac{-2\pi}{3}, 0, \frac{2\pi}{3}$

  (iv) $\theta = \frac{-3\pi}{4}, \frac{-\pi}{4}, \frac{\pi}{4}, \frac{3\pi}{4}$

  (v) $\theta = \frac{-11\pi}{12}, \frac{-3\pi}{4}, \frac{-7\pi}{12}, \frac{-\pi}{4}, \frac{\pi}{12},$
  $\frac{\pi}{4}, \frac{5\pi}{12}, \frac{3\pi}{4}$

3 $3\sin\theta - 4\sin^3\theta$,
$\theta = 0, \frac{\pi}{4}, \frac{3\pi}{4}, \pi, \frac{5\pi}{4}, \frac{7\pi}{4}, 2\pi$

4 $\theta = 51°, 309°$

5 $\cot\theta$

6 $\dfrac{\tan\theta\,(3 - \tan^2\theta)}{1 - 3\tan^2\theta}$

8 (ii) $\theta = 63.4°$

9 (i)

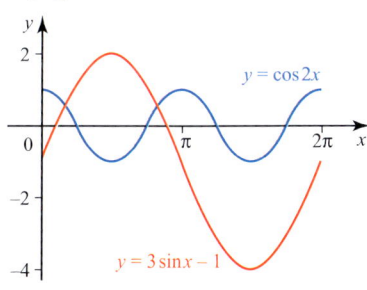

  (iii) $x = \frac{\pi}{6}, \frac{5\pi}{6}$

10 (ii) $\theta = 27.2°, 152.8°, 207.2°, 332.8°$

11 (ii) $\theta = 26.6°, 206.6°$

12 (i) $\frac{1}{10}(4\sqrt{3} - 3)$

  (ii) $\tan 2\alpha = -\frac{24}{7}, \tan 3\alpha = -\frac{44}{117}$

Exercise 3D (Page 70)

1 (i) $\sqrt{2}\cos(\theta - 45°)$

  (ii) $29\cos(\theta - 46.4°)$

  (iii) $2\cos(\theta - 60°)$

  (iv) $3\cos(\theta - 41.8°)$

2 (i) $\sqrt{2}\cos\left(\theta + \frac{\pi}{4}\right)$

  (ii) $2\cos\left(\theta + \frac{\pi}{6}\right)$

3 (i) $\sqrt{5}\sin(\theta + 63.4°)$

  (ii) $3\sin(\theta + 48.2°)$

4 (i) $\sqrt{2}\sin\left(\theta - \frac{\pi}{4}\right)$

  (ii) $3\sin(\theta - 0.49 \text{ rad})$

5 (i) $2\cos(\theta - (-60°))$

  (ii) $4\cos(\theta - (-45°))$

  (iii) $2\cos(\theta - 30°)$

  (iv) $13\cos(\theta - 22.6°)$

  (v) $2\cos(\theta - 150°)$

  (vi) $2\cos(\theta - 135°)$

6 (i) $13\cos(\theta + 67.4°)$

  (ii) Max 13, min -13

  (iii)

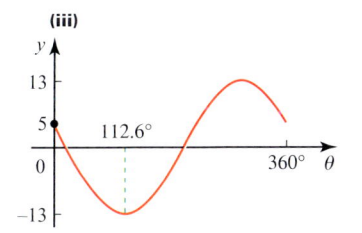

  (iv) $\theta = 4.7°, 220.5°$

7 (i) $2\sqrt{3}\sin\left(\theta - \frac{\pi}{6}\right)$

  (ii) Max $2\sqrt{3}, \theta = \frac{2\pi}{3}$;
  min $-2\sqrt{3}, \theta = \frac{5\pi}{3}$

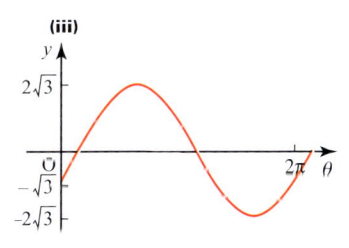

  (iii)

  (iv) $\theta = \frac{\pi}{3}, \pi$

8 (i) $\sqrt{13}\sin(2\theta + 56.3°)$

  (ii) Max $\sqrt{13}, \theta = 16.8°$;
  min $-\sqrt{13}, \theta = 106.8°$

  (iii)

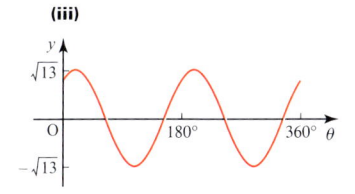

  (iv) $\theta = 53.8°, 159.9°, 233.8°, 339.9°$

9 (i) $\sqrt{3}\cos(\theta - 54.7°)$

  (ii) Max $\sqrt{3}, \theta = 54.7°$;
  min $-\sqrt{3}, \theta = 234.7°$

  (iii)

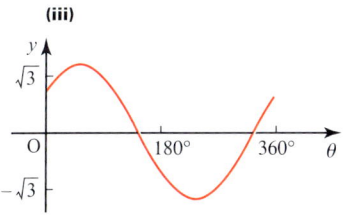

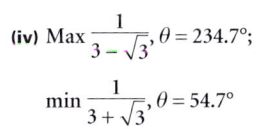

  (iv) Max $\dfrac{1}{3 - \sqrt{3}}, \theta = 234.7°$;
  min $\dfrac{1}{3 + \sqrt{3}}, \theta = 54.7°$

10 (ii) $\theta = 30.6°, 82.0°$

11 (i) $\cos x \cos\alpha - \sin x \sin\alpha$

  (ii) $r = \sqrt{29}, \alpha = 68.2°$

  (iii) Max $\sqrt{29}$ when $x = 291.8°$,
  min $-\sqrt{29}$ when $x = 111.8°$

  (iv) $x = 235.7°, 347.9°$

12 (i) $30.96°$

  (ii) $x = 15.7°, 282.4°$

  (iii) $x = 7.9°, 141.2°, 187.9°, 321.2°$

13 (i) $R = 10, \alpha = 53.13°$

  (ii)

  (iii) $x = 119.55°, 346.71°$

  (iv) $\theta = 103.29°, 330.45°$

P2 Answers

14 (i) $c = \sqrt{a^2 + b^2}$

(ii) $\tan a = \dfrac{b}{a}$

(iii) $a = 36.87°$

(iv) $\theta = 103.29°, 330.45°$

15 (i) $5\cos(x - 53.13°)$

(ii) $x = 27.29°, 78.97°$

16 (i) $\sqrt{26}\cos(\theta + 11.31°)$

(ii) $\theta = 27.02°, 310.36°$

17 (i) $25\cos(\theta - 73.74°)$

(ii) $\theta = 20.6°, 126.9°$

18 $\theta = 81.3°, 172.4°$

Investigation (Page 74)

The total current is

$I = A_1 \sin\omega t + A_2 \sin(\omega t + a)$
(where $\omega = 2\pi f$).

$I = A_1 \sin\omega t + A_2 \sin\omega t \cos a$
$\quad + A_2 \cos\omega t \sin a$

$= (A_1 + A_2 \cos a)\sin\omega t$
$\quad + (A_2 \sin a)\cos\omega t$

Let $A_1 + A_2 \cos a = P$ and $A_2 \sin a = Q$

so $\quad I = P\sin\omega t + Q\cos\omega t$

$= \sqrt{P^2 + Q^2} \sin(\omega t + \varepsilon)$

where $\varepsilon = \tan^{-1}\left(\dfrac{Q}{P}\right)$.

This is a sine wave with the same frequency but a greater amplitude. The phase angle ε is between 0 and a.

Exercise 3E (Page 74)

1 (i) $\sin 6\theta$

(ii) $\cos 6\theta$

(iii) 1

(iv) $\cos\theta$

(v) $\sin\theta$

(vi) $\tfrac{3}{2}\sin 2\theta$

(vii) $\cos\theta$

(viii) -1

2 (i) $1 - \sin 2x$

(ii) $\cos 2x$

(iii) $\tfrac{1}{2}(5\cos 2x - 1)$

4 (i) $\theta = 4.4°, 95.6°$

(ii) $\theta = 199.5°, 340.5°$

(iii) $\theta = \dfrac{-\pi}{6}, \dfrac{\pi}{2}$

(iv) $\theta = -15.9°, 164.1°$

(v) $\theta = \dfrac{\pi}{2}, \dfrac{\pi}{6}, \dfrac{5\pi}{6}$

(vi) $\theta = 20.8°, 122.3°$

(vii) $\theta = 76.0°, 135°$

Chapter 4

Activity 4.1 (Page 81)

$y = \dfrac{u}{v}$ where $u = x^{10}$ and $v = x^7$

gives $\dfrac{du}{dx} = 10x^9$ and $\dfrac{dv}{dx} = 7x^6$

Using the quotient rule,

$\dfrac{dy}{dx} = \dfrac{v\dfrac{du}{dx} - u\dfrac{dv}{dx}}{v^2}$

$= \dfrac{x^7 \times 10x^9 - x^{10} \times 7x^6}{x^{14}}$

$= \dfrac{10x^{16} - 7x^{16}}{x^{14}} = 3x^2$

$y = \dfrac{u}{v} = \dfrac{x^{10}}{x^7} = x^3 \Rightarrow \dfrac{dy}{dx} = 3x^2$

Exercise 4A (Page 82)

1 (i) $x(5x^3 - 3x + 6)$

(ii) $x^4(21x^2 + 24x - 35)$

(iii) $2x(6x + 1)(2x + 1)^3$

(iv) $-\dfrac{2}{(3x - 1)^2}$

(v) $\dfrac{x^2(x^2 + 3)}{(x^2 + 1)^2}$

(vi) $2(2x + 1)(12x^2 + 3x - 8)$

(vii) $\dfrac{2(1 + 6x - 2x^2)}{(2x^2 + 1)^2}$

(viii) $\dfrac{7 - x}{(x + 3)^3}$

(ix) $\dfrac{3x - 1}{2\sqrt{x - 1}}$

2 (i) $-\dfrac{1}{(x - 1)^2}$

(ii) $-1; y = -x$

(iii) $-1; y = -x + 4$

(iv) The two tangents are parallel.

3 (i) $3x(x - 2)$

(ii) $(0, 4)$, maximum; $(2, 0)$, minimum

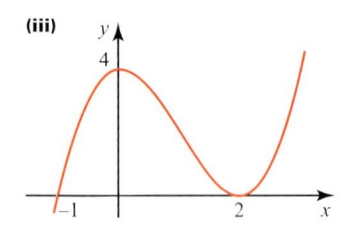

(iii)

4 (i) $-\dfrac{1}{(x - 4)^2}$

(ii) $4y + x = 12$

(iii) $y = x - 3$

(iv) $\dfrac{dy}{dx} \neq 0$ for any value of x

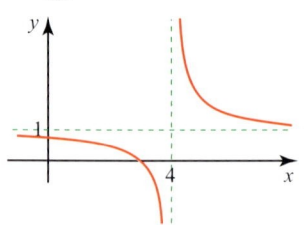

5 (i) $\dfrac{\sqrt{x} - 2}{(\sqrt{x} - 1)^2}$

(ii) $\dfrac{1}{4}$

(iii) $(4, 8)$

(iv) Tangent: $y = 8$; normal: $x = 4$

(v) (a) $Q\left(\dfrac{37}{4}, 8\right)$

(b) $R(4, 29)$

6 (i) $\dfrac{2(x + 1)(x + 2)}{(2x + 3)^2}$

(ii) $(-1, -2); (-2, -3)$

(iii) $(-1, -2)$, minimum; $(-2, -3)$, maximum

7 (i) $\dfrac{2x(x + 1)}{(2x + 1)^2}$; $(0, 0)$ and $(-1, -1)$

(ii) $(0, 0)$ minimum; $(-1, -1)$ maximum

8 (i) $\dfrac{3\sqrt{2}}{2}$

(iii) $\tfrac{3}{2}$; 3; gradient $= \infty$

316

? (Page 85)

$\frac{d}{dx}(f(x))$ is a polynomial of order $(n-1)$ so it has no term in x^n.

? (Page 87)

$y = \ln(3x)$ is a translation of $y = \ln(x)$ through $\begin{pmatrix} 0 \\ 3 \end{pmatrix}$.

The curves have the same shape.

The gradient function is valid for $x > 0$.

Exercise 4B (Page 89)

1. (i) $\frac{3}{x}$
   (ii) $\frac{1}{x}$
   (iii) $\frac{2}{x}$
   (iv) $\frac{2x}{x^2+1}$
   (v) $-\frac{1}{x}$
   (vi) $1 + \ln x$
   (vii) $x(1 + 2\ln(4x))$
   (viii) $-\frac{1}{x(x+1)}$
   (ix) $\frac{x}{x^2-1}$
   (x) $\frac{1-2\ln x}{x^3}$

2. (i) $3e^x$
   (ii) $2e^{2x}$
   (iii) $2xe^{x^2}$
   (iv) $2(x+1)e^{(x+1)^2}$
   (v) $e^{4x}(1+4x)$
   (vi) $2x^2e^{-x}(3-x)$
   (vii) $\frac{1-x}{e^x}$
   (viii) $6e^{2x}(e^{2x}+1)^2$

3. (i) $0.108e^{0.9t}$
   (ii) $0.108\,\text{m h}^{-1}$; $0.266\,\text{m h}^{-1}$; $0.653\,\text{m h}^{-1}$; $1.61\,\text{m h}^{-1}$

4. (i) $\frac{dy}{dx} = (1+x)e^x$;
   $\frac{d^2y}{dx^2} = (2+x)e^x$

   (ii) $\left(-1, -\frac{1}{e}\right)$

5. (i) Rotation symmetry, centre $(0, 0)$ of order 2. $f(x)$ is an odd function since $f(-x) = -f(x)$.
   (ii) $f'(x) = 2 + \ln(x^2)$; $f''(x) = \frac{2}{x}$
   (iii) $\left(-\frac{1}{e}, \frac{2}{e}\right)$, maximum; $\left(\frac{1}{e}, -\frac{2}{e}\right)$, minimum

6. (i) $\frac{e^x(x-1)}{x^2}$
   (ii) $(1, e)$, minimum
   (iii)

7. (i) $y = \ln x \Rightarrow \frac{dy}{dx} = \frac{1}{x}$;
   $y = x \ln x \Rightarrow \frac{dy}{dx} = 1 + \ln x$

8. (i) $(1-x)e^{-x}$
   (ii) $\left(1, \frac{1}{e}\right)$

9. (i) 1
   (ii) $f'(x) = \frac{1-\ln x}{x^2}$;
   $f''(x) = \frac{2\ln x - 3}{x^3}$
   (iii) $\frac{1}{e}$; $-\frac{1}{e^3}$

10. $(4, 4e^{-2})$

11. (i) $(1, -e)$
    (ii) Minimum

12. (ii) $ey - 2x + 1 = 0$

Activity 4.2 (Page 92)

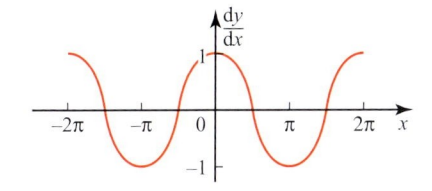

When $y = \sin x$ the graph of $\frac{dy}{dx}$ against x looks like the graph of $\cos x$.

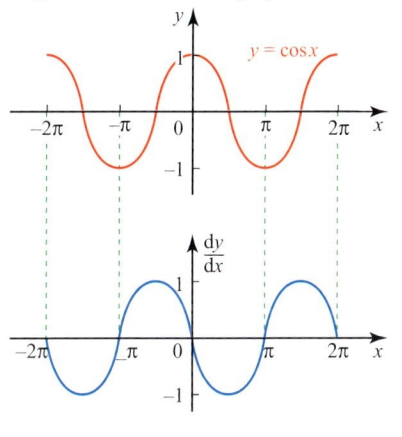

? (Page 93)

No. You can see this if you draw both graphs.

? (Page 93)

This is a demonstration but 'looking like' is not the same as proof.

Activity 4.3 (Page 93)

$y = \tan x = \frac{\sin x}{\cos x}$

$\frac{dy}{dx} = \frac{\cos x(\cos x) - \sin x(-\sin x)}{\cos^2 x}$

$= \frac{\cos^2 x + \sin^2 x}{\cos^2 x} = \frac{1}{\cos^2 x}$

$= \sec^2 x$

Exercise 4C (Page 96)

1. (i) $-2\sin x + \cos x$
   (ii) $\sec^2 x$
   (iii) $\cos x + \sin x$

2. (i) $x\sec^2 x + \tan x$
   (ii) $\cos^2 x - \sin^2 x = \cos 2x$
   (iii) $e^x(\sin x + \cos x)$

3. (i) $\frac{x\cos x - \sin x}{x^2}$
   (ii) $e^x(\cos x + \sin x)\sec^2 x$
   (iii) $\frac{\sin x(1-\sin x) - \cos x(x+\cos x)}{\sin^2 x}$

4 (i) $2x\sec^2(x^2+1)$

(ii) $2\cos 2x$

(iii) $\dfrac{1}{\tan x}$

5 (i) $-\dfrac{\sin x}{2\sqrt{\cos x}}$

(ii) $e^x(\tan x + \sec^2 x)$

(iii) $8x\cos 4x^2$

(iv) $-2\sin 2x e^{\cos 2x}$

(v) $\dfrac{1}{1+\cos x}$

(vi) $\dfrac{1}{\sin x \cos x}$

6 (i) $\cos x - x\sin x$

(ii) -1

(iii) $y = -x$

(iv) $y = x - 2\pi$

7 $\dfrac{dy}{dx} = e^x \cos 3x - 3e^x \sin 3x$

$\dfrac{d^2y}{dx^2} = -6e^x \sin 3x - 8e^x \cos 3x$

8 (i) $e^{-x}(\cos x - \sin x)$

(iii) $(0.79, 0.32), (-2.4, -7.5)$

(iv) Differentiate with respect to x again and evaluate the second derivative at the stationary points.

9 Maximum at $x = \tfrac{1}{6}\pi$,

minimum at $x = \tfrac{5}{6}\pi$

10 Maximum at $x = \tfrac{1}{12}\pi$,

minimum at $x = \tfrac{5}{12}\pi$

11 (i) $\tfrac{1}{4}\pi$

(ii) Maximum

12 $-\tfrac{1}{4}\pi$

? (Page 99)

The mapping is one-to-many.

Exercise 4D (Page 102)

1 (i) $4y^3 \dfrac{dy}{dx}$

(ii) $2x + 3y^2 \dfrac{dy}{dx}$

(iii) $x\dfrac{dy}{dx} + y + 1 + \dfrac{dy}{dx}$

(iv) $-\sin y \dfrac{dy}{dx}$

(v) $e^{(y+2)} \dfrac{dy}{dx}$

(vi) $y^3 + 3xy^2 \dfrac{dy}{dx}$

(vii) $4xy^5 + 10x^2y^4 \dfrac{dy}{dx}$

(viii) $1 + \dfrac{1}{y}\dfrac{dy}{dx}$

(ix) $xe^y \dfrac{dy}{dx} + e^y + \sin y \dfrac{dy}{dx}$

(x) $\dfrac{x^2}{y}\dfrac{dy}{dx} + 2x \ln y$

(xi) $e^{\sin y} + x\cos y e^{\sin y}\dfrac{dy}{dx}$

(xii) $\tan y + x\sec^2 y \dfrac{dy}{dx}$

$- (\tan x \dfrac{dy}{dx} + y\sec^2 x)$

2 $\tfrac{1}{5}$

3 0

4 (i) 0

(ii) $y = -1$

5 $(1, -2)$ and $(-1, 2)$

6 (i) $\dfrac{y+4}{6-x}$

(ii) $x - 2y - 11 = 0$

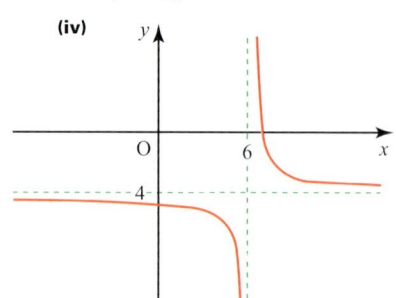

(iii) $\left(2, -4\tfrac{1}{2}\right)$

(iv)

Asymptotes $x = 6$, $y = -4$

7 (i) $\ln y = x \ln x$

(ii) $\dfrac{1}{y}\dfrac{dy}{dx} = 1 + \ln x$

(iii) $(0.368, 0.692)$

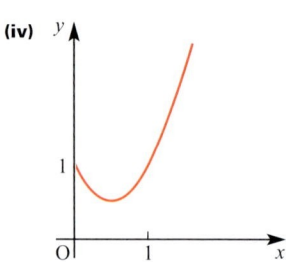

(iv)

8 (ii) $(1, -3), (-1, 3)$

9 (ii) $4x - 5y = -12$

10 (ii) $(2, 1), (-2, -1)$

11 (i) $\dfrac{3x^2 - 2xy}{x^2 + 3y^2}$

(ii) $8x - 7y - 9 = 0$

12 $(a, -2a)$

? (Page 105)

At points where the rate of change of gradient is greatest.

Exercise 4E (Page 112)

1 (i) t

(ii) $\dfrac{1 + \cos\theta}{1 + \sin\theta}$

(iii) $\dfrac{t^2 + 1}{t^2 - 1}$

(iv) $-\tfrac{2}{3}\cot\theta$

(v) $\dfrac{t - 1}{t + 1}$

(vi) $-\tan\theta$

(vii) $\dfrac{1}{2e^t}$

(viii) $\dfrac{(1+t)^2}{(1-t)^2}$

2 (i) 6

(ii) $y = 6x - \sqrt{3}$

(iii) $3x + 18y - 19\sqrt{3} = 0$

3 (i) $\left(\tfrac{1}{4}, 0\right)$

(ii) 2

(iii) $y = 2x - \tfrac{1}{2}$

(iv) $\left(0, -\tfrac{1}{2}\right)$

4 (i) $x - ty + at^2 = 0$

(ii) $tx + y = at^3 + 2at$

(iii) $(at^2 + 2a, 0), (0, at^3 + 2at)$

6 (i) $-\dfrac{b}{at^2}$

(ii) $at^2y + bx = 2abt$

(iii) $X(2at, 0), Y\left(0, \dfrac{2b}{t}\right)$

(iv) Area $= 2ab$

7 (ii) $y = tx - 2t^2$

(iii) $[2(t_1 + t_2), 2t_1 t_2]$

(iv) $x = 4$

8 (i) $t = 1$

(iii) $x + y = 3$

(v) $(-8, -5)$

9 (i) $t = -2$

(iii) $y = 2x - 6$

(iv) $(-5, 9)$

10 (i) $-\dfrac{3\cos t}{4\sin t}$

(ii) $3x\cos t + 4y\sin t = 12$

(iii) $t = 0.6435 + n\pi$

11 (iii) $\left(1 + \dfrac{1}{\sqrt{3}}, \dfrac{2}{\sqrt{3}}\right)$

12 (i) $\dfrac{2t(t-1)}{3t - 2}$

(ii) $(6, 5)$

13 $2\sin\theta$

14 (ii) $\dfrac{\ln 3}{2}$

16 (i) $-\tan t$

Chapter 5

Activity 5.1 (Page 119)

1 The areas of the two shaded regions are equal since $y = \dfrac{1}{x}$ is an odd function.

(P) (Page 120)

The polynomial $p_2(x)$ can take the value zero.

$x^2 - 2x + 3 = (x - 1)^2 + 2$ so is defined for all values of x and is always greater than or equal to 2.

Exercise 5A (Page 120)

1 (i) $3\ln|x| + c$

(ii) $\dfrac{1}{4}\ln|x| + c$

(iii) $\ln|x - 5| + c$

(iv) $\dfrac{1}{2}\ln|2x - 9| + c$

2 (i) $\dfrac{1}{3}e^{3x} + c$

(ii) $-\dfrac{1}{4}e^{-4x} + c$

(iii) $-3e^{-x/3} + c$

(iv) $-\dfrac{2}{e^{5x}} + c$

(v) $e^x - 2e^{-2x} + c$

3 (i) $2(e^8 - 1) = 5960$

(ii) $\ln\dfrac{49}{9} = 1.69$

(iii) 4.70

(iv) 0.906

4 (i) $P(2, 4); Q(-2, -4)$

(ii) 8.77; 14.2 (to 3 s.f.)

5 (i) $4; 5\ln 5 - 4$

(ii) Reflection in $y = x$

(iv) (a) $3(5\ln 5 - 4)$

(b) $4\ln 3 + 5\ln 5 - 4$

6 (i) $\dfrac{2}{2x + 3}$

(iii) Quotient $= 2x + 1$, remainder $= -3$

7 (i) $y = \dfrac{1}{2}e^{2x} + 2e^{-x} - \dfrac{3}{2}$

(ii) Minimum when $x = 0.231$

8 (i) $y = 3x - 3$

(ii) (a) 4

9 $\dfrac{1}{2}(e^2 + 1)$

Investigations (Page 123)

A series for e^x

$a_0 = 1$

$a_1 = 1$

$a_2 = \dfrac{1}{2!}$

$a_3 = \dfrac{1}{3!}$

$a_4 = \dfrac{1}{4!}$

$e = 2.71828183$ (8 d.p.)

Compound interest

Scheme B: $R = 2.594$

Scheme C: $R = 2.653$

1000 instalments: $R = 2.717$

10^4 instalments: $R = 2.718$

10^6 instalments: R agrees with the value of e to 5 d.p.

Exercise 5B (Page 126)

1 (i) $-\cos x - 2\sin x + c$

(ii) $3\sin x - 2\cos x + c$

(iii) $-5\cos x + 4\sin x + c$

(iv) $4\tan x + c$

(v) $-\dfrac{1}{2}\cos(2x + 1) + c$

(vi) $\dfrac{1}{5}\sin(5x - \pi) + c$

(vii) $3\tan 2x + c$

(viii) $\tan 3x + \dfrac{1}{2}\cos 2x + c$

(ix) $4\tan x - \dfrac{1}{2}\sin 2x + c$

2 (i) $\dfrac{1}{2}$

(ii) 1

(iii) $\dfrac{\sqrt{3} - 1}{2}$

(iv) $\dfrac{3}{4}$

(v) $\dfrac{1}{3}$

(vi) $\dfrac{\sqrt{3} - 1}{2}$

(vii) 0

(viii) 1

(ix) $\dfrac{3}{4}$

3 (ii) $\dfrac{3}{8}$

4 (i) (a) $\dfrac{1}{2}x + \dfrac{1}{4}\sin 2x + c$

(b) $\dfrac{\pi}{4}$

(ii) (a) $\dfrac{1}{2}x - \dfrac{1}{4}\sin 2x + c$

(b) $\dfrac{\pi}{6} - \dfrac{\sqrt{3}}{8}$

6 (ii) $\frac{\pi}{12}, \frac{5\pi}{12}$

(iii) $\frac{\pi}{2}$

7 (i) $\frac{1}{2} + \frac{1}{2}\cos 2x$

(iii) $\frac{1}{6}\pi - \frac{1}{8}\sqrt{3}$

8 (ii) $\frac{1}{4}(5\pi - 2)$

9 (ii) $2\sqrt{3} - \frac{\pi}{2}$

Activity 5.2 (Page 130)

For example

32 strips: 8.398

50 strips: 8.409

100 strips: 8.416

1000 strips: 8.420

❓ (Page 130)

The curve is part of the circle centre $\left(2\frac{1}{2}, 0\right)$, radius $2\frac{1}{2}$.
Area required is half a major segment = 8.4197 units².
Error from 16-strip estimate is about 0.7%.

❓ (Page 131)

(i) Underestimates – all trapezia below the curve

(ii) Impossible to tell

(iii) Overestimates – all trapezia above the curve

Exercise 5C (Page 131)

1 (i) 458 m

(ii) A curve is approximated by a straight line. The speeds are not given to a high level of accuracy.

2 (i) 3.1349...

(ii) 3.1399..., 3.1411...

(iii) 3.14

3 (i) 7.3

(ii) Overestimate

4 (i)

x	y
2	2
3	2.2361
4	2.4495
5	2.6458
6	2.8284
7	3

(ii) 12.6598; too small

(iii) $2\frac{1}{3}$ square units

(iv) $12\frac{2}{3}$ square units, 0.054%

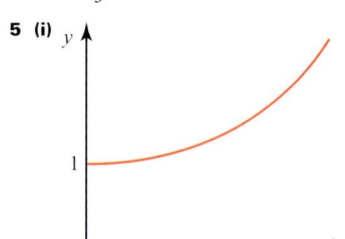

5 (i)

(ii) 2.179 218, 2.145 242, 2.136 756, 2.134 635

(iii) 2.13

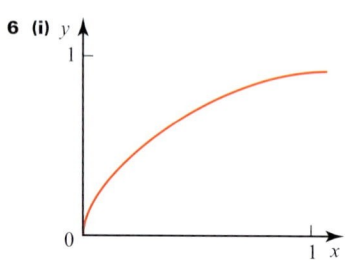

6 (i)

(ii) 0.458 658, 0.575 532, 0.618 518, 0.634 173, 0.639 825

(iii) 0.64

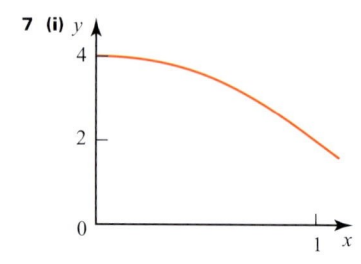

7 (i)

(ii) 3, 3.1, 3.131 176, 3.138 988

(iii) 3.14 (This actually converges to π.)

8 (i) 2, 2, 4, 4

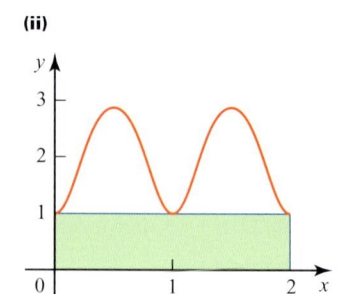

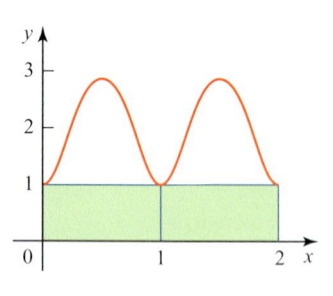

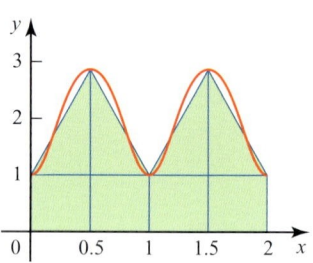

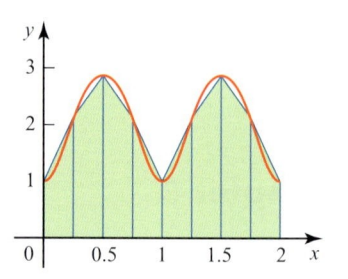

(ii)

(iii) 4

9 (i) $(1, 0)$

(ii) $\frac{1}{e}$

(iii) 0.89

(iv) Underestimate

10 (i) $(0, 1)$

(ii) $\frac{1}{4}\pi$

(iii) 1.77

(iv) Underestimate

11 (i) 2

(iii) 0.95

12 (i) 1.23

(ii) One of the intervals gives an overestimate and the other gives an underestimate.

Chapter 6

? (Page 136)

(i), (ii) and (iv) can be solved algebraically; (iii) and (v) cannot.

? (Page 138)

0.012 takes 5 steps

0.385 takes 18 steps

0.989 takes 28 steps

In general 0.abc takes (a + b + c + 2) steps.

Activity 6.1 (Page 139)

For 1 d.p., an interval length of < 0.05 is usually necessary, requiring $n = 5$. However, it depends on the position of the end points of the interval.

For example, the interval [0.25, 0.3125] obtained in 4 steps gives 0.3 (1 d.p.) but the interval [0.3125, 0.375] obtained in 4 steps is inconclusive. As are the interval [0.34375, 0.375] obtained in 5 steps, the interval [0.34375, 0.359375] obtained in 6 steps, the interval [0.34375, 0.3515625] obtained in 7 steps, etc.

In cases like this, 2 and 3 d.p. accuracy is obtained very quickly after 1 d.p.

The expected number of steps for 2 d.p., requiring an interval of length < 0.005, is 8 steps.

Exercise 6A (Page 141)

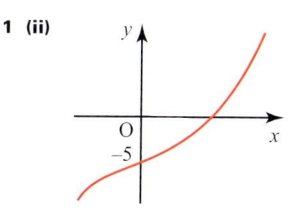

1 (ii)

(iii) 1.154

2 (i) 2

(ii) [0, 1]; [1, 2]

(iii) 0.62, 1.51

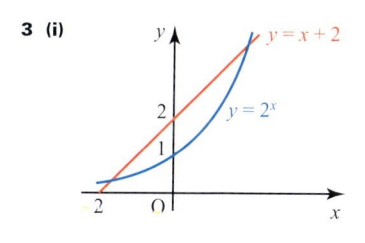

3 (i)

(ii) 2 roots

(iii) 2, −1.690

4 −1.88, 0.35, 1.53

5 1.62, 1.28

6 (i) [−2, −1]; [1, 2]; [4, 5]

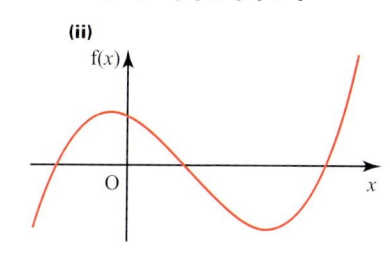

(ii)

(iii) −1.51, 1.24, 4.26

(iv) $a = 1.51171875$, $n = 8$
    $a = 1.244384766$, $n = 12$
    $a = 4.262695313$, $n = 10$

7 (i) [1, 2]; [4, 5]

(ii) 1.857, 4.536

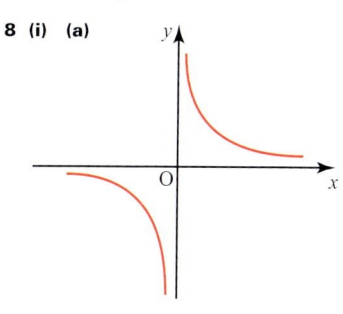

8 (i) (a)

(b) No root

(c) Convergence to a non-existent root

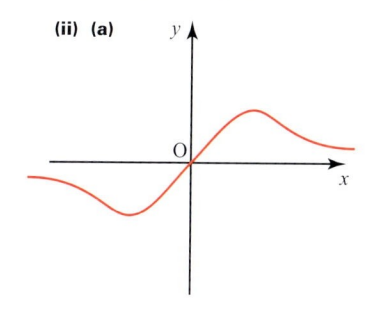

(ii) (a)

(b) $x = 0$

(c) Success

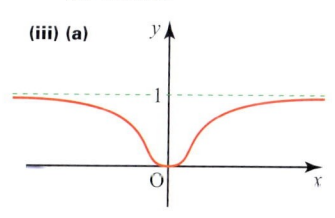

(iii) (a)

(b) $x = 0$

(c) Failure to find root

Investigation (Page 142)

(i) Converges to 0.7391 (to 4 d.p.) since $\cos 0.7391 = 0.7391$ (to 4 d.p.).

(ii) Converges to 1.
$\sqrt{x} < x$ for $x > 1$, $\sqrt{x} > x$ for $x < 1$ and $\sqrt{1} = 1$

(iii) Converges to 1.6180 (to 4 d.p.) since this is the solution of $x = \sqrt{x+1}$ (i.e. the positive solution of $x^2 - x - 1 = 0$).

? (Page 144)

Writing $x^5 - 5x + 3 = 0$
as $x^5 - 4x + 3 = x$
gives $g(x) = x^5 - 4x + 3$

Generalising this to
$$x^5 + (n-5)x + 3 = nx$$
gives $g(x) = \dfrac{x^5 + (n-5)x + 3}{n}$

and indicates that infinitely many rearrangements are possible.

P2 Answers

ⓟ (Page 146)

Bounds for the root have now been established.

Activity 6.2 (Page 148)

$x_0 = -2$ gives divergence to $-\infty$

$x_0 = -1$ gives convergence to 0.618

$x_0 = 1$ gives convergence to 0.618

$x_0 = 2$ gives divergence to $+\infty$.

- Between $x = -1$ and $x = 1$, $-1 <$ gradient < 1.
- Gradient is just greater than zero here.
- At this root gradient > 1 so the root is not found.
- $x = -1$, gradient = 1.
- $x = 1$, gradient = 1.
- At this root $-1 <$ gradient < 1 so the root is found.
- At this root gradient > 1 so the root is not found.

Exercise 6B (Page 148)

1 (ii) 1.521

2 (ii) 2.120

3 (iii) 1.503

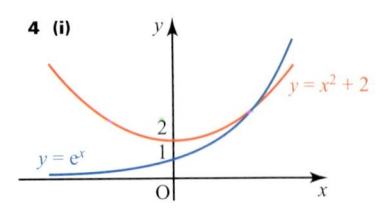

4 (i) [graph of $y = e^x$ and $y = x^2 + 2$]

 (ii) Only one point of intersection

 (iii) $F(x) = \ln(x^2 + 2)$ is possible.

 (iv) 1.319

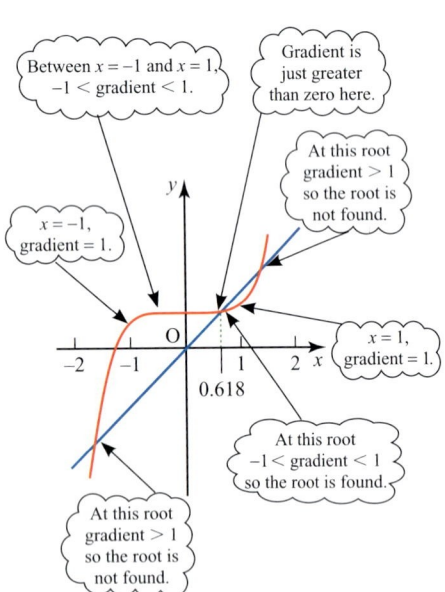

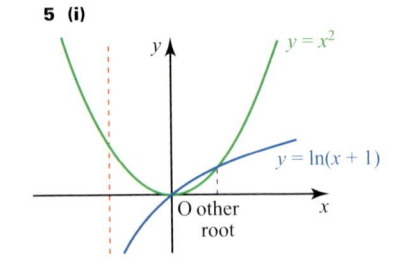

5 (i) [graph of $y = x^2$ and $y = \ln(x+1)$, O other root]

 (ii) 0.747

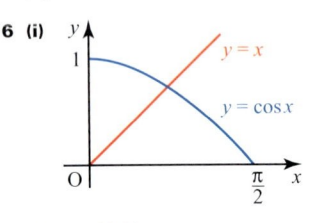

6 (i) [graph of $y = x$ and $y = \cos x$]

 (ii) 0.739 09

7 (i) 1.68

 (ii) $x = \dfrac{3x}{4} + \dfrac{2}{x^3}$; $\alpha = \sqrt[4]{8}$

8 (i) 2.29

 (ii) $x = \dfrac{2x}{3} + \dfrac{4}{x^2}$; $\alpha = \sqrt[3]{12}$

9 (i) [graph of $y = 2 - 2x$ and $y = \cos x$]

 (iv) 0.58

10 (iii) 1.08

11 (i) $\left(-\dfrac{1}{2}, -\dfrac{1}{2e}\right)$

 (iii) 1.35

12 (i) [graph of $y = 2 - x^2$ and $y = \ln x$]

 (iv) $x = 1.31$

13 (i) 3 and 4

 (ii) 3.43

14 (iii) 1.77

Chapter 7

Investigation (Page 154)

1.01, 1.02, 1.03

$\sqrt{1+x} \approx 1 + \dfrac{1}{2}x$ or $\sqrt{x} \approx \dfrac{1}{2}(1+x)$

$k = \dfrac{1}{2}$

0.20

ⓟ (Page 156)

$(1+x)^{1/2} = 3$ but substituting $x = 8$ into the expansion gives successive approximations of 1, 5, −3, 29, −131, ... and these are getting further from 3 rather than closer to it.

Investigation (Page 157)

$-0.19 < x < 0.60$

$-0.08 < x < 0.07$

Activity 7.1 (Page 157)

For $|x| < 1$ the sum of the geometric series is $\dfrac{1}{1+x}$ which is the same as $(1+x)^{-1}$.

Investigation (Page 159)

$(1-x)^{-3} = 1 + 3x + 6x^2 + 10x^3 \ldots$

The coefficients of x are the triangular numbers.

❓ (Page 160)

$\sqrt{101} = \sqrt{100 \times 1.01}$

$= 10\sqrt{1.01}$

$= 10(1 + 0.01)^{\frac{1}{2}}$

$= 10[1 + \dfrac{1}{2}(0.01)$

$+ \dfrac{\left(\dfrac{1}{2}\right)\left(-\dfrac{1}{2}\right)}{2!}(0.01)^2 + \ldots]$

$= 10.050$ (3 d.p.)

322

? (Page 162)

$\sqrt{x-1}$ is only defined for $x > 1$.
A possible rearrangement is
$\sqrt{x\left(1-\frac{1}{x}\right)} = \sqrt{x}\left(1-\frac{1}{x}\right)^{\frac{1}{2}}$.
Since $x > 1 \Rightarrow 0 < \frac{1}{x} < 1$
the binomial expansion could be used but the resulting expansion would not be a series of positive powers of x.

Exercise 7A (Page 162)

1 (i) (a) $1 - 2x + 3x^2$
   (b) $|x| < 1$
   (c) 0.43%

 (ii) (a) $1 - 2x + 4x^2$
   (b) $|x| < \frac{1}{2}$
   (c) 0.8%

 (iii) (a) $1 - \frac{x^2}{2} - \frac{x^4}{8}$
   (b) $|x| < 1$
   (c) 0.0000063%

 (iv) (a) $1 + 4x + 8x^2$
   (b) $|x| < \frac{1}{2}$
   (c) 1.3%

 (v) (a) $\frac{1}{3} - \frac{x}{9} + \frac{x^2}{27}$
   (b) $|x| < 3$
   (c) 0.0037%

 (vi) (a) $2 - \frac{7x}{4} - \frac{17x^2}{64}$
   (b) $|x| < 4$
   (c) 0.00095%

 (vii) (a) $-\frac{2}{3} - \frac{5x}{9} - \frac{5x^2}{27}$
   (b) $|x| < 3$
   (c) 0.0088%

 (viii) (a) $\frac{1}{2} - \frac{3x}{16} + \frac{27x^2}{256}$
   (b) $|x| < \frac{4}{3}$
   (c) 0.013%

 (ix) (a) $1 + 6x + 20x^2$
   (b) $|x| < \frac{1}{2}$
   (c) 4%

 (x) (a) $1 + 2x^2 + 2x^4$
   (b) $|x| < 1$
   (c) 0.00020%

 (xi) (a) $1 + \frac{2x^2}{3} - \frac{4x^4}{9}$
   (b) $|x| < \frac{1}{\sqrt{2}}$
   (c) 0.000048%

 (xii) (a) $1 - 3x + 7x^2$
   (b) $|x| < \frac{1}{2}$
   (c) 1.64%

2 (i) $1 + 3x + 3x^2 + x^3$
  (ii) $1 + 4x + 10x^2 + 20x^3$ for $|x| < 1$
  (iii) $a = 25$, $b = 63$

3 (i) $16 - 32x + 24x^2 - 8x^3 + x^4$
  (ii) $1 - 6x + 21x^2 - 80x^3$ for $|x| < \frac{1}{2}$
  (iii) $a = -128$, $b = 600$

4 (i) $1 + x + x^2 + x^3$ for $|x| < 1$
  (ii) $1 - 4x + 12x^2 - 32x^3$ for $|x| < \frac{1}{2}$
  (iii) $1 - 3x + 9x^2 - 23x^3$ for $|x| < \frac{1}{2}$

5 (ii) $1 + \frac{x}{8} + \frac{3x^2}{128}$ for $|x| < 4$
  (iii) $1 + \frac{9x}{8} + \frac{19x^2}{128}$

6 (i) $1 - y + y^2 - y^3 \ldots$
  (ii) $1 - \frac{2}{x} + \frac{4}{x^2} - \frac{8}{x^3}$
  (iv) $\frac{x}{2} - \frac{x^2}{4} + \frac{x^3}{8} - \frac{x^4}{16}$
  (v) $x < -2$ or $x > 2$; $-2 < x < 2$; no overlap in range of validity.

7 $\frac{1}{4} - \frac{3}{4}x + \frac{27}{16}x^2$

8 $1 - \frac{3}{2}x^2$

9 (i) -3
  (ii) $-\frac{10}{3}x^3$

Exercise 7B (Page 166)

1 $\frac{2a^2}{3b^3}$

2 $\frac{1}{9y}$

3 $\frac{x+3}{x-6}$

4 $\frac{x+3}{x+1}$

5 $\frac{2x-5}{2x+5}$

6 $\frac{3(a+4)}{20}$

7 $\frac{x(2x+3)}{(x+1)}$

8 $\frac{2}{5(p-2)}$

9 $\frac{a-b}{2a-b}$

10 $\frac{(x+4)(x-1)}{x(x+3)}$

11 $\frac{9}{20x}$

12 $\frac{x-3}{12}$

13 $\frac{a^2+1}{a^2-1}$

14 $\frac{5x-13}{(x-3)(x-2)}$

15 $\frac{2}{(x+2)(x-2)}$

16 $\frac{2p^2}{(p^2-1)(p^2+1)}$

17 $\frac{a^2-a+2}{(a+1)(a^2+1)}$

18 $\frac{-2(y^2+4y+8)}{(y+2)^2(y+4)}$

19 $\frac{x^2+x+1}{x+1}$

20 $-\frac{(3b+1)}{(b+1)^2}$

21 $\frac{13x-5}{6(x-1)(x+1)}$

22 $\frac{4(3-x)}{5(x+2)^2}$

23 $\frac{3a-4}{(a+2)(2a-3)}$

24 $\frac{3x^2-4}{x(x-2)(x+2)}$

? (Page 168)

The identity is true for all values of x. Once a particular value of x is substituted you have an equation. Equating constant terms is equivalent to substituting $x = 0$.

Exercise 7C (Page 170)

1. $\dfrac{1}{(x-2)} - \dfrac{1}{(x+3)}$

2. $\dfrac{1}{x} - \dfrac{1}{(x+1)}$

3. $\dfrac{2}{(x-4)} - \dfrac{2}{(x-1)}$

4. $\dfrac{2}{(x-1)} - \dfrac{1}{(x+2)}$

5. $\dfrac{1}{(x+1)} + \dfrac{1}{(2x-1)}$

6. $\dfrac{2}{(x-2)} - \dfrac{2}{x}$

7. $\dfrac{1}{(x-1)} - \dfrac{3}{(3x-1)}$

8. $\dfrac{3}{5(x-4)} + \dfrac{2}{5(x+1)}$

9. $\dfrac{5}{(2x-1)} - \dfrac{2}{x}$

10. $\dfrac{2}{(2x-3)} - \dfrac{1}{(x+2)}$

11. $\dfrac{8}{13(2x-5)} + \dfrac{9}{13(x+4)}$

12. $\dfrac{19}{24(3x-2)} - \dfrac{11}{24(3x+2)}$

13. $\dfrac{1}{(x+1)} + \dfrac{2}{(x+2)} + \dfrac{3}{(x+3)}$

14. $\dfrac{4}{(x-1)} + \dfrac{3}{(3-x)} + \dfrac{2}{(2x+1)}$

15. $\dfrac{1}{(2+x)} - \dfrac{2}{(2-x)} - \dfrac{1}{(2x+3)}$

Exercise 7D (Page 172)

1. (i) $\dfrac{9}{(1-3x)} - \dfrac{3}{(1-x)} - \dfrac{2}{(1-x)^2}$

   (ii) $\dfrac{4}{(2x-1)} - \dfrac{2x}{(x^2+1)}$

   (iii) $\dfrac{1}{(x-1)^2} - \dfrac{1}{(x-1)} + \dfrac{1}{(x+2)}$

   (iv) $\dfrac{5}{8(x-2)} + \dfrac{6-5x}{8(x^2+4)}$

   (v) $\dfrac{5-2x}{(2x^2-3)} + \dfrac{2}{(x+2)}$

   Can be taken further using surds.

   (vi) $\dfrac{2}{x} - \dfrac{1}{x^2} - \dfrac{3}{(2x+1)}$

   (vii) $\dfrac{10x}{(3x^2-1)} - \dfrac{3}{x}$

   Can be taken further using surds.

   (viii) $\dfrac{1}{(2x^2+1)} + \dfrac{1}{(x+1)}$

   (ix) $\dfrac{8}{(2x-1)} - \dfrac{4}{(2x-1)^2} - \dfrac{3}{x}$

2. $A=1,\ B=0,\ C=1$

3. $A=1,\ B=0,\ C=-4$

Investigation (Page 174)

The binomial expansion is

$1 - x + 3x^2$.

The expansion is valid when $|x| < \tfrac{1}{2}$.

Which method is preferred is a matter of personal preference for (a) and (b) but for (c) must be (iii).

Exercise 7E (Page 174)

1. (i) $4 + 20x + 72x^2$

   (ii) $-4 - 10x - 16x^2$

   (iii) $\dfrac{5}{2} + \dfrac{11x}{4} + \dfrac{33x^2}{8}$

   (iv) $-\dfrac{1}{8} - \dfrac{5x}{16} - \dfrac{x^2}{8}$

2. (i) $\dfrac{2}{(2x-1)} - \dfrac{3}{(x+2)}$

   (ii) $1 + 2x + 4x^2 + \ldots$
   $a=1, b=2, c=4$, for $|x| < \tfrac{1}{2}$

   (iii) $\dfrac{1}{2} - \dfrac{x}{4} + \dfrac{x^2}{8}$ for $|x| < 2$

   (iv) $-\dfrac{7}{2} - \dfrac{13x}{4} - \dfrac{67x^2}{8}$; 0.505%

3. (i) $2 + x - x^2$
   $\dfrac{2}{(2-x)} - \dfrac{1}{(1+x)}$

   (ii) $|x| < 1$

4. (i) $\dfrac{1}{(1-x)} - \dfrac{9}{(3-x)}$

   (ii) $0, 1\tfrac{1}{2}$

   (iii) $\dfrac{4x}{3} + \dfrac{8x^2}{3}$

5. (i) $\dfrac{2}{(2-x)} + \dfrac{2x+4}{(1+x^2)}$

   (ii) $5 + \tfrac{5}{2}x - \tfrac{15}{4}x^2 - \tfrac{15}{8}x^3$

6. (i) $\dfrac{2}{(2+x)} + \dfrac{x-1}{(x^2+1)}$

   (ii) $\tfrac{1}{2}x + \tfrac{5}{4}x^2 - \tfrac{9}{8}x^3$

7. (i) $\dfrac{1}{(1-x)} + \dfrac{2}{(1+2x)} - \dfrac{4}{(2+x)}$

   (ii) $1 - 2x + \tfrac{17}{2}x^2$

Chapter 8

? (Page 177)

It is the same as

$\displaystyle\int_1^4 \sqrt{x}\,dx.$

? (Page 179)

Yes: Using the chain rule

$\dfrac{dy}{dx} = \dfrac{dy}{du} \times \dfrac{du}{dx}$

Integrating both sides with respect to x

$y = \displaystyle\int\left(\dfrac{dy}{du} \times \dfrac{du}{dx}\right)dx = \int\left(\dfrac{dy}{du}\right)du$

Activity 8.1 (Page 181)

$\tfrac{2}{5}(x-2)^{5/2} + \tfrac{4}{3}(x-2)^{3/2} + c$

$= \tfrac{2}{15}(x-2)^{3/2}\,[3(x-2) + 10] + c$

$= \tfrac{2}{15}(3x+4)(x-2)^{3/2} + c$

Exercise 8A (Page 181)

1. (i) $\tfrac{1}{8}(x^3+1)^8 + c$

   (ii) $\tfrac{1}{6}(x^2+1)^6 + c$

   (iii) $\tfrac{1}{5}(x^3-2)^5 + c$

   (iv) $\tfrac{1}{6}(2x^2-5)^{3/2} + c$

   (v) $\tfrac{1}{15}(2x+1)^{3/2}(3x-1) + c$

   (vi) $\tfrac{2}{3}(x+9)^{1/2}(x-18) + c$

2. (i) $222\,000$

   (ii) 586

   (iii) 18.1

3. (i) $22\tfrac{1}{2}$

   (ii) $1\tfrac{1}{9}$

4. (i) $A(-1, 0)$, $x \geqslant -1$

5. (i) (a) $\dfrac{(1+x)^4}{4} + c$

   (b) $2\tfrac{2}{5}$

   (ii) $\tfrac{1}{3}(2\sqrt{2} - 1) \approx 0.609$

6 (i) (a) $8\sqrt{x} - \dfrac{3}{2x^2} + c$

   (b) $2(1 + x^2)^{3/2} + c$

   (ii) $k = 2, a = 1, b = 2; 32.5$

Exercise 8B (Page 184)

1 (i) $\ln|x^2 + 1| + c$

   (ii) $\tfrac{1}{3}\ln|3x^2 + 9x - 1| + c$

   (iii) $4e^{x^3} + c$

2 (i) 0.018

   (ii) 0

3 (i) $\tfrac{1}{2}(e - 1)$

   (ii) $\tfrac{1}{2}(e^4 - 1)$

   (iii) $\tfrac{1}{2}(e + e^4) - 1 = 27.7$ (to 3 s.f.)

4 $0.490; 0.314$

5 (i) $-(x + 2)e^{-x}$

   (ii) $-(x + 3)e^{-x}$

   (iii) $(-2, e^2)$

   (iv) $-e^2$; max. at $x = -2$

   (vi) $3 - \dfrac{4}{e}$

6 (i) $\tfrac{1}{5}(2x - 3)^{5/2} + (2x - 3)^{3/2} + c$

   (ii) $\dfrac{\ln x + 2}{2\sqrt{x}}; 2\sqrt{x}\ln x + c$

   (iii) (a) $-2xe^{-x^2}$

        (b) $3x^2 e^{-x^6}$

7 (i) (a) $\tfrac{1}{2}\ln 3$

        (b) $\sqrt{9 + x^2} + c$

   (ii) (b) $\left(\dfrac{1}{\sqrt{2}}, \dfrac{1}{\sqrt{2}}e^{-1/2}\right)$ and $\left(-\dfrac{1}{\sqrt{2}}, -\dfrac{1}{\sqrt{2}}e^{-1/2}\right)$

        (c) 0.074

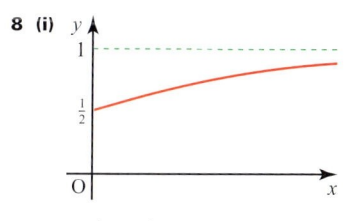

8 (i)

   (ii) $\ln\left(\dfrac{e^2 + 1}{2}\right) \approx 1.434$

   (iii) $\ln\left(\dfrac{e^2 + 1}{2}\right) \approx 1.434$

   (iv) The same. The substitution $e^x = t^2$ transforms the integral in part (ii) into that in part (iii).

9 (i) (a) $-4xe^{-2x^2}$

        (b) $e^{-2x^2} - 4x^2 e^{-2x^2}$

   (ii) $\tfrac{1}{4}(1 - e^{-2k^2})$

   (iv) Max. at $\left(\dfrac{1}{2}, \dfrac{1}{2}e^{-1/2}\right)$

10 (i) 1

   (ii) $\tfrac{1}{2}\ln\left(p^2 + 1\right)$

   (iii) 2.53

Exercise 8C (Page 189)

1 (i) $\tfrac{1}{3}\sin 3x + c$

   (ii) $\cos(1 - x) + c$

   (iii) $-\tfrac{1}{4}\cos^4 x + c$

   (iv) $\ln|2 - \cos x| + c$

   (v) $-\ln|\cos x| + c$

   (vi) $-\tfrac{1}{6}(\cos 2x + 1)^3 + c$

2 (i) $-\cos(x^2) + c$

   (ii) $e^{\sin x} + c$

   (iii) $\tfrac{1}{2}\tan^2 x + c$

   (iv) $\dfrac{-1}{\sin x} + c$

3 (i) 1

   (ii) $\dfrac{1}{16}$

   (iii) 1

   (iv) $e - 1$

   (v) $\ln 2$

4 (ii) $\tfrac{1}{2}$

5 (i) $2\cos\left(\theta - \tfrac{1}{3}\pi\right)$

6 (ii) $\tfrac{1}{6}\pi - \tfrac{1}{4}\sqrt{3}$

❓ (Page 190)

Substitution using $u = x^2 - 1$ needs $2x$ in the numerator. Not a product, not suitable for integration by parts.

Exercise 8D (Page 193)

1 (i) $\ln\left|\dfrac{3x - 2}{1 - x}\right| + c$

   (ii) $\dfrac{1}{1 - x} + \ln\left|\dfrac{x - 1}{2x + 3}\right| + c$

   (iii) $\ln\left|\dfrac{x - 1}{\sqrt{x^2 + 1}}\right| + c$

   (iv) $\ln\left|\dfrac{(x + 1)^2}{\sqrt{2x + 1}}\right| + c$

   (v) $\ln\left|\dfrac{x}{1 - x}\right| - \dfrac{1}{x} + c$

   (vi) $\tfrac{1}{2}\ln\left|\dfrac{x + 1}{x + 3}\right| + c$

   (vii) $\ln\left|\dfrac{\sqrt{x^2 + 4}}{x + 2}\right| + c$

   (viii) $\ln\left|\dfrac{2x + 1}{x + 2}\right| + \dfrac{1}{2(2x + 1)} + c$

2 $-\dfrac{x}{x^2 + 4} + \dfrac{1}{x - 3}, \ln\left(\dfrac{\sqrt{2}}{6}\right)$

3 $\dfrac{1}{x^2} - \dfrac{2}{x} + \dfrac{4}{2x + 1}$

4 (i) (a) $\dfrac{2}{1 - 2x} + \dfrac{1}{1 + x}$

        (b) $\ln\left(\tfrac{11}{8}\right) = 0.31845$

   (ii) (a) $3 + 3x + 9x^2 + \ldots$

        (b) 0.31800

        (c) 0.14%

5 (i) $A = 1, B = 3, C = -2$

   (ii) $2 + \ln\left(\tfrac{125}{3}\right) = 5.73$

6 (i) $B - 1, C - 16$

   (ii) $\tfrac{55}{2}\ln 2$

   (iii) $8 + 5x + 2x^2 + \dfrac{x^4}{2}$ for $|x| < 1$

7 (i) $A = 1, B = 2, C = 1, D = -3$

8 (i) $1 + \dfrac{1}{2(x + 1)} - \dfrac{3}{2(x + 3)}$

Activity 8.2 (Page 195)

(i) (a) $\dfrac{d}{dx}(x\cos x) = -x\sin x + \cos x$

   (b) $\Rightarrow x\cos x = \int -x\sin x\, dx + \int \cos x\, dx$

   $\Rightarrow \int x\sin x\, dx = -x\cos x + \int \cos x\, dx$

(c) $\Rightarrow \int x \sin x \, dx$
$= -x \cos x + \sin x + c$

(ii) (a) $\dfrac{d}{dx}(xe^{2x}) = x \times 2e^{2x} + e^{2x}$

(b) $\Rightarrow xe^{2x} = \int 2xe^{2x} \, dx + \int e^{2x} \, dx$
$\Rightarrow \int 2xe^{2x} \, dx = xe^{2x} - \int e^{2x} \, dx$

(c) $\Rightarrow \int 2xe^{2x} \, dx = xe^{2x} - \tfrac{1}{2}e^{2x} + c$

? (Page 195)

Each of the integrals in Activity 8.2 is of the form $\int x \dfrac{dv}{dx} \, dx$ and is found by starting with the product xv.

Exercise 8E (Page 199)

1 (i) (a) $u = x, \dfrac{dv}{dx} = e^x$
(b) $xe^x - e^x + c$

(ii) (a) $u = x, \dfrac{dv}{dx} = \cos 3x$
(b) $\tfrac{1}{3}x \sin 3x + \tfrac{1}{9} \cos 3x + c$

(iii) (a) $u = 2x + 1, \dfrac{dv}{dx} = \cos x$
(b) $(2x + 1)\sin x + 2\cos x + c$

(iv) (a) $u = x, \dfrac{dv}{dx} = e^{-2x}$
(b) $-\tfrac{1}{2}xe^{-2x} - \tfrac{1}{4}e^{-2x} + c$

(v) (a) $u = x, \dfrac{dv}{dx} = e^{-x}$
(b) $-xe^{-x} - e^{-x} + c$

(vi) (a) $u = x, \dfrac{dv}{dx} = \sin 2x$
(b) $-\tfrac{1}{2}x \cos 2x + \tfrac{1}{4} \sin 2x + c$

2 (i) $\tfrac{1}{4}x^4 \ln x - \tfrac{1}{16}x^4 + c$

(ii) $xe^{3x} - \tfrac{1}{3}e^{3x} + c$

(iii) $x \sin 2x + \tfrac{1}{2} \cos 2x + c$

(iv) $\tfrac{1}{3}x^3 \ln 2x - \tfrac{1}{9}x^3 + c$

3 $\tfrac{2}{15}(1 + x)^{3/2}(3x - 2) + c$

4 $\tfrac{1}{15}(x - 2)^5(5x + 2) + c$

5 (i) $x \ln x - x + c$
(ii) $x \ln 3x - x + c$
(iii) $x \ln px - x + c$

6 $x^2 e^x - 2xe^x + 2e^x + c$

7 $(2 - x)^2 \sin x - 2(2 - x)\cos x - 2 \sin x + c$

Exercise 8F (Page 201)

1 (i) $\tfrac{2}{9}e^3 + \tfrac{1}{9}$
(ii) -2
(iii) $2e^2$
(iv) $3 \ln 2 - 1$
(v) $\dfrac{\pi}{4}$
(vi) $\tfrac{64}{3} \ln 4 - 7$

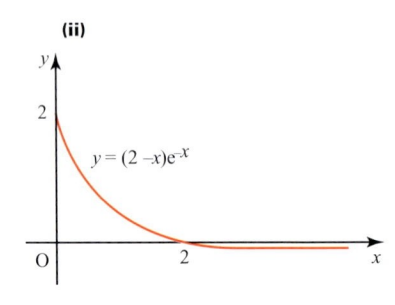

2 (i) $(2, 0), (0, 2)$

(ii)

$y = (2 - x)e^{-x}$

(iii) $e^{-2} + 1$

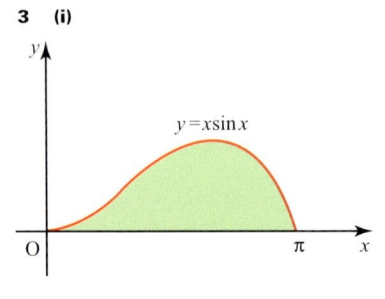

3 (i)

$y = x \sin x$

(ii) π

4 $5 \ln 5 - 4$

5 $\dfrac{\pi}{2} - 1$

6 $-\tfrac{4}{13}$ so area $= \tfrac{4}{15}$ square units

7 $x = 0.5$; area $= 0.134$ square units

8 The curve is below the trapezia.

9 (i) $\tfrac{1}{k}x \sin kx + \tfrac{1}{k^2} \cos kx + c$
(ii) $\cos 2x - \cos 8x$

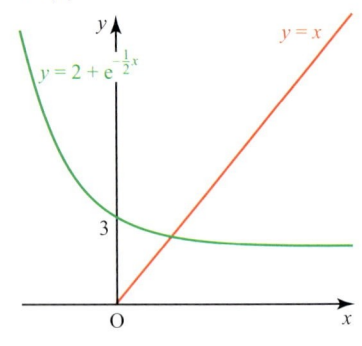

11 (ii)

(iv) 2.31

12 (i) $\tfrac{1}{2}$
(ii) $\pi\left(2\sqrt{e} - 3\right)$

? (Page 204)

You will return to these integrals in Activity 8.3.

Activity 8.3 (Page 205)

(i) This is a quotient. The derivative of the expression on the bottom is not related to the expression on the top, so you cannot use substitution. However, as the expression on the bottom can be factorised, you can write it as partial fractions.

$\int \dfrac{x - 5}{x^2 + 2x - 3} \, dx$

$= \int \dfrac{2}{(x + 3)} \, dx - \int \dfrac{1}{(x - 1)} \, dx$

$= 2 \ln|x + 3| - \ln|x - 1| + c$

(ii) The derivative of the expression on the bottom line is $2x + 2$, which is twice the expression on the top line. So the integral is of the form

$k \int \dfrac{f'(x)}{f(x)} \, dx = k \ln|f(x)| + c.$

This integral can also be found using partial fractions, but using logarithms is quicker.

$\int \dfrac{x + 1}{x^2 + 2x - 3} \, dx$

$= \tfrac{1}{2} \int \dfrac{2x + 2}{x^2 + 2x - 3} \, dx$

$= \tfrac{1}{2} \ln|x^2 + 2x - 3| + c$

(iii) This is a product of x and e^x. There is no relationship between one expression and the derivative of the other, so you cannot use substitution. As one of the expressions is x, you can use integration by parts.
$$\int xe^x\,dx = xe^x - \int e^x\,dx$$
$$= xe^x - e^x + c$$

(iv) This is also a product, this time of x and e^{x^2}. e^{x^2} is a function of x^2, and $2x$ is the derivative of x^2, so you can use the substitution $u = x^2$.

Using $u = x^2$
$$\int xe^{x^2}\,dx = \int \tfrac{1}{2}e^u\,du$$
$$= \tfrac{1}{2}e^u + c$$
$$= \tfrac{1}{2}e^{x^2} + c$$

(v) In this case the numerator is the differential of the denominator and so the integral is the natural logarithm of the modulus of the denominator.
$$\int \frac{2x + \cos x}{x^2 + \sin x}\,dx$$
Since $\dfrac{d}{dx}(x^2 + \sin x) = 2x + \cos x$ the integral is $\ln|x^2 + \sin x| + c$.

(vi) This is a product: $\sin^2 x$ is a function of $\sin x$, and $\cos x$ is the derivative of $\sin x$, so you can use the substitution $u = \sin x$.

Using $u = \sin x$
$$\int \cos x \sin^2 x\,dx = \int u^2\,du$$
$$= \tfrac{1}{3}u^3 + c$$
$$= \tfrac{1}{3}\sin^3 x + c$$

Exercise 8G (Page 206)

1 (i) $\tfrac{1}{3}\sin(3x-1) + c$

(ii) $\dfrac{-1}{(x^2 + x - 1)} + c$

(iii) $-e^{1-x} + c$

(iv) $\tfrac{1}{2}\sin 2x + c$

(v) $x\ln 2x - x + c$

(vi) $\dfrac{-1}{4(x^2 - 1)^2} + c$

(vii) $\tfrac{1}{3}(2x-3)^{2/3} + c$

(viii) $\ln\left|\dfrac{x-1}{x+2}\right| - \dfrac{1}{x-1} + c$

(ix) $\tfrac{1}{4}x^4\ln x - \tfrac{1}{16}x^4 + c$

(x) $\ln\left|\dfrac{x-3}{2x-1}\right| + c$

(xi) $\tfrac{1}{2}e^{x^2+2x} + c$

(xii) $-\ln(\sin x + \cos x) + c$

(xiii) $-\tfrac{1}{2}x^2\cos 2x + \tfrac{1}{2}x\sin 2x + \tfrac{1}{4}\cos 2x + c$

(xiv) $-\tfrac{1}{2}\cos 2x + \tfrac{1}{6}\cos^3 2x + c$

2 (i) $\tfrac{8}{3}$

(ii) $\tfrac{1}{3}\ln 4$

(iii) $48 + 8\ln 4$

(iv) $\tfrac{2}{3}$

(v) $\tfrac{8}{3}\ln 2 - \tfrac{7}{9}$

3 $\tfrac{4}{3}$

4 $\tfrac{1}{3}(2\sqrt{2} - 1)$

5 0.24

6 $\tfrac{1}{8}\pi - \tfrac{1}{4}$

7 (i) $-\tfrac{1}{2}xe^{-2x} - \tfrac{1}{4}e^{-2x} + c$

(ii) 0.112

8 (i) $-\tfrac{1}{2}\cos(2x - 3) + c$

(ii) $\tfrac{3}{4}e^4 + \tfrac{1}{4}$

(iii) $\tfrac{1}{2}\ln|x^2 - 9| + c$

9 (i) $\tfrac{38}{9}$

(ii) $\tfrac{1}{4} - \dfrac{3}{4e^2}$

10 $\tfrac{1}{4}, \tfrac{1}{4} - \dfrac{3}{4e^2}$

Chapter 9

Exercise 9A (Page 212)

1 $\dfrac{dv}{dt}$ is the rate of change of velocity with respect to time, i.e. the acceleration.

The differential equation tells you that the acceleration is proportional to the square of the velocity.

2 $\dfrac{ds}{dt} = \dfrac{k}{s^2}$

3 $\dfrac{dh}{dt} = k\ln(H - h)$

4 $\dfrac{dm}{dt} = \dfrac{k}{m}$

5 $\dfrac{dP}{dt} = k\sqrt{P}$

6 $\dfrac{de}{d\theta} = k\theta$

7 $\dfrac{d\theta}{dt} = -\dfrac{(\theta - 15)}{160}$

8 $\dfrac{dN}{dt} = \dfrac{N}{20}$

9 $\dfrac{dv}{dt} = \dfrac{4}{\sqrt{v}}$

10 $\dfrac{dA}{dt} = \dfrac{2k\sqrt{\pi}}{\sqrt{A}} = \dfrac{k'}{\sqrt{A}}$

11 $\dfrac{d\theta}{ds} = -\dfrac{s}{4}$

12 $\dfrac{dV}{dt} = -\dfrac{2V}{1125\pi}$

13 $\dfrac{dh}{dt} = \dfrac{(2 - k\sqrt{h})}{100}$

Investigation (Page 214)

H is about $(70°\text{ N}, 35°\text{ W})$ and L is about $(62°\text{ N}, 5°\text{ W})$ so they are separated by $30°$ in longitude at a mean latitude of $66°$. Reference to the scale shows this to be about 900 nautical miles.

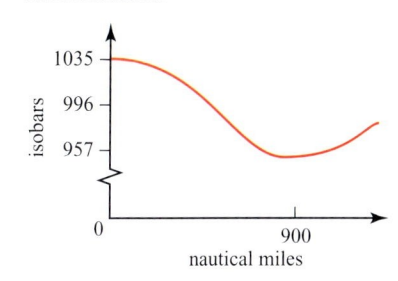

The mean level is 996 and the amplitude 39 so a model is

$$p = 996 + 39 \cos\left(\frac{\pi x}{900}\right)$$

and $\dfrac{dp}{dx} = \dfrac{-39\pi}{900} \sin\left(\dfrac{\pi x}{900}\right)$

or $\dfrac{dp}{dx} = -a \sin bx$

with $a = 0.136$ and $b = 0.0035$.

The model covers the main features of the situation.

❓ (Page 215)

$\ln|y| + c_1 = \tfrac{1}{2}x^2 + c_2$

can be rewritten as

$\ln|y| = \tfrac{1}{2}x^2 + (c_2 - c_1)$.

Exercise 9B (Page 217)

1 (i) $y = \tfrac{1}{3}x^3 + c$

(ii) $y = \sin x + c$

(iii) $y = e^x + c$

(iv) $y = \tfrac{2}{3}x^{3/2} + c$

2 (i) $y = -\dfrac{2}{(x^2 + c)}$

(ii) $y^2 = \tfrac{2}{3}x^3 + c$

(iii) $y = Ae^x$

(iv) $y = \ln|e^x + c|$

(v) $y = Ax$

(vi) $y = \left(\tfrac{1}{4}x^2 + c\right)^2$

(vii) $y = -\dfrac{1}{(\sin x + c)}$

(viii) $y^2 = A(x^2 + 1) - 1$

(ix) $y = -\ln\left(c - \tfrac{1}{2}x^2\right)$

(x) $y^3 = \tfrac{3}{2}x^2 \ln x - \tfrac{3}{4}x^2 + c$

Exercise 9C (Page 221)

1 (i) $y = \tfrac{1}{3}x^3 - x - 4$

(ii) $y = e^{x^3/3}$

(iii) $y = \ln\left(\tfrac{1}{2}x^2 + 1\right)$

(iv) $y = \dfrac{1}{(2-x)}$

(v) $y = e^{(x^2-1)/2} - 1$

(vi) $y = \sec x$

2 (i) $\theta = 20 - Ae^{-2t}$

(ii) $\theta = 20 - 15e^{-2t}$

(iii) $t = 1.01$ hours

3 (i) $N = Ae^t$

(ii) $N = 10e^t$

(iii) N tends to ∞, which would never be realised because of the combined effects of food shortage, predators and human controls.

4 $\dfrac{ds}{dt} = \dfrac{2}{s}$; $s = \sqrt{4t + c}$

5 (i) $\dfrac{1}{3y} + \dfrac{1}{3(3-y)}$

(ii) $\tfrac{1}{3}\ln\left|\dfrac{y}{3-y}\right| + c$ or $\tfrac{1}{3}\ln\left|\dfrac{Ay}{3-y}\right|$

(iii) $y = \dfrac{3x^3}{(4 + x^3)}$ $(x \in \mathbb{R})$

6 $y = 2 + e^{-kt}$

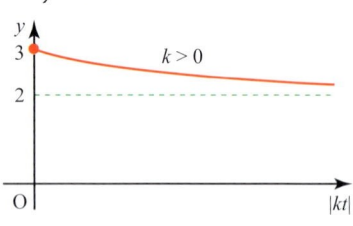

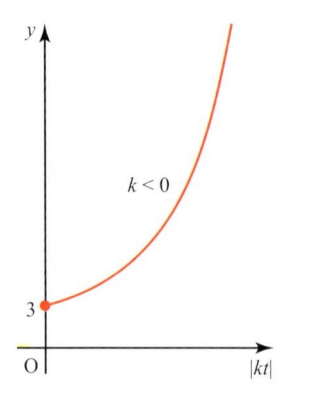

$k > 0$

$k < 0$

7 (i) $N = 1500e^{0.0347t}$
    $= 1500 \times 2^{t/20}$

(ii) $N = 24\,000$

(iii) 11 hours 42 minutes

8 (ii) $\dfrac{1}{x-1} - \dfrac{1}{x+1}$

(iii) $y = \dfrac{(x+1)}{2(x-1)} e^{3-x}$ $(x \neq \pm 1)$

9 (i) $\dfrac{dr}{dt} = \dfrac{k}{r^2}$

(ii) $k = 5000$; 141 m (3 s.f.)

(iii) $\dfrac{dr}{dt} = \dfrac{k_1}{r^2(2+t)}$; $k_1 = 10\,000$

(iv) 104 m (3 s.f.)

10 (i) $\dfrac{1}{3(2-x)} + \dfrac{1}{3(1+x)}$

(ii) $\tfrac{1}{3}$

(iv) 1.18 hours (2 d.p.)

(v) 0.728 kg

11 (i) $2x \sin 2x + \cos 2x + c$

(iii) $y^2 = 4x^2 + 4x \sin 2x + 2 \cos 2x + 1$

12 (i) $\dfrac{1}{1+x} - \dfrac{x}{1+x^2}$

(iii) $1 - \dfrac{x^2}{2} + \dfrac{3x^4}{8}$;

$1 + x - \dfrac{x^2}{2} - \dfrac{x^3}{2} + \dfrac{3x^4}{8} + \dfrac{3x^5}{8}$

13 (i) $\dfrac{3}{(3x-1)} - \dfrac{1}{x}$

(iii) $t = 1.967$ (3 d.p.)

(iv) 500 and 3550

14 (ii) $\cot x$; $\ln(\sin x) + c$

(iii) $y = 0.185$ (3 s.f.); minimum

15 (i) $\tfrac{1}{4}\ln y - \tfrac{1}{4}\ln(4 - y)$

(ii) $\dfrac{4}{3e^{-4x} + 1}$

(iii) The value of y tends to 4.

16 (i) $\theta = A(1 + 3e^{-kt})$

(iii) $\dfrac{7A}{3}$

17 (i) $\tan^{-1}\left(\tfrac{1}{2} - \tfrac{1}{2}e^{-2t}\right)$

(ii) The value of x tends to $\tan^{-1}\tfrac{1}{2}$.

(iii) As $\tfrac{1}{2} - \tfrac{1}{2}e^{-2t}$ increases so does $\tan^{-1}\left(\tfrac{1}{2} - \tfrac{1}{2}e^{-2t}\right)$.

18 (iii) $100 \ln\left(\dfrac{10+h}{10-h}\right) - 20h$

Investigation (Page 226)

Using the assumptions in Exercise 9A, question 7: the rate of cooling is proportional to the temperature of the tea above the surrounding air. The initial temperature is 95°C and the cooling rate is 0.5°Cs^{-1}. So

$$\theta = 15 + 80e^{-t/160}.$$

Adding 10% milk at 5°C gives

$$\theta = 15 + 71e^{-t/160}.$$

The final temperature is lower if the milk is added at the end.

Chapter 10

(Page 228)

$\overrightarrow{OP} = \overrightarrow{OA} + \lambda(\overrightarrow{OB} - \overrightarrow{OA})$
$= (1-\lambda)\overrightarrow{OA} + \lambda\overrightarrow{OB}$

Activity 10.1 (Page 229)

(ii) $\begin{pmatrix} -2 \\ -9 \end{pmatrix}, \begin{pmatrix} 0 \\ -5 \end{pmatrix}, \begin{pmatrix} 2 \\ -1 \end{pmatrix}, \begin{pmatrix} 3 \\ 1 \end{pmatrix},$
$\begin{pmatrix} 3\frac{1}{2} \\ 2 \end{pmatrix}, \begin{pmatrix} 4 \\ 3 \end{pmatrix}, \begin{pmatrix} 8 \\ 11 \end{pmatrix}$

(iv) $0, 1, \frac{1}{2}, \frac{3}{4}$

(v) (a) It lies between A and B.
(b) It lies beyond B.
(c) It lies beyond A.

Activity 10.2 (Page 231)

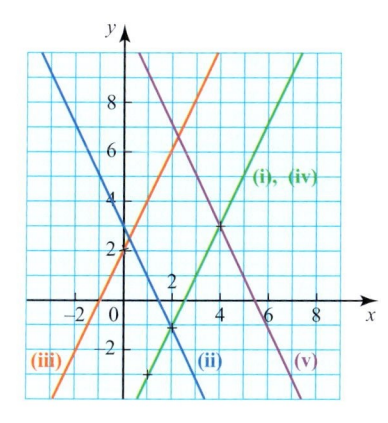

(graph showing lines labelled (i), (iv), (ii), (iii), (v) on xy-axes)

(i) and (iv) are the same since putting $\lambda = -1$ in (i) gives $\begin{pmatrix} 1 \\ -3 \end{pmatrix}$

and $\begin{pmatrix} 1 \\ 2 \end{pmatrix}$ is parallel to $\begin{pmatrix} 3 \\ 6 \end{pmatrix}$.

(iii) is parallel to (i) since the direction vector is the same.

(iv) is parallel to (ii) since

$\begin{pmatrix} -1 \\ 2 \end{pmatrix} = -\begin{pmatrix} 1 \\ -2 \end{pmatrix}.$

Exercise 10A (Page 232)

1 (i) (a) $2\mathbf{i} + 8\mathbf{j}$
(b) $\sqrt{68}$
(c) $3\mathbf{i} + 7\mathbf{j}$

(ii) (a) $-4\mathbf{i} - 3\mathbf{j}$
(b) 5
(c) $2\mathbf{i} + 1.5\mathbf{j}$

(iii) (a) $6\mathbf{i} + 8\mathbf{j}$
(b) 10
(c) $\mathbf{i} + 3\mathbf{j}$

(iv) (a) $6\mathbf{i} - 8\mathbf{j}$
(b) 10
(c) 0

(v) (a) $5\mathbf{i} + 12\mathbf{j}$
(b) 13
(c) $-7.5\mathbf{i} - 2\mathbf{j}$

2 Note: These answers are not unique.

(i) $\mathbf{r} = \begin{pmatrix} 2 \\ 1 \end{pmatrix} + \lambda \begin{pmatrix} 1 \\ 2 \end{pmatrix}$

(ii) $\mathbf{r} = \begin{pmatrix} 3 \\ 5 \end{pmatrix} + \lambda \begin{pmatrix} -1 \\ 1 \end{pmatrix}$

(iii) $\mathbf{r} = \begin{pmatrix} -6 \\ -6 \end{pmatrix} + \lambda \begin{pmatrix} 1 \\ 1 \end{pmatrix}$

(iv) $\mathbf{r} = \begin{pmatrix} 5 \\ 3 \end{pmatrix} + \lambda \begin{pmatrix} 1 \\ 1 \end{pmatrix}$

(v) $\mathbf{r} = \lambda \begin{pmatrix} 2 \\ 1 \end{pmatrix}$

(vi) $\mathbf{r} = \lambda \begin{pmatrix} -1 \\ 4 \end{pmatrix}$

(vii) $\mathbf{r} = \lambda \begin{pmatrix} -1 \\ 4 \end{pmatrix}$

(viii) $\mathbf{r} = \begin{pmatrix} 3 \\ -12 \end{pmatrix} + \lambda \begin{pmatrix} -1 \\ 4 \end{pmatrix}$

3 Note: These answers are not unique.

(i) $\mathbf{r} = \begin{pmatrix} 2 \\ 4 \\ -1 \end{pmatrix} + \lambda \begin{pmatrix} 3 \\ 6 \\ 4 \end{pmatrix}$

(ii) $\mathbf{r} = \begin{pmatrix} 1 \\ 0 \\ -1 \end{pmatrix} + \lambda \begin{pmatrix} 1 \\ 0 \\ 0 \end{pmatrix}$

(iii) $\mathbf{r} = \begin{pmatrix} 1 \\ 0 \\ 4 \end{pmatrix} + \lambda \begin{pmatrix} 5 \\ 3 \\ -6 \end{pmatrix}$

(iv) $\mathbf{r} = \begin{pmatrix} 0 \\ 0 \\ 1 \end{pmatrix} + \lambda \begin{pmatrix} 2 \\ 1 \\ 3 \end{pmatrix}$

(v) $\mathbf{r} = \lambda \begin{pmatrix} 1 \\ 2 \\ 3 \end{pmatrix}$

4 (i) Yes, $\lambda = 2$
(ii) Yes, $\lambda = -1$
(iii) No
(iv) No
(v) Yes, $\lambda = -5$

5 (i) $\mathbf{r} = \begin{pmatrix} -1 \\ -2 \\ 1 \end{pmatrix} + \lambda \begin{pmatrix} -1 \\ 3 \\ -3 \end{pmatrix}$

or $\mathbf{r} = \begin{pmatrix} -1 \\ -2 \\ 1 \end{pmatrix} + \lambda \begin{pmatrix} -2 \\ 6 \\ -6 \end{pmatrix}$

(ii) $(-2, 1, 2)$

(iii) $\mathbf{r} = \begin{pmatrix} -2 \\ 1 \\ -2 \end{pmatrix} + \lambda \begin{pmatrix} 0 \\ 1 \\ 0 \end{pmatrix}$

Exercise 10B (Page 238)

1 (i) $\begin{pmatrix} 4 \\ 1 \end{pmatrix}$

(ii) $\begin{pmatrix} 5 \\ 5 \end{pmatrix}$

(iii) $\begin{pmatrix} 12 \\ 17 \end{pmatrix}$

(iv) $\begin{pmatrix} -5 \\ 6 \end{pmatrix}$

P3 Answers

(v) $\begin{pmatrix} 6 \\ 3 \end{pmatrix}$

2 (i) Intersect at $(3, -2, 5)$
  (ii) Parallel
  (iii) Intersect at $(3, 2, -13)$
  (iv) Intersect at $(1, 2, 7)$
  (v) Skew
  (vi) Intersect at $(4, -7, 11)$
  (vii) Skew

3 (i) 12.8 km
  (ii) 20 km h^{-1}, 5 km h^{-1}
  (iii) After 40 minutes there is a collision.

4 (i) $\overrightarrow{OL} = \begin{pmatrix} 10 \\ 4.5 \end{pmatrix}$; $\overrightarrow{OM} = \begin{pmatrix} 7 \\ 3.5 \end{pmatrix}$;
  $\overrightarrow{ON} = \begin{pmatrix} 4 \\ 1 \end{pmatrix}$

  (ii) AL: $\mathbf{r} = \begin{pmatrix} 1 \\ 0 \end{pmatrix} + \lambda \begin{pmatrix} 2 \\ 1 \end{pmatrix}$;
  BM: $\mathbf{r} = \begin{pmatrix} 7 \\ 2 \end{pmatrix} + \mu \begin{pmatrix} 0 \\ 1 \end{pmatrix}$;
  CN: $\mathbf{r} = \begin{pmatrix} 13 \\ 7 \end{pmatrix} + \nu \begin{pmatrix} 3 \\ 2 \end{pmatrix}$

  (iii) (a) $(7, 3)$
  (b) $(7, 3)$

  (iv) The lines AL, BM and CN are concurrent. (They are the medians of the triangle, and this result holds for the medians of any triangle.)

5 $(-2, -6, -1)$; 30 units

6 No

7 6 units, 9 units, $\sqrt{77}$ units

8 (i) $\begin{pmatrix} -0.25 \\ 0 \\ 0 \end{pmatrix}$

  (ii) $(0, 0.05, 1.1)$

  (iii) DE: $\mathbf{r} = \begin{pmatrix} 0 \\ 0 \\ 1 \end{pmatrix} + \lambda \begin{pmatrix} 1 \\ 0 \\ 0 \end{pmatrix}$

  EF: $\mathbf{r} = \begin{pmatrix} 0.25 \\ 0 \\ 1 \end{pmatrix} + \lambda \begin{pmatrix} 0 \\ 1 \\ 2 \end{pmatrix}$

Exercise 10C (Page 242)

1 53.6°
2 81.8°
3 8.72°
4 35.3°
5 61.0°
6 (i) $A(4, 0, 0)$, $F(4, 0, 3)$
  (ii) 114.1°, 109.5°
  (iii) They touch but are not perpendicular.
7 (ii) $5\mathbf{i} + 3\mathbf{j} + 4\mathbf{k}$

Exercise 10D (page 245)

1 (i) (a) $(-2, 6, 7)$
  (b) $\sqrt{29}$ units
  (ii) (a) $(3, -1, 7)$
  (b) $\sqrt{17}$ units
  (iii) (a) $(2, 7, -3)$
  (b) 7 units

2 $2\sqrt{10}$ units

3 $\sqrt{35}$ units

4 (i) $(0, 4, 3)$
  (ii) $\begin{pmatrix} -5 \\ 4 \\ 3 \end{pmatrix}$, $\sqrt{50}$
  (iii) $\mathbf{r} = \begin{pmatrix} 5 \\ 0 \\ 0 \end{pmatrix} + \lambda \begin{pmatrix} -5 \\ 4 \\ 3 \end{pmatrix}$
  (iv) $\begin{pmatrix} 3\frac{3}{4} \\ 1 \\ \frac{3}{4} \end{pmatrix}$, 63.4°
  (v) Spider is then at $P(2.5, 2, 1.5)$ and $\overrightarrow{OP} \cdot \overrightarrow{AG} = 0$, $|\overrightarrow{OP}| = 3.54$

5 (i) $(1, 0.5, 0)$
  (ii) 41.8°
  (iii) 027°
  (iv) $(2, 2.5, 2)$
  (v) $t = 2$, $\sqrt{5}$ km

? (Page 247)

A three-legged stool is the more stable. Three points, such as the ends of the legs, define a plane but a fourth will not, in general, be in the same plane. So the ends of the legs of a three-legged stool lie in a plane but those of a four-legged stool need not. The four-legged stool will rest on three legs but could rock on to a different three.

? (Page 250)

(i) 90° with all lines.
(ii) No, so long as the pencil remains perpendicular to the table.

Activity 10.3 (Page 255)

Repeat the work in Example 10.13 replacing $(7, 5, 3)$ by (α, β, γ), so 7 by α, 5 by β and 3 by γ; and $(3, 2, 1)$ by (n_1, n_2, n_3) and 6 by d.

Exercise 10E (Page 257)

1 (i) Parallel, line in plane
  (ii) Parallel, line not in plane
  (iii) Not parallel
  (iv) Parallel, line in plane
  (v) Not parallel
  (vi) Parallel, line not in plane

2 (i) $\overrightarrow{LM} = \begin{pmatrix} 2 \\ 2 \\ -2 \end{pmatrix}$; $\overrightarrow{LN} = \begin{pmatrix} 5 \\ 2 \\ -1 \end{pmatrix}$
  (iii) $x - 4y - 3z = -2$

3 (iii) B

4 (iii) Three points define a plane.
  (iv) $(1, 0, -1)$

5 (i) $(0, 1, 3)$
  (ii) $(1, 1, 1)$
  (iii) $(8, 4, 2)$
  (iv) $(0, 0, 0)$
  (v) $(11, 19, -10)$

6 (i) (a) $\mathbf{r} = \begin{pmatrix} 2 \\ 2 \\ 3 \end{pmatrix} + \lambda \begin{pmatrix} 1 \\ -1 \\ 2 \end{pmatrix}$

(b) (1, 3, 1)

(c) $\sqrt{6}$

(ii) (a) $\mathbf{r} = \begin{pmatrix} 2 \\ 3 \\ 0 \end{pmatrix} + \lambda \begin{pmatrix} 2 \\ 5 \\ 3 \end{pmatrix}$

(b) (1, 0.5, −1.5)

(c) 3.08

(iii) (a) $\mathbf{r} = \begin{pmatrix} 3 \\ 1 \\ 3 \end{pmatrix} + \lambda \begin{pmatrix} 1 \\ 0 \\ 0 \end{pmatrix}$

(b) (0, 1, 3)

(c) 3

(iv) (a) $\mathbf{r} = \begin{pmatrix} 2 \\ 1 \\ 0 \end{pmatrix} + \lambda \begin{pmatrix} 3 \\ -4 \\ 1 \end{pmatrix}$

(b) (2, 1, 0): A is in the plane

(c) 0

(v) (a) $\mathbf{r} = \lambda \begin{pmatrix} 1 \\ 1 \\ 1 \end{pmatrix}$

(b) (2, 2, 2)

(c) $\sqrt{12}$

7 (i) $x + 2y + 3z = 25$

(ii) $206 = 150 + 56$

(iii) W is in the plane; $\overrightarrow{UW} \cdot \overrightarrow{UV} = 0$

8 (i) $\mathbf{r} = \begin{pmatrix} 13 \\ 5 \\ 0 \end{pmatrix} + \lambda \begin{pmatrix} 3 \\ 1 \\ -2 \end{pmatrix}$

(ii) (4, 2, 6)

(iii) 11.2

9 (i) 4.1°

(ii) 32.3°

(iii) 35.6°

10 (ii) $\overrightarrow{AB} = \begin{pmatrix} -1 \\ 2 \\ 1 \end{pmatrix}; \overrightarrow{AC} = \begin{pmatrix} 8 \\ -4 \\ 1 \end{pmatrix};$ in both cases the scalar product = 0

(iii) 132.9°

(iv) 8.08

11 (i) 5, $\sqrt{89}$

(ii) 62.2°

(iii) 20.9

(iv) (4, 6, −3)

12 (i) PQ: $\mathbf{r} = \begin{pmatrix} 2 \\ 2 \\ 4 \end{pmatrix} + \lambda \begin{pmatrix} -1 \\ 2 \\ 2 \end{pmatrix}$;

XY: $\mathbf{r} = \begin{pmatrix} -2 \\ -2 \\ -3 \end{pmatrix} + \mu \begin{pmatrix} 1 \\ 2 \\ 3 \end{pmatrix}$

(iii) Yes

(iv) Yes, (1, 4, 6)

13 (ii) $\begin{pmatrix} 2 \\ -1 \\ 3 \end{pmatrix}$

(iii) (10, −5, 15)

(iv) OA: $\mathbf{r} = \lambda \begin{pmatrix} 5 \\ -12 \\ 16 \end{pmatrix}$;

AB: $\mathbf{r} = \begin{pmatrix} 5 \\ -12 \\ 16 \end{pmatrix} + \mu \begin{pmatrix} 1 \\ 5 \\ 1 \end{pmatrix}$

(v) 69°

14 (i) $\begin{pmatrix} 2 \\ -3 \\ 4 \end{pmatrix}$

(ii) $\mathbf{r} = \begin{pmatrix} 3 \\ -8 \\ 12 \end{pmatrix} + \lambda \begin{pmatrix} 2 \\ -3 \\ 4 \end{pmatrix}$;

(−1, −2, 4)

(iii) (0, −3.5, 6)

(iv) 15.6° (1 d.p.)

15 (i) $2x - 3y + 7z = -5$

(ii) $\mathbf{r} = (130\mathbf{i} - 40\mathbf{j} + 20\mathbf{k}) + \lambda(8\mathbf{i} - 4\mathbf{j} + \mathbf{k})$

(iii) $10\mathbf{i} + 20\mathbf{j} + 5\mathbf{k}$

(iv) 135 m

16 (i) $\mathbf{r} = \begin{pmatrix} 2 \\ 3 \\ 5 \end{pmatrix} + \lambda \begin{pmatrix} 1 \\ 1 \\ -0.5 \end{pmatrix}$

(ii) (12, 13, 0)

(iii) 109.5° (1 d.p.)

(iv) 25 m

17 (i) (3, 1, 0)

(ii) 63.4°

(iv) $\mathbf{r} = \begin{pmatrix} 1 \\ 1 \\ 1 \end{pmatrix} + \lambda \begin{pmatrix} 1 \\ 2 \\ 2 \end{pmatrix}$

(v) $\left(\dfrac{5}{3}, \dfrac{7}{3}, \dfrac{7}{3}\right)$ or $\left(\dfrac{1}{3}, -\dfrac{1}{3}, -\dfrac{1}{3}\right)$

18 (i) $b = -2, c = 3$

19 (ii) $6x + y - 8z = 6$

20 (i) $\mathbf{r} = \begin{pmatrix} -1 \\ 3 \\ 5 \end{pmatrix} + \lambda \begin{pmatrix} 3 \\ -1 \\ -4 \end{pmatrix}$

(ii) $\mathbf{r} = \begin{pmatrix} 5 \\ 1 \\ -3 \end{pmatrix}$

(iii) $7x - 11y + 8z = 0$

21 (i) $3\mathbf{i} + 2\mathbf{j} + \mathbf{k}$

(ii) 72.2°

(iii) $\mathbf{r} = 3\mathbf{i} + 2\mathbf{j} + \mathbf{k} + \lambda(6\mathbf{i} + 2\mathbf{j} - \mathbf{k})$

❓ (Page 265)

π_3 is parallel to π_1 and π_2 (the common line is at infinity).

Exercise 10F (Page 265)

1 (i) $\mathbf{r} = \begin{pmatrix} 3 \\ 1 \\ 0 \end{pmatrix} + \lambda \begin{pmatrix} 15 \\ 27 \\ 7 \end{pmatrix}$

(ii) $\mathbf{r} = \begin{pmatrix} 0 \\ -3 \\ 5 \end{pmatrix} + \lambda \begin{pmatrix} 0 \\ 0 \\ -4 \end{pmatrix}$

(iii) $\mathbf{r} = \begin{pmatrix} 0 \\ 1 \\ -1 \end{pmatrix} + \lambda \begin{pmatrix} 16 \\ 15 \\ 13 \end{pmatrix}$

(iv) $\mathbf{r} = \begin{pmatrix} 2 \\ 0 \\ 4 \end{pmatrix} + \lambda \begin{pmatrix} 11 \\ 4 \\ 21 \end{pmatrix}$

2 (i) 56.5°

(ii) 80.0°

(iii) 24.9°

(iv) 63.5°

3 (i) $\mathbf{r} = \begin{pmatrix} -2 \\ 3 \\ 5 \end{pmatrix} + \lambda \begin{pmatrix} 0 \\ 0 \\ -1 \end{pmatrix}$

(ii) $\mathbf{r} = \begin{pmatrix} 4 \\ -3 \\ 2 \end{pmatrix} + \lambda \begin{pmatrix} 3 \\ 2 \\ -6 \end{pmatrix}$

4 $41x - 19y + 26z = 33$

5 $x + 3y - z = -8$

6 $\mathbf{r} = \begin{pmatrix} 4 \\ -2 \\ -7 \end{pmatrix} + \lambda \begin{pmatrix} 21 \\ 4 \\ 11 \end{pmatrix}$

7 $60x + 11y + 100z = 900$;
$60x - 11y - 100z = -300$;
$\mathbf{r} = \begin{pmatrix} 5 \\ 0 \\ 6 \end{pmatrix} + t \begin{pmatrix} 0 \\ 100 \\ -11 \end{pmatrix}$; $6.3°$

8 (i) $x + 3z = -800$

  (ii) Normal is approx. 18.4° to the horizontal

  (iii) $14x - 15y + 3450z = 15950$

  (iv) $x = 15\lambda$,
$y = -1136\lambda - 62396.7$,
$z = -5\lambda - 266.7$

  (v) 62 km (assuming seam is sufficiently extensive)

9 (i) $\mathbf{r} = (2\mathbf{i} + 3\mathbf{j} + 5\mathbf{k}) + \lambda(3\mathbf{i} + \mathbf{j} - 2\mathbf{k})$

  (ii) $\lambda = 1$; (5, 4, 3)

  (iii) (9.5, 5.5, 0)

  (iv) (6.5, 4.5, 2); 1.87 (3 s.f.)

  (v) $\mathbf{i} + 2\mathbf{j} = 3\mathbf{k}$; 38.2° (1 d.p.)

10 (i) $\begin{pmatrix} a \\ b \\ 1 \end{pmatrix}$

  (ii) $\overrightarrow{AB} = \begin{pmatrix} 2 \\ -3 \\ 0 \end{pmatrix}$; $\overrightarrow{AC} = \begin{pmatrix} 3 \\ -5 \\ 1 \end{pmatrix}$

  (iii) $2a - 3b = 0$; $3a - 5b + 1 = 0$

  (iv) $3x + 2y + z = 6$

  (v) 36.7° (1 d.p.)

  (vi) $\left(3\tfrac{2}{3}, -3\tfrac{2}{3}, 2\tfrac{1}{3}\right)$

11 (i) (6, 4.5, 3)

  (iii) $x - 2z = 0$

  (iv) AOBC: $\begin{pmatrix} 0 \\ 2 \\ -3 \end{pmatrix}$; DOBE: $\begin{pmatrix} 1 \\ 0 \\ -2 \end{pmatrix}$;
41.9° (1 d.p.); 138.1°

12 (i) $a = -2$

  (ii) 3

13 (i) $4x + 2y + z = 8$

  (ii) 77.4°

14 (i) 57.7°

  (ii) $\mathbf{r} = 2\mathbf{i} - \mathbf{k} + \lambda(4\mathbf{i} - 7\mathbf{j} + 5\mathbf{k})$

15 (i) $2x - 3y + 6z = 2$

  (ii) 2

  (iii) $\mathbf{r} = \lambda(6\mathbf{i} + 2\mathbf{j} - \mathbf{k})$

Chapter 11

Activity 11.1 (Page 272)

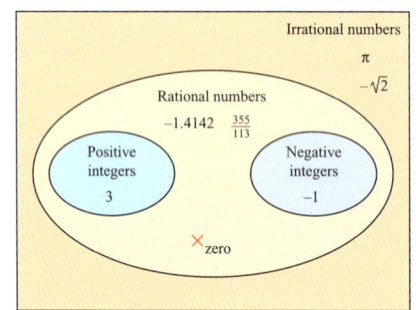

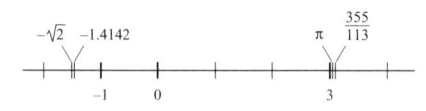

Activity 11.2 (Page 272)

(i) Positive integer

(ii) Rational number

(iii) Irrational number

(iv) Negative integer

(v) Zero, negative integer

(vi) No real number is possible

Activity 11.3 (Page 273)

$z = 3 - 7i$

$\Rightarrow z^2 - 6z + 58$

$= (3 - 7i)^2 - 6(3 - 7i) + 58$

$= 9 - 42i + 49i^2 - 18 + 42i + 58$

$= 9 - 42i - 49 - 18 + 42i + 58$

$= 0$

? (Page 274)

$i^3 = -i$, $i^4 = 1$, $i^5 = i$

All numbers of the form

- i^{4n} are equal to 1
- i^{4n+1} are equal to i
- i^{4n+2} are equal to -1
- i^{4n+3} are equal to $-i$.

Activity 11.4 (Page 275)

(i) (a) 6

  (b) 2

  (c) 34

  (d) 5

They are all real.

(ii) $z + z^* = (x + iy) + (x - iy) = 2x$

$zz^* = (x + iy)(x - iy)$

$= x^2 - ixy + ixy - i^2 y^2$

$= x^2 + y^2$

These are real for any real values of x and y.

Exercise 11A (Page 275)

1 (i) $14 + 10i$

  (ii) $5 + 2i$

  (iii) $-3 + 4i$

  (iv) $-1 + i$

  (v) 21

  (vi) $12 + 21i$

  (vii) $3 + 29i$

  (viii) $14 + 5i$

  (ix) $40 + 42i$

  (x) 100

  (xi) $43 + 76i$

  (xii) $-9 + 46i$

2 (i) $-1 \pm i$

  (ii) $1 \pm 2i$

  (iii) $2 \pm 3i$

  (iv) $-3 \pm 5i$

(v) $\frac{1}{2} \pm 2i$

(vi) $-2 \pm \sqrt{2}\,i$

3 (i) $2i$

(ii) $5i$ and $-3i$

(iii) $1 + i$ and $-1 + i$

(iv) $2 - 3i$ and $-2 - 3i$

(v) $-1 - 4i$ and $1 - 4i$

(vi) $-3i$ and $2i$

4 (i) 2

(ii) -4

(iii) $2 - 3i$

(iv) $6 + 4i$

(v) $8 + i$

(vi) $-4 - 7i$

(vii) 0

(viii) 0

(ix) -39

(x) $-46 - 9i$

(xi) $-46 - 9i$

(xii) $52i$

❓ (Page 276)

Yes, for example $\frac{2}{3} = \frac{4}{6}$, although $2 \neq 4$ and $3 \neq 6$.

Activity 11.5 (Page 277)

$\frac{1}{x + iy} = p + iq$

$\Rightarrow (p + iq)(x + iy) = 1$

$\Rightarrow px + ipy + iqx + iqy^2 = 1$

$\Rightarrow (px - qy) + i(py + qx) = 1$

$px - qy = 1$ and $py + qx = 0$

Solving simultaneously gives

$p = \dfrac{x}{x^2 + y^2}$, $q = \dfrac{-y}{x^2 + y^2}$

so $\dfrac{1}{x + iy} = \dfrac{x - iy}{x^2 + y^2}$

❓ (Page 279)

$\frac{1}{i} = -i$, $\frac{1}{i^2} = -1$, $\frac{1}{i^3} = i$

All numbers of the form

- $\dfrac{1}{i^{4n}}$ are equal to 1
- $\dfrac{1}{i^{4n+1}}$ are equal to $-i$
- $\dfrac{1}{i^{4n+2}}$ are equal to -1
- $\dfrac{1}{i^{4n+3}}$ are equal to i.

Exercise 11B (Page 279)

1 (i) $\frac{3}{10} - \frac{1}{10}i$

(ii) $\frac{6}{37} + \frac{1}{37}i$

(iii) $-\frac{1}{4} + \frac{3}{4}i$

(iv) $\frac{4}{5} + \frac{11}{10}i$

(v) $\frac{5}{2} - \frac{1}{2}i$

(vi) $7 - 5i$

(vii) $-i$

(viii) $\frac{11}{25} - \frac{27}{25}i$

(ix) $\frac{7}{29} + \frac{32}{29}i$

(x) $-1 - \frac{3}{2}i$

2 (i) $a = 5$, $b = 2$

(ii) $a = 3$, $b = -7$

(iii) $a = 2$, $b = -3$

(iv) $a = 4$, $b = 5$

(v) $a = \frac{5}{4}$, $b = -\frac{3}{4}$

(vi) $a = \dfrac{1}{\sqrt{2}}$, $b = \dfrac{1}{\sqrt{2}}$

3 $a = 2$, $b = 2$

4 (i) $z = 2 - i$

(ii) $z = 3 + i$

(iii) $z = 11 - 10i$

(iv) $z = \dfrac{-35 + 149i}{34}$

5 $0, 2, -1 \pm \sqrt{3}\,i$

6 $\dfrac{2x}{x^2 + y^2}$

8 (i) $a^3 - 3ab^2 + (3a^2b - b^3)i$

(iii) $z = 1, -\frac{1}{2} \pm \frac{1}{2}\sqrt{3}\,i$

9 (i) $(z - \alpha)(z - \beta)$
$= z^2 - (\alpha + \beta)z + \alpha\beta$

(ii) (a) $z^2 - 14z + 65 = 0$

(b) $9z^2 + 25 = 0$

(c) $z^2 + 4z + 12 = 0$

(d) $z^2 - (5 + 3i)z + 4 + 7i = 0$

10 (i) $3i$ and $-3i$

(ii) $2 + i$ and $-2 - i$

(iii) $3 + 5i$ and $-3 - 5i$

(iv) $3 - 4i$ and $-3 + 4i$

(v) $5 - 2i$ and $-5 + 2i$

(vi) $2 - 3i$ and $-2 + 3i$

Activity 11.6 (Page 281)

(i) Rotation through 180° about the origin

(ii) Reflection in the real axis

❓ (Page 281)

z and $-z^*$ (or $-z$ and z^*) are reflections of each other in the imaginary axis.

Activity 11.7 (Page 283)

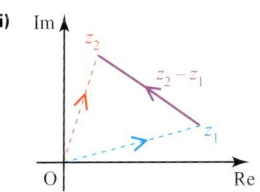

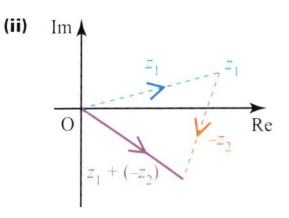

(i) [Argand diagram showing z_1, z_2, and $z_2 - z_1$]

(ii) [Argand diagram showing z_1, $-z_2$, and $z_1 + (-z_2)$]

Exercise 11C (Page 283)

1

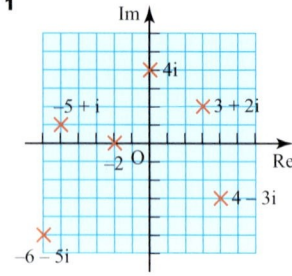

(i) $\sqrt{13}$

(ii) 4

(iii) $\sqrt{26}$

(iv) 2

(v) $\sqrt{61}$

(vi) 5

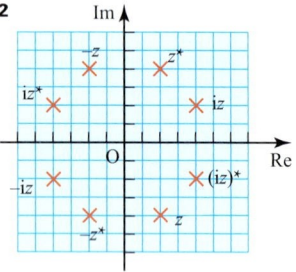

2

3 Points:

(i) $10 + 5i$

(ii) $1 + 2i$

(iii) $11 + 7i$

(iv) $9 + 3i$

(v) $-9 - 3i$

4 (i) 5

(ii) 13

(iii) 65

(iv) $\frac{5}{13}$

(v) $\frac{13}{5}$

$|zw| = |z||w|, \left|\dfrac{z}{w}\right| = \dfrac{|z|}{|w|}, \left|\dfrac{w}{z}\right| = \dfrac{|w|}{|z|}$

5 (i) $z^{-1} = \frac{1}{2} - \frac{1}{2}i, |z^{-1}| = \frac{1}{\sqrt{2}}$

$z^0 = 1, |z^0| = 1$

$z^1 = 1 + i, |z^1| = \sqrt{2}$

$z^2 = 2i, |z^2| = 2$

$z^3 = -2 + 2i, |z^3| = 2\sqrt{2}$

$z^4 = -4, |z^4| = 4$

$z^5 = -4 - 4i, |z^5| = 4\sqrt{2}$

(ii)

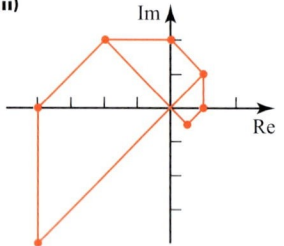

(iii) The half-squares formed are enlarged by $\sqrt{2}$ and rotated through $\frac{\pi}{4}$ each time.

6 Half a turn about O followed by reflection in the x axis is the same as reflection in the x axis followed by half a turn about O.

❓ (Page 284)

$|z_2 - z_1|$ is the distance between the points representing z_1 and z_2 in the Argand diagram.

❓ (Page 285)

(i)

(ii)

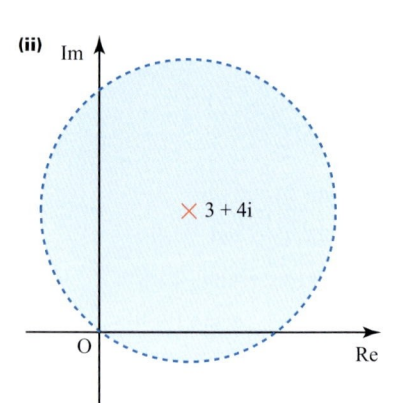

(iii)

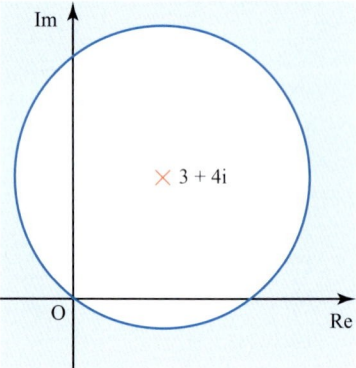

❓ (Page 286)

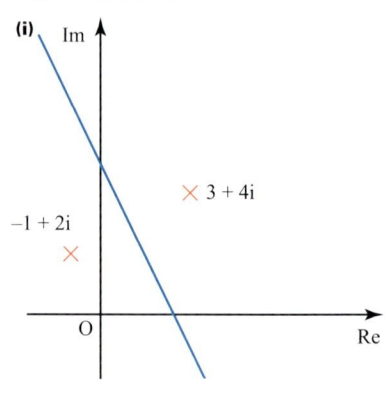

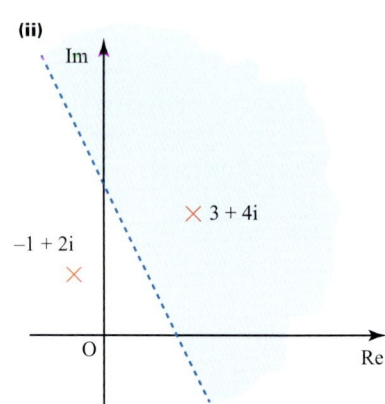

(i)

(ii)

(iii), 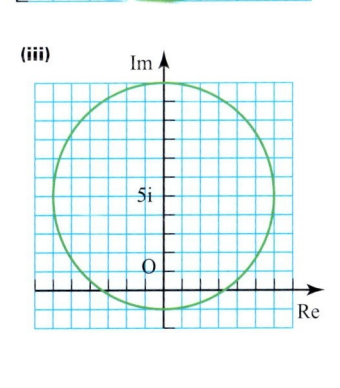 (iv) [diagrams]

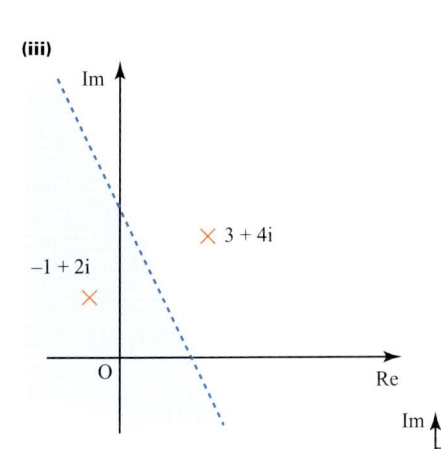

2 [diagram]

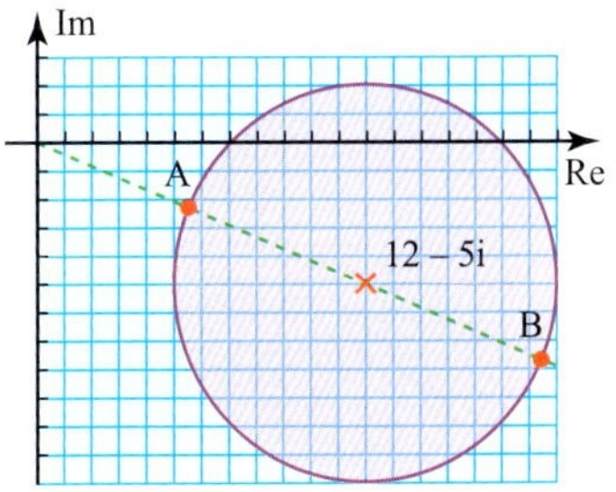

|z| is least at A and greatest at B.

$|12 - 5i| = \sqrt{144 + 25} = 13$

At A, $|z| = 13 - 7 = 6$

At B, $|z| = 13 + 7 = 20$

Exercise 11D (Page 286)

1 (i), (ii), (iii), (iv), (v), (vi), (vii), (viii) [diagrams]

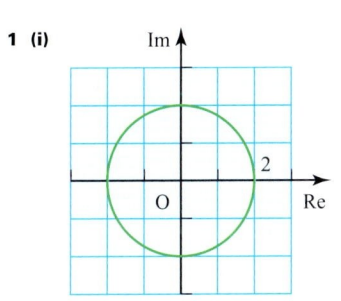

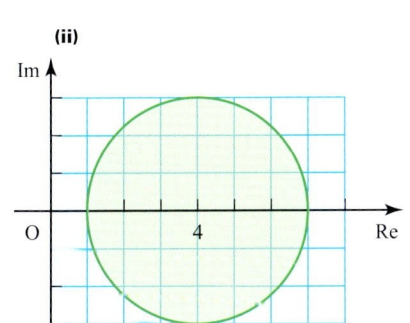

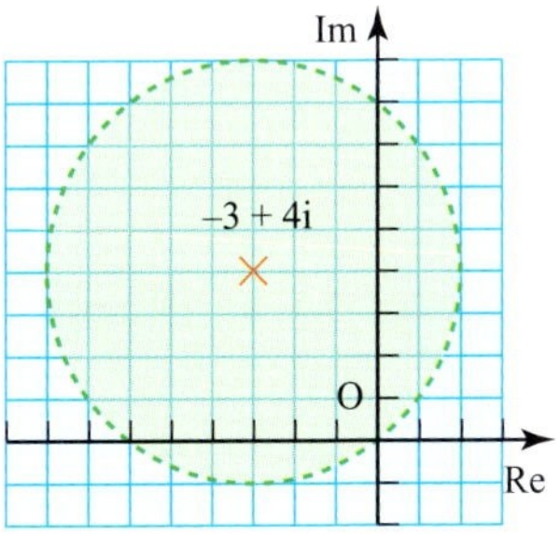

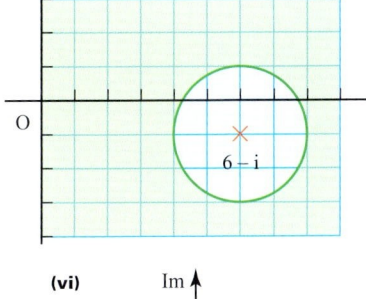

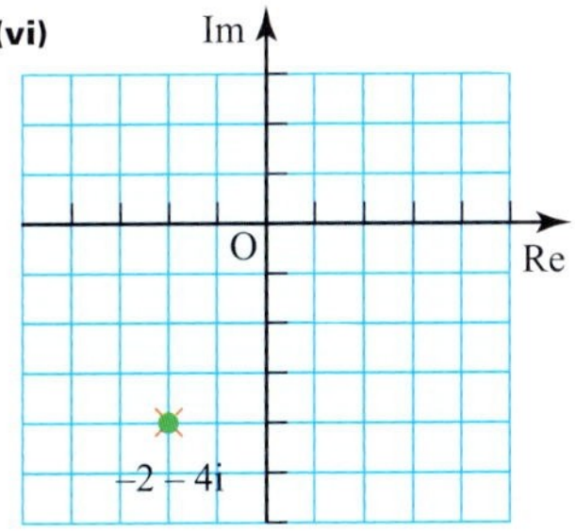

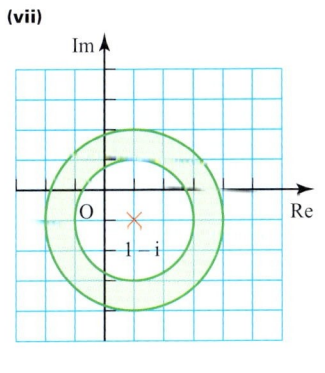

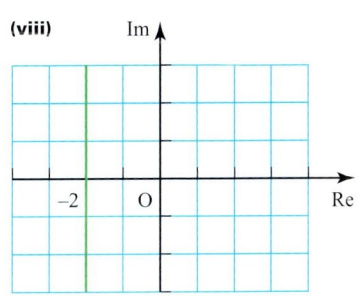

3 (i) [diagram]

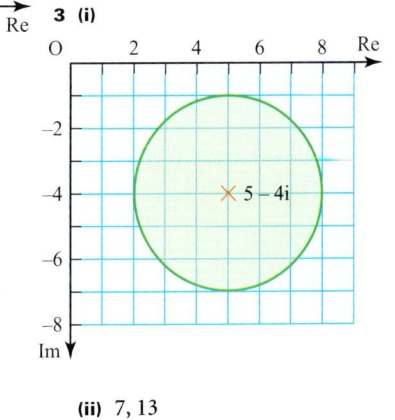

(ii) 7, 13

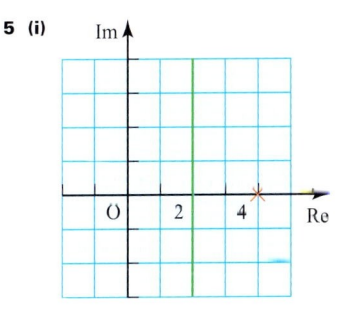

4 Not possible

5 (i), (ii) [diagrams]

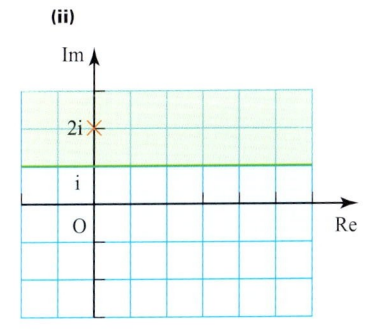

(iii)

Im axis plot showing $-1+i$ and $1-i$

(iv)

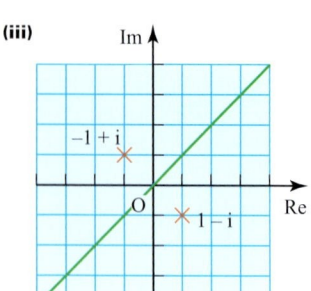

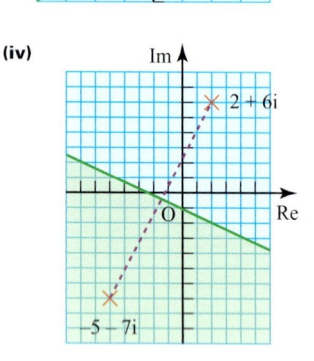

Im axis plot showing $2+6i$ and $-5-7i$

? (Page 288)

(i) $\dfrac{\pi}{2}$

(ii) $-\dfrac{3\pi}{4}$

(iii) $-\dfrac{\pi}{4}$

Activity 11.8 (Page 288)

(i) (a) 45°
(b) 63.4°
(c) 89.4°
(d) −63.4°
(e) −88.9°
(f) −89.7°

$-90° < \tan^{-1} x < 90°$

(ii) $-\dfrac{\pi}{2} < \tan^{-1} x < \dfrac{\pi}{2}$

Activity 11.9 (Page 289)

$\arg(1+i) = \dfrac{\pi}{4}$,

$\arg(1-i) = -\dfrac{\pi}{4}$,

$\arg(-1+i) = \dfrac{3\pi}{4}$,

$\arg(-1-i) = -\dfrac{3\pi}{4}$

? (Page 290)

(i) $2(\cos(\pi - \alpha) + i \sin(\pi - \alpha))$

(ii) $2(\cos(-\alpha) + i \sin(-\alpha))$

Activity 11.11 (Page 290)

	$\dfrac{\pi}{4}$	$\dfrac{\pi}{6}$	$\dfrac{\pi}{3}$
tan	1	$\dfrac{1}{\sqrt{3}}$	$\sqrt{3}$
sin	$\dfrac{1}{\sqrt{2}}$	$\dfrac{1}{2}$	$\dfrac{\sqrt{3}}{2}$
cos	$\dfrac{1}{\sqrt{2}}$	$\dfrac{\sqrt{3}}{2}$	$\dfrac{1}{2}$

Exercise 11E (Page 291)

1 (i) $r = 8, \theta = \dfrac{\pi}{5}$

(ii) $r = \dfrac{1}{4}, \theta = 2.3$

(iii) $r = 4, \theta = -\dfrac{\pi}{3}$

(iv) $r = 3, \theta = \pi - 3$

2 (i) $r = 1, \theta = 0$,
$z = 1(\cos 0 + i \sin 0)$

(ii) $r = 2, \theta = \pi$,
$z = 2(\cos \pi + i \sin \pi)$

(iii) $r = 3, \theta = \dfrac{\pi}{2}$,
$z = 3\left(\cos \dfrac{\pi}{2} + i \sin \dfrac{\pi}{2}\right)$

(iv) $r = 4, \theta = -\dfrac{\pi}{2}$,
$z = 4\left(\cos\left(-\dfrac{\pi}{2}\right) + i \sin\left(-\dfrac{\pi}{2}\right)\right)$

(v) $r = \sqrt{2}, \theta = \dfrac{\pi}{4}$,
$z = \sqrt{2}\left(\cos \dfrac{\pi}{4} + i \sin \dfrac{\pi}{4}\right)$

(vi) $r = 5\sqrt{2}, \theta = -\dfrac{3\pi}{4}$,
$z = 5\sqrt{2}\left(\cos\left(-\dfrac{3\pi}{4}\right) + i \sin\left(-\dfrac{3\pi}{4}\right)\right)$

(vii) $r = 2, \theta = -\dfrac{\pi}{3}$,
$z = 2\left(\cos\left(-\dfrac{\pi}{3}\right) + i \sin\left(-\dfrac{\pi}{3}\right)\right)$

(viii) $r = 12, \theta = \dfrac{\pi}{6}$,
$z = 12\left(\cos \dfrac{\pi}{6} + i \sin \dfrac{\pi}{6}\right)$

(ix) $r = 5, \theta = -0.927$,
$z = 5(\cos(-0.927) + i \sin(-0.927))$

(x) $r = 13, \theta = 2.747$,
$z = 13(\cos 2.747 + i \sin 2.747)$

(xi) $r = \sqrt{65}, \theta = 1.052$,
$z = \sqrt{65}(\cos 1.052 + i \sin 1.052)$

(xii) $r = \sqrt{12013}, \theta = -2.128$,
$z = \sqrt{12013}(\cos(-2.128) + i \sin(-2.128))$

3 (i) $z = 2i$

(ii) $z = \dfrac{3}{2} + \dfrac{3\sqrt{3}}{2}i$

(iii) $z = -\dfrac{7\sqrt{3}}{2} + \dfrac{7}{2}i$

(iv) $z = \dfrac{1}{\sqrt{2}} - \dfrac{1}{\sqrt{2}}i$

(v) $z = -\dfrac{5}{2} - \dfrac{5\sqrt{3}}{2}i$

(vi) $z = -2.497 - 5.456i$

4 (i) $\alpha - \pi$

(ii) $-\alpha$

(iii) $\pi - \alpha$

(iv) $\dfrac{\pi}{2} - \alpha$

(v) $\dfrac{\pi}{2} + \alpha$

5 (ii) Real part $= \dfrac{1}{2}$

6 (i) Argand diagram showing circle centred at approximately $1+i$ passing through origin and with radius giving the shown circle

(ii) Real part $= \dfrac{1}{4}$

7 (i) (a) $2 + i$

(b) $r = \sqrt{5}, \theta = 0.464$

(ii) $-3 + 2i$ and $3 - 2i$

8 (i)

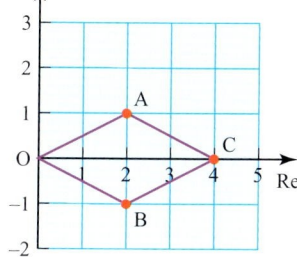

OACB is a rhombus.

(ii) $\frac{3}{5} + \frac{4}{5}i$

? (Page 293)

$\arg(z_1 - z_2)$ is the angle between the line joining z_1 and z_2 and a line parallel to the real axis.

Exercise 11F (Page 294)

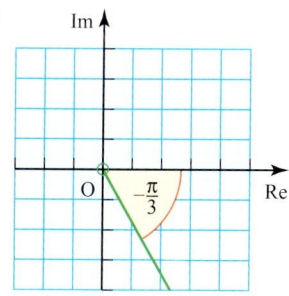

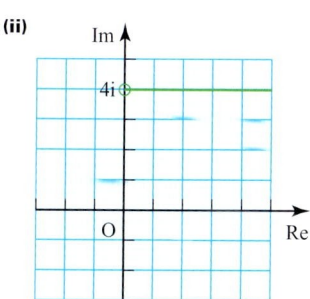

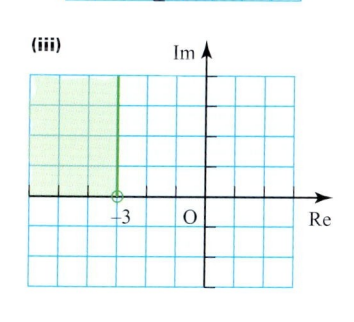

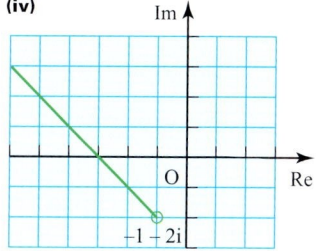

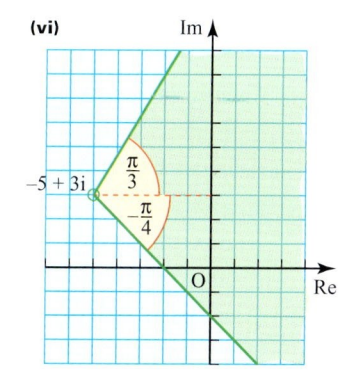

1 (i), **(ii)**, **(iii)**, **(iv)**, **(v)**, **(vi)**

2 $\frac{\pi}{3}, \frac{2\pi}{3}$

3 (i) $\frac{2\pi}{3}, 2$

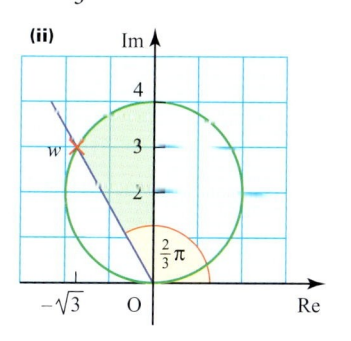

(ii)

(iii) $\sqrt{12}$

4 (i) $r = 1, \theta = \frac{2}{3}\pi$

(ii) wz: modulus $= R$, argument $= \theta + \frac{2}{3}\pi$

$\frac{z}{w}$: modulus $= R$, argument $= \theta - \frac{2}{3}\pi$

(iii) The three points are the same distance from the origin and separated by equal angles of $\frac{2\pi}{3}$ (i.e. 120°).

(iv) $-(2 + \sqrt{3}) + (2\sqrt{3} - 1)i$
$-(2 - \sqrt{3}) - (2\sqrt{3} + 1)i$

5 (i) $2 + i$ and $-2 + i$

(ii) $2 + i$: $r = 2.24$, $\theta = 0.464$ radians

$-2 + i$: $r = 2.24$, $\theta = 2.68$ radians

(iii)

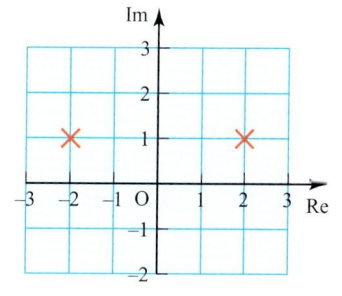

6 (i) $1 - \sqrt{3}i$, $-1 - \sqrt{3}i$

(ii)

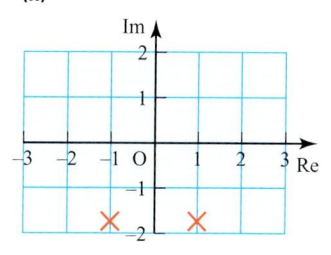

(iii) $1 - \sqrt{3}i$: $r = 2, \theta = -\frac{\pi}{3}$

$-1 - \sqrt{3}i$: $r = 2, \theta = -\frac{2\pi}{3}$

(iv) The three points are the same distance from the origin and separated by equal angles of $\frac{2\pi}{3}$ (i.e. 120°).

7 (i) (a) $1 + 2i$

(b) $-\frac{1}{2} + \frac{1}{2}i$

(ii) $\frac{3\pi}{4}$

(iv) OA = BC and OA and BC are parallel

8 (i) $u: r = \sqrt{2}, \theta = -\frac{3}{4}\pi$

$u^2: r = 2, \theta = \frac{1}{2}\pi$

(ii)

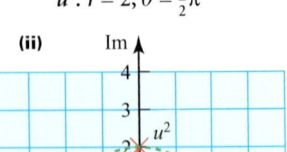

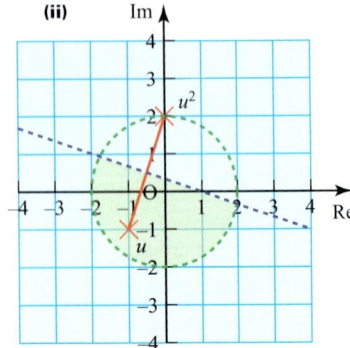

Activity 11.12 (Page 296)

(i) Rotation of vector z through $+\frac{\pi}{2}$

(ii) Half turn of vector z
(= two successive $\frac{\pi}{2}$ rotations: $-1 = i \times i$)

? (Page 299)

$3 + i$ and $-3 - i$

Exercise 11G (Page 301)

1 (i) $32(\cos 0.6 + i \sin 0.6)$

(ii) $2(\cos(-0.2) + i \sin(-0.2))$

(iii) $12\left(\cos\frac{\pi}{2} + i \sin\frac{\pi}{2}\right)$

(iv) $3\left(\cos\frac{\pi}{6} + i \sin\frac{\pi}{6}\right)$

(v) $24\left(\cos\frac{5\pi}{4} + i \sin\frac{5\pi}{4}\right)$

(vi) $6\left(\cos\frac{3\pi}{4} + i \sin\frac{3\pi}{4}\right)$

2 (i) $6\left(\cos\frac{7\pi}{12} + i \sin\frac{7\pi}{12}\right)$

(ii) $\frac{3}{2}\left(\cos\frac{\pi}{12} + i \sin\frac{\pi}{12}\right)$

(iii) $\frac{2}{3}\left[\cos\left(-\frac{\pi}{12}\right) + i \sin\left(-\frac{\pi}{12}\right)\right]$

(iv) $\frac{1}{2}\left[\cos\left(-\frac{\pi}{4}\right) + i \sin\left(-\frac{\pi}{4}\right)\right]$

(v) $9\left(\cos\frac{2\pi}{3} + i \sin\frac{2\pi}{3}\right)$

(vi) $32\left[\cos\left(-\frac{3\pi}{4}\right) + i \sin\left(-\frac{3\pi}{4}\right)\right]$

(vii) $432(\cos 0 + i \sin 0)$

(viii) $10\left(\cos\frac{3\pi}{4} + i \sin\frac{3\pi}{4}\right)$

(ix) $3\sqrt{2}\left(\cos\frac{7\pi}{12} + i \sin\frac{7\pi}{12}\right)$

3 Exceptions

(i) if $z = 0$ then $\frac{1}{z}$ does not exist

(iii) if z = real and negative then $\arg\left(\frac{1}{z}\right) = \arg z$

4 (i) Enlarge from O ×3

(ii) Enlarge from O ×2 and rotate $+\frac{\pi}{2}$

(iii) Complete the parallelogram $3z, 0, 2iz$

(iv) Reflect in the real axis

(v) Find where the circle with centre O through z meets the positive real axis

(vi) Complete the similar triangles $0, 1, z$ and $0, z, z^2$

5 $\frac{\sqrt{3}-1}{4}, \frac{\sqrt{3}+1}{4}$;

$\sqrt{2}\left(\cos\frac{3\pi}{4} + i \sin\frac{3\pi}{4}\right)$,

$\sqrt{10}\left(\cos\frac{\pi}{3} + i \sin\frac{\pi}{3}\right); \frac{\sqrt{3}+1}{2\sqrt{2}}$

6 (ii) $\frac{\pi}{4}, \frac{5\pi}{6}$

(iii) $8, -\frac{11\pi}{12}$

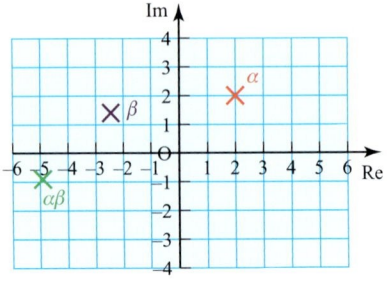

(iv) Perpendicular bisector of line from α to β

(v) $\frac{13\pi}{24}$

7 (i) -1

(ii) $\frac{1+i}{\sqrt{2}}$

(iii) $-1.209 + 0.698i$

(iv) $-13.129 + 15.201i$

8 (i) (a) $10e^i$

(b) 4

(c) $6e^{8i}$

(d) $3e^i$

(e) $3e^{3i}$

(f) $4e^{-i}$

(ii) (a) $2(\cos 3 + i \sin 3)$
$\times 5(\cos(-2) + i \sin(-2))$
$= 10(\cos 1 + i \sin 1)$

(b) $8(\cos 5 + i \sin 5)$
$\div 2(\cos 5 + i \sin 5)$
$= 4$

(c) $3(\cos 7 + i \sin 7)$
$\times 2(\cos 1 + i \sin 1)$
$= 6(\cos 8 + i \sin 8)$

(d) $12(\cos 5 + i \sin 5)$
$\div 4(\cos 4 + i \sin 4)$
$= 3(\cos 1 + i \sin 1)$

(e) $3(\cos 2 + i \sin 2)$
$\times (\cos 1 + i \sin 1)$
$= 3(\cos 3 + i \sin 3)$

(f) $8(\cos 3 + i \sin 3)$
$\div 2(\cos 4 + i \sin 4)$
$= 4(\cos(-1) + i \sin(-1))$

Exercise 11H (Page 304)

1 $2 - i, -3$

2 $z = 7, 4 \pm 2i$

3 $p = 4, q = -10$, other roots $1 + i, -6$

4 $z = 3 \pm 2i, 2 \pm i$

5 $z = \pm 3i, 4 \pm \sqrt{5}$

6 (i) $w^2 = -2i, w^3 = -2 - 2i$, $w^4 = -4$

(ii) $p = -4, q = 2$

(iii) two of $1 - i, 1 + i, -1, -4$

7 (i) $a^2 = -3 - 4i$, $a^3 = 11 - 2i$

(ii) $-1 - 2i$, -5

(iii) $|-5| = 5$, $\arg(-5) = \pi$
$|-1 + 2i| = \sqrt{5}$,
$\arg(-1 + 2i) = 2.03$
$|-1 + 2i| = \sqrt{5}$,
$\arg(-1 + 2i) = 2.03$

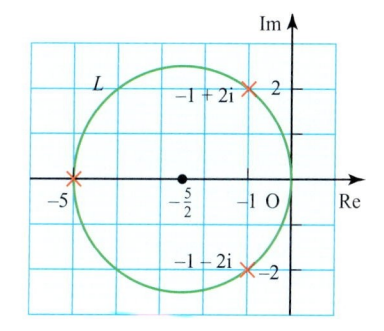

8 (i) $\beta = -1 + \sqrt{3}i$, $\gamma = -1 - \sqrt{3}i$

(ii) $\dfrac{1}{\beta} = -\dfrac{1}{4} - \dfrac{\sqrt{3}}{4}i$, $\dfrac{1}{\gamma} = -\dfrac{1}{4} + \dfrac{\sqrt{3}}{4}i$

(iii) $|\alpha| = 4$, $\arg \alpha = \pi$

$|\beta| = 2$, $\arg \beta = \dfrac{2\pi}{3}$

$|\gamma| = 2$, $\arg \gamma = -\dfrac{2\pi}{3}$

(iv)

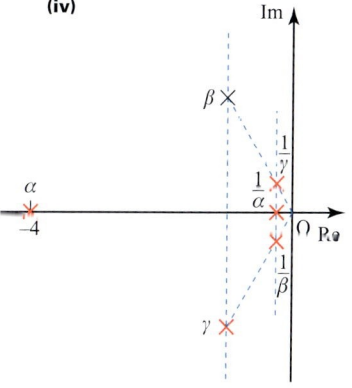

9 (i) $\alpha^2 = -15 + 8i$,
$\alpha^3 = -47 - 52i$

(ii) $k = 3$

(iii) -7, $1 - 4i$
$\arg(1 + 4i) = 1.326$
$\arg(-7) = \pi$
$\arg(1 - 4i) = -1.326$

(iv) $c = 5$

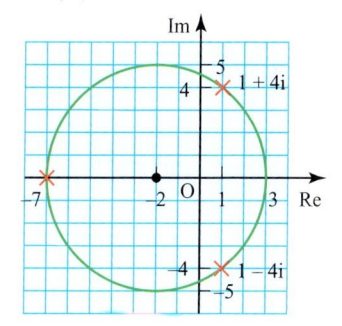

10 (i) $\alpha^2 = -8 - 6i$, $\alpha^3 = 26 - 18i$

(ii) $\mu = 20$

(iii) $z = -\dfrac{2}{3}, -1 \pm 3i$

$\left|-\dfrac{2}{3}\right| = \dfrac{2}{3}$, $\arg\left(-\dfrac{2}{3}\right) = \pi$

$|-1 + 3i| = \sqrt{10}$,
$\arg(-1 + 3i) = 1.893$
$|-1 + 3i| = \sqrt{10}$,
$\arg(-1 + 3i) = 1.893$

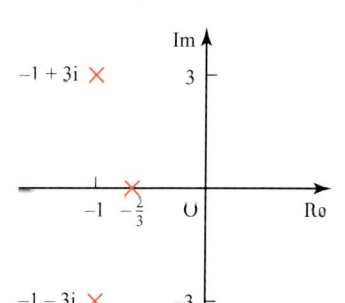

11 (i) $\beta = -2 + 2\sqrt{3}i$,
$\gamma = -2 - 2\sqrt{3}i$

(ii) $|\alpha| = 3$, $\arg \alpha = 0$

$|\beta| = 4$, $\arg \beta = \dfrac{2\pi}{3}$

$|\gamma| = 4$, $\arg \gamma = \dfrac{2\pi}{3}$

$\left|\dfrac{\beta}{\gamma}\right| = 1$, $\arg\left(\dfrac{\beta}{\gamma}\right) = -\dfrac{2\pi}{3}$

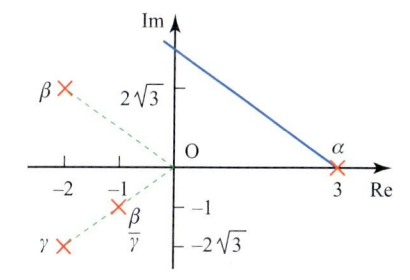

12 (ii) $1 - 2i$

(iii)

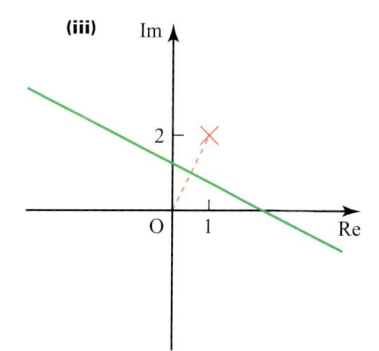

Index

Page numbers in black are in *Pure Mathematics 2*. Page numbers in blue are in *Pure Mathematics 3*.

acceleration 209
addition
  of complex numbers 274
  of fractions 165
  of polynomials 3
algebraic fractions 164
  expressing in partial fractions 167
alternating current 74
angle
  between a line and a plane 256
  between two lines 240–242
  between two planes 264
Argand, Jean-Robert 281
Argand diagram 281, 284–286

binomial coefficients 155–156
binomial expansion 155–162
  use of partial fractions 173–174
Brahmagupta 271
Bürgi, Jolst 43

cartesian equations 227, 249
chain rule 80, 93, 97, 108, 111
  reverse 179
change-of-sign methods 137–141
co-ordinate geometry 227
cobweb diagram 145, 146
complex conjugates 274–275
complex exponents 299–300
complex numbers
  addition 274
  conjugate 302
  division 276–278, 296–297
  equality 276
  and equations 302
  geometrical representation 281
  historical development 272
  modulus 283
  modulus–argument (polar) form 287–291, 293–294, 296–299
  multiplication 274, 296–297
  notation 273
  real and imaginary parts 273

  square root 278–279, 298–299
  subtraction 274
  sum and difference 281–282
  vector representation 281–282
complex plane 281
compound interest 123
compound-angle formulae 55–58, 296
cosecant (cosec) 52
cosine graph 76
cotangent (cot) 52
Cotes, Roger 299
counting numbers 271
cubic equations 302
cubic expressions 3
curves, modelling 30–35

decimal search 138–139
Devi, Shakuntala 154
differential equations
  forming 209–212
  general solution 214–215
  order 209
  particular solution 217–220
differentiation
  of implicit functions 97–102
  of natural logarithms and exponentials 85–89
  parametric 108–111
  product rule 78–80
  quotient rule 80–82
  of trigonometrical functions 92–95
distance
  between a point and a line 244–245
  of a point from a plane 254–255
division
  by zero 279
  of complex numbers 276–278, 296–297
  of fractions 164

  of logarithms 25
  of polynomials 5–6
double-angle formulae 61–65

e (base of natural logarithms) 32, 41–42
equations
  numerical solution 136–151
  rearranging 143, 146–147
  of a straight line 30–31, 32
  of the tangent to the curve 108
error (or solution) bounds 139
Euler, Leonhard 299
existence theorem 303
experimental data, mathematical relationships 30
exponential curves 43
exponential functions 28, 43
  differentiation 85–89
  infinite series 123
  integrals involving 117, 183–184
exponential growth and decay 43–44
exponential relationships 33–35
exponents, complex 299–300

factor theorem 9
factorisation, of algebraic expressions 164, 165
fixed-point iteration 142–147
fractions
  addition and subtraction 165
  multiplication and division 164
  simplifying 164
Fundamental Theorem of Algebra 302–303

Gauss, Carl Friedrich 272, 302
general binomial theorem 155–158
geometric progression 157
Girard, Albert 302
Golden Ratio 143
gradient function 78, 87, 93

graphs
   of the exponential function 43
   and logarithms 27–28, 31–33
   of the natural logarithm function 42, 43
   of parametric equations 105–107
Greek mathematicians 271

Hardy, G.H. 2, 78

i, square root of −1 273
identities
   in partial fraction methods 167, 168
   Pythagorean 61
   trigonometrical 56, 125
implicit functions 97–102
indices, and logarithms 25–26
inequalities, involving the modulus sign 19–21
integrals
   indefinite 120, 180–181
   involving exponentials and logarithms 117–118, 183–184
   involving trigonometrical functions 124–126, 187–189
   standard 203
integration
   by parts 194–201
   by substitution (by change of variable) 177–182
   choice of methods 203–205
   general 203
   numerical 128–131
   use of partial fractions 190–193
intersection
   of a line and a plane 252–253
   of two lines 234–237
   of two planes 262
interval bisection 139
intervals
   estimation 137–141
   notation 136
inverse functions 43
irrational numbers 271
isobars 214
iteration, fixed-point 142–147
iterative process 42

line of intersection, of two planes 262

lines
   angle between 240–242
   angle to a plane 256
   cartesian and vector equations 227
   direction and location 230
   intersection 234–240
   intersection with a plane 252–253
   parallel, intersecting or skew 234–236
liquid, cooling rate 208, 226
logarithmic scales 24
logarithms
   base 23, 24, 27
   discovery 43
   and graphs 27–28, 31–33
   and indices 25–26
   integrals involving 117–120, 183–184
   laws of 25–27
   multiplication and division 25
   natural 32, 39–42, 85–89, 183–184
   power zero 25
   reciprocals 27
   and roots 26
   to the base 10 24
lowest common multiple 165

mental arithmetic 154
modelling
   curves 30–35
   pressure gradient 214
   use of double-angle formula 61
   waves, by trigonometrical functions 51, 55, 58
modulus, of a complex number 283
modulus function 17–18
modulus–argument (polar) form of complex numbers 287–291, 293–294, 296–299
multiplication
   of complex numbers 274
   of fractions 164
   of logarithms 25
   of polynomials 4

Napier, John 43
natural logarithm function 39–42, 117–118
   see also logarithms, natural

negative numbers 271
   square roots 272
Newton's law of cooling 208–209
number system, historical development 271–272

oscillations see waves

parameters, eliminating 107
parametric equations 104–116
partial fractions 166–172
   with a quadratic factor in the denominator 192–193
   with a repeated factor in the denominator 191–192
   use with the binomial expansion 173–174
   use in integration 190–193
Pascal's triangle 156
perpendicular distance from a point to a line 244–245
perpendicular (normal) to a plane 250–252
planes
   angle between 264
   distance to a point 254–255
   equation 247–248, 250–252
   intersection 262–263
   sheaf 264
point of intersection, co-ordinates 234–236
polar form see modulus–argument (polar) form of complex numbers
polynomial equations
   factorisation 9–14
   roots 8, 9, 11–12
   solution 8–14
      by means of complex numbers 302–304
polynomials
   addition 3
   division 5–6
   multiplication 4
   order 3
   subtraction 3–4
position vector 227, 228
   of point of intersection 238
pressure gradient, modelling 214
principal argument 287

principal value, of trigonometrical functions 75
product rule 78–80
Pythagoras' theorem, trigonometrical forms 53

quadratic equations 3
 with no real roots 272, 302
quadratic expressions 3
quadratic formula 8, 273
quartic and quintic expressions 3
quotient 5, 6
quotient rule 80–82

$r\cos(\theta \pm \alpha)$, $r\sin(\theta \pm \alpha)$ formulae 66–70
Ramanujan, Srinivasa 2
rate of change, and differential equation 209–212
rational function 164
rational numbers 271
real and imaginary axes 281
real numbers 155, 271–272
reciprocals, logarithms 27
remainder theorem 12–13
roots, and logarithms 26

scalar product 240, 241, 244
secant (sec) 52
separation of variables 215–216
series, infinite 157–158
sets of points
 in an Argand diagram 284–286
 using the polar form 293–294
sheaf of planes 264
sine graph 76, 92
skew lines 234, 241
spiral dilatation 297
square root
 of a complex number 278–279, 298–299
 of −1 (i) 273
 of a negative number 272
staircase diagram 145, 146
stationary points 80, 99, 100–101
 of a parametric curve 111
subtraction
 of complex numbers 274
 of fractions 165
 of polynomials 3–4

temperature–time graph 208
trapezium rule 128–131

trigonometrical equations, general solutions 75
trigonometrical functions
 differentiation 92–95
 integrals involving 124–126, 187–189
 principal value 75
 reciprocal 52–54
trigonometrical identities, using in integration 125–126

vector equations
 of a line 227, 228–232
 of a plane 247, 248
vectors
 ease of use 227
 joining two points 227–228
 one-dimensional 209
velocity 209

waves
 combining 70, 74
 modelling 51, 55, 58
Wessel, Caspar 281